INFRARED AND MILLIMETER WAVES

VOLUME 11   MILLIMETER COMPONENTS
                        AND TECHNIQUES, PART III

CONTRIBUTORS

D. BÖHME

YAT MAN CHOI

I. G. EDDISON

V. A. FLYAGIN

H. FREUND

L. GENZEL

A. L. GOL'DENBERG

A. GOVER

V. L. GRANATSTEIN

DOUGLAS J. HARRIS

F. KREMER

FREDERICK LABAAR

ALGIE L. LANCE

J. H. MCADOO

G. S. NUSINOVICH

A. POGLITSCH

WENDELL D. SEAL

CHA-MEI TANG

TSUKASA YONEYAMA

INFRARED AND MILLIMETER WAVES

VOLUME 11   MILLIMETER COMPONENTS AND TECHNIQUES, PART III

Edited by   *KENNETH J. BUTTON*

NATIONAL MAGNET LABORATORY
MASSACHUSETTS INSTITUTE OF TECHNOLOGY
CAMBRIDGE, MASSACHUSETTS

1984

ACADEMIC PRESS, INC.

(Harcourt Brace Jovanovich, Publishers)

Orlando   San Diego   New York   London
Toronto   Montreal   Sydney   Tokyo

ACADEMIC PRESS, INC.
Orlando, Florida 32887

United Kingdom Edition published by
ACADEMIC PRESS, INC. (LONDON) LTD.
24/28 Oval Road, London NW1 7DX

Library of Congress Cataloging in Publication Data

Main entry under title:

Infrared and millimeter waves.

Includes bibliographies and indexes.
Contents: v. 1. Sources of radiation. -- v. 2.
Instrumentation. -- [etc.] -- v. 11. Millimeter
components and techniques, part III.
1. Infra-red apparatus and appliances. 2. Millimeter
wave devices. I. Button, Kenneth J.
TA1570.I52   621.36'2   79-6949
ISBN 0-12-147711-8 (V.11)

PRINTED IN THE UNITED STATES OF AMERICA

84 85 86 87   9 8 7 6 5 4 3 2 1

CONTENTS

Chapter 4   **The Application of Oversized Cavities
for Millimeter-Wave Spectroscopy**
F. Kremer, A. Poglitsch, D. Böhme, and L. Genzel

Chapter 5   **Powerful Gyrotrons**
V. A. Flyagin, A. L. Gol'denberg, and G. S. Nusinovich

Chapter 6   **Some Perspectives on Operating Frequency
Increase in Gyrotrons**
G. S. Nusinovich

Chapter 7   **Phase Noise and AM Noise Measurements
in the Frequency Domain**
Algie L. Lance, Wendell D. Seal, and Frederick Labaar

Chapter 8 Basic Design Considerations for Free-Electron Lasers Driven by Electron Beams from rf Accelerators

A. Gover, H. Freund, V. L. Granatstein, J. H. McAdoo, and Cha-Mei Tang

LIST OF CONTRIBUTORS

Numbers in parentheses indicate the pages on which the authors' contributions begin.

D. Böhme (141), *Max-Planck-Institut für Festkörperforschung, 7000 Stuttgart 80, Federal Republic of Germany*

Yat Man Choi[1] (99), *Department of Physics, Electronics, and Electrical Engineering, University of Wales Institute of Science and Technology, Cardiff, United Kingdom*

I. G. Eddison (1), *Plessey Research (Caswell) Limited, Allen Clark Research Centre, Caswell, Towcester, Northants, England NN12 8EQ*

V. A. Flyagin (179), *Institute of Applied Physics, Academy of Sciences of the USSR, Gorky, USSR*

H. Freund[2] (291), *Naval Research Laboratory, Washington, D.C.*

L. Genzel (141), *Max-Planck-Institut für Festkörperforschung, 7000 Stuttgart 80, Federal Republic of Germany*

A. L. Gol'denberg (179), *Institute of Applied Physics, Academy of Sciences of the USSR, Gorky, USSR*

A. Gover[3] (291), *Naval Research Laboratory, Washington, D.C.*

V. L. Granatstein[4] (291), *Naval Research Laboratory, Washington, D.C.*

Douglas J. Harris (99), *Department of Physics, Electronics, and Electrical Engineering, University of Wales Institute of Science and Technology, Cardiff, United Kingdom*

F. Kremer (141), *Max-Planck-Institut für Festkörperforschung, 7000 Stuttgart 80, Federal Republic of Germany*

Frederick Labaar (239), *TRW Operations and Support Group, Redondo Beach, California 90278*

---

[1] Present address: Department of Electronic Engineering, Hong Kong Polytechnic, Kowloon, Hong Kong.

[2] Present address: Science Applications, Inc., Mclean, Virginia 22102.

[3] Present address: Science Applications, Inc., Mclean, Virginia 22102, and Tel Aviv University, Faculty of Engineering, Tel Aviv, Israel.

[4] Present address: Electrical Engineering Department, University of Maryland, College Park, Maryland 20742.

Algie L. Lance (239), *TRW Operations and Support Group, Redondo Beach, California 90278*

J. H. McAdoo[5] (291), *Naval Research Laboratory, Washington, D. C. 20375*

G. S. Nusinovich (179, 227), *Institute of Applied Physics, Academy of Sciences of the USSR, Gorky, USSR*

A. Poglitsch (141), *Max-Planck-Institut für Festkörperforschung, 7000 Stuttgart 80, Federal Republic of Germany*

Wendell D. Seal (239), *TRW Operations and Support Group, Redondo Beach, California 90278*

Cha-Mei Tang (291), *Naval Research Laboratory, Washington, D. C. 20375*

Tsukasa Yoneyama[6] (61), *Research Institute of Electrical Communication, Tohoku University, Sendai, Japan*

[5] Present address: Electrical Engineering Department, University of Maryland, College Park, Maryland 20742.

[6] Present address: Department of Electronics and Information Engineering, Ryukyu University, Nishiharacho, Okinawa 903-01, Japan.

PREFACE

This is the third of several books in this treatise that will adhere closely to the theme ''Millimeter Components and Techniques.'' Parts I and II on this topic have already been published as Volumes 9 and 10; the contents of Volumes 9 and 10 appear in ''Contents of Other Volumes.'' In each of these books we try to provide a few chapters on millimeter-wave hardware and practical applications. Then we give coverage to the latest developments in gyrotron technology, because gyrotrons are an important new high-power millimeter-wave source of radiation and it is rarely possible to find comprehensive explanations elsewhere in the literature. Finally, each book contains a few reviews of current millimeter-wave developments, which we classify as ''techniques.'' In the present book, this last category is represented by Chapter 7, ''Phase Noise and AM Noise Measurements in the Frequency Domain,'' by Algie L. Lance, Wendell D. Seal, and Frederick Labaar. Another new ''technique'' is reviewed in Chapter 4, in which F. Kremer, A. Poglitsch, D. Böhme, and L. Genzel describe the very recent revival of ''Oversized Cavities for Millimeter-Wave Spectroscopy.'' This is an excellent example of the thorough treatment of a subject that cannot be found elsewhere. Kremer and colleagues have been describing their unique results at the annual ''Conference on Infrared and Millimeter Waves,'' but here they have a chance to describe the theory of the procedure and experimental technique, make comparisons with other methods of measurement, and provide a summary of their own results.

On the other hand, millimeter-wave hardware and devices are very well represented here by the first three chapters. We are extremely fortunate to have I. G. Eddison prepare his review of ''Indium Phosphide and Gallium Arsenide Transferred-Electron Devices,'' which appears as Chapter 1. It is then very appropriate that we have Chapter 2, ''Nonradiative Dielectric Waveguide,'' by Tsukasa Yoneyama and Chapter 3, ''Groove Guide for Short Millimetric Waveguide Systems,'' by Yat Man Choi and Douglas J. Harris. This almost suggests that millimeter-wave systems have finally arrived, but we shall have to reserve our decision on that question until we see more chapters on millimeter-wave integrated circuits in "Millimeter Components and Techniques, Part IV" (Volume 13).

To be specific, we now expect manuscripts from A. G. Cardiasmenos on "Millimeter-Wave Hybrid Integrated Circuit Techniques," Jeffrey Paul on "Quasi-Optical Planar Mixer Techniques," K. Sigfrid Yngvesson and colleagues on "Near-Millimeter Imaging with Integrated Planar Receptors," Naresh Deo on "Dielectric-Based Active and Passive Millimeter-Wave Components," and Wolfgang Menzel on "Integrated Fin-Line Components for Communication, Radar, and Radiometer Applications." In the meantime, we are still developing the subseries titled "Electromagnetic Waves in Matter." Part I was published as Volume 8 and Part II will appear as Volume 12. The contents of these books can be found in "Contents of Other Volumes"; Part III is in preparation.

CONTENTS OF OTHER VOLUMES

Volume 13: Millimeter Components and Techniques, Part IV
(In Press)

CHAPTER 1

Indium Phosphide and Gallium Arsenide Transferred-Electron Devices

I. G. Eddison

Plessey Research (Caswell) Limited
Allen Clark Research Centre
Caswell, Towcester, Northants, England

I. Introduction

There is growing recognition that the 30–300-GHz frequency range offers distinct advantages for radar and communication purposes. With this recognition has come a move from simple systems to smaller, more complex millimeter-wave systems having improved range, frequency agility, and ruggedness. Inevitably, these demands call for major performance improvements from solid-state amplifiers and oscillators such as the transferred-electron device (TED).

The TED or, as it is more popularly known, the Gunn diode is a low-power, low-efficiency device that offers very good overall noise performance. It is therefore a very popular device for local-oscillator applications in both the microwave and millimeter-wave frequency ranges. At millimeter-wave frequencies the second generation system designs now call for TEDs with improved output power and higher operating frequency capabilities. These

1

new performance demands are necessary to provide local oscillators capable of driving multiple mixer systems as well as to provide reflection amplifiers with high gain and high added power. Other system aims call for frequency agility, wide bandwidths, low noise output, and improved voltage and temperature stabilities allied to device ruggedness. Equally crucial is the question of the overall cost of the device because the eventual success of working at millimeter wavelengths will require volume production at a sensible system cost. These considerations and the need for small size are pushing technology toward complete system integration, and of course the transferred-electron oscillator (TEO) and transferred-electron amplifier (TEA) must develop accordingly.

These recent requirements are placing heavy demands on the performance parameters of the TED, and in many cases the physical limitations of the device are being approached. It is therefore the aim of this chapter to review the internal physical processes involved in TEDs to identify those parameters critical to the transferred-electron (Gunn) effect at millimeter-wave frequencies. Emphasis is placed throughout on the relation of theoretical device performance to existing practical device behavior. This theoretical work is also used to highlight those areas of internal device physics that are likely to provide the improved performance characteristics demanded by system designers.

The TED will be approached in this way both as an oscillator and as a reflection amplifier device. Practical equipment results are quoted to illustrate the potential of the TED before consideration is given to the future prospect of this class of device in the vitally important field of millimeter-wave technology.

II.   Transferred-Electron Oscillators

A.   ELECTRON TRANSFER AT MILLIMETER WAVELENGTHS

As its name suggests, the transferred-electron effect in many-valleyed semiconductors is characterized by the transfer of conduction-band electrons from the central Γ valley (a high-mobility, low-energy state) into the satellite L valley (a low-mobility, high-energy state). This process gives the electrons in the semiconductor the unusual velocity–field characteristics shown in Fig. 1. At low applied fields the electrons occupy the central valley and exhibit the high-mobility relationship between the electron velocity v and the applied field ε. However, as the field approaches a threshold value determined by the intervalley energy gap the electrons transfer into the low-mobility state. There is a consequent drop in electron velocity to the satellite valley's saturated drift velocity (v_v). This region of negative differential mobility

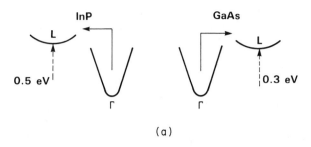

(a)

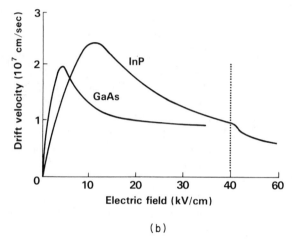

(b)

Fig. 1   (a) Schematic of band structure and (b) electron drift velocity versus electric field in GaAs and InP. The intervalley energy gap determines the threshold field value where electrons transfer to a low-mobility state. · · ·, change of scale.

(NDM) gives rise to the device's bulk negative resistance and hence its ability to convert dc energy into an rf output. As might be expected the NDM region has a crucial effect on overall device performance. In particular it can be shown that the device dc–rf conversion efficiency, and thus its output power capability, is directly related to the peak-to-valley electron velocity ratio (v_p/v_v).

When different TED materials are compared their respective electron velocity–field characteristics can provide useful indications of the possible rf performances of the device. Figure 1 illustrates the electron behavior in both indium phosphide and gallium arsenide and shows some important differences between the materials. It is immediately clear that indium phosphide exhibits a higher threshold field as well as a higher peak-to-valley electron velocity ratio. Further, the electron velocity–field behavior is less temperature dependent in indium phosphide (Fawcett and Hill, 1975), a fact attributable to the greater energy separation between the central and satellite valleys for electrons in indium phosphide. Thus the important electron peak-to-valley velocity ratio has a lower reduction with temperature in indium phosphide compared with the fairly rapid fall seen in gallium arsenide.

Consideration of these factors, especially the high peak-to-valley velocity ratio, suggests that indium phosphide would make the basis of an excellent high-efficiency, high-peak-power pulsed device at microwave frequencies (Smith and Tebbenham, 1979). However, the real benefits of indium phosphide as an improved high-frequency material become apparent only when the limitations of electron transfer speed of the two materials are examined. Theoretical analysis of electron transfer shows that the dominant speed limitation is the rate at which electrons gain or lose energy in the central Γ valley rather than the much faster intervalley scattering rate. This effect is illustrated in Fig. 2, which considers an electron just scattered out of the satellite valley with a velocity v_0 (point A). Now, depending on the direction of this velocity with respect to the applied field, the electron will either be decelerated (A' to B) and then reaccelerated (B to C), or simply reaccelerated before transferring back into the L satellite valley (A or C to D).

This rethermalization process has an associated time constant Υ_0 that is inversely proportional to the applied field, because the larger the field acting on the electron the faster it will gain energy.

To understand the effect of this time constant Υ_0 at higher frequencies, it is necessary to consider again the electron velocity–field characteristics of

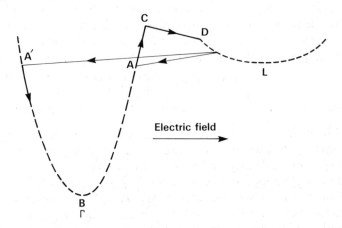

FIG. 2   Momentum space representation of intervalley scattering showing energy relaxation effects.

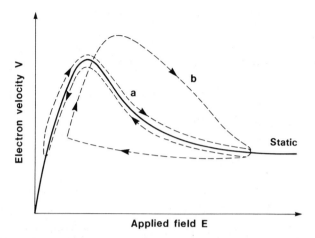

FIG. 3   Dynamic electron velocity V versus field characteristics E showing energy relaxation effects: curve a, at 10 GHz; curve b, at 50 GHz.

Fig. 3. At low oscillation frequencies Υ_0 has minimal effect, and during any rf cycle the electrons closely follow the static v–ε curve as shown by curve a. As higher frequencies are approached, the time delay due to Υ_0 introduces a hysteresis effect in the dynamic v–ε curve (curve b), thereby lowering the effective electron peak-to-valley velocity ratio. Consequently there is a steady reduction in conversion efficiency and output power as the operating frequency is increased.

A further property of the v–ε characteristics of the two materials is that the higher peak-to-valley ratio and the higher threshold velocity of indium phosphide lead to a higher effective electron velocity ratio. This fact gives the advantage of longer active-layer lengths for a given operating frequency but at the expense of an increased current density, i.e.,

$$l \propto (v_{\text{eff}}/f_0), \tag{1}$$

$$J = nqv_{\text{eff}}, \tag{2}$$

where l is the active layer length, v_{eff} the effective electron velocity in the active layer, f_0 the oscillation frequency, J the device current density, n the active-layer doping density, and q the electron charge.

These considerations suggest that the higher threshold field of indium phosphide permits faster electron transfer and hence higher oscillation frequencies than are possible with gallium arsenide. Analysis shows that indium phosphide should operate to frequencies approximately a factor of two higher than gallium arsenide (Kroemer, 1978). Besides this important advantage, indium phosphide's higher peak-to-valley ratio promises better dc–rf conversion efficiencies with the attendant improved output power levels. Also, the less temperature-dependent electron-velocity behavior in indium phosphide suggests a greater rf power stability with temperature from the final oscillator device. Further, the faster effective electron velocity promises longer active-layer lengths, thereby easing material growth constraints.

Among all these advantages it should be recognized that nature demands a penalty. For indium phosphide the high threshold field and high effective electron velocities conspire to give thermal limitations that are a serious disadvantage at frequencies below 50 GHz. At higher frequencies these difficulties are eased by the shorter active-layer lengths, but thermal design is a critical factor in indium phosphide device operation, as will be shown later.

Displayed in Table I is a summary of the important properties of the two TED semiconductor materials. It is revealed in Section II.F that the practical behavior of the TED materials agrees well with the theoretical predictions described earlier.

TABLE I

COMPARISON OF GaAs AND InP AS SEMICONDUCTOR MATERIALS FOR
TRANSFERRED-ELECTRON DEVICES

Property	GaAs	InP
Electron peak-to-valley velocity ratio	2.2	3.5
Threshold field (kV/cm)	3.2	10.0
Temperature dependence of peak electron velocity (%/°C)	−0.15	−0.1
Inertial energy time constant τ_0 (psec)	1.5	0.75
Diffusion coefficient–mobility ratio $D(E)/\mu(E)$(cm^2/sec), where $E = 2E_{th}$	142	72

B. COMPUTER SIMULATIONS OF OSCILLATING DEVICES

Computer modeling is a valuable means of gaining an insight into TED operation, and simulation results are available describing the internal physical behavior of many devices (Jones and Rees, 1972, 1975). A recent simulation of the transferred-electron effect is particularly relevant to the millimeter-wave device (Friscourt et al., 1983). Here a single-electron gas model is used as the basis of a numerical solution of the Boltzmann transport equations. The results of this detailed simulation confirm the previous predictions and reveal very good agreement with the practical behavior of both gallium arsenide and indium phosphide devices.

The main operating features of a high-frequency TEO are illustrated by the n^+–n–n^+ structure of Fig. 4. The simulation reveals that there are two

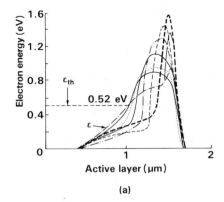

(a)

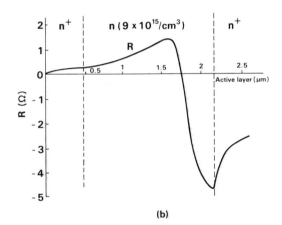

(b)

FIG. 4 (a) Evolution of electron energy and (b) diode resistance as a function of device active-layer length: $F = 100$ GHz, $m = 10\%$, $V_0 = 4.7$ V, $I_0 = 760$ mA. The graphs in part (a) show the electron energy profile at varying times during one rf cycle: beginning of cycle, $--$; $--$, $-\cdots-$, $-\cdot-$, $\underline{\quad\quad}$; $\cdots$, end of cycle. (Courtesy University of Lille.)

very distinct regions in the active layer of the device, as can be seen from the energy profile of electrons in an indium phosphide 100-GHz TED (Fig. 4a). The first region adjacent to the cathode contact behaves as a heating (or thermalization) zone and the second region adjacent to the anode is the active zone where electrons have sufficient energy for intervalley transfer. Thus only 40 % of the device n layer supports the propagation of the accumulation layers that are responsible for rf power generation, and the remaining 60 % is therefore a dead-space region. The difficulties that this causes are emphasized when the available resistance is calculated as a function of active-layer lengths, as shown in Fig. 4b. Here it is clear that only a small part of the layer is actually generating the necessary negative resistance, and the rest of the layer is simply a parasitic loss region. Not only does this effect seriously limit the power generated in the device but also the consequent low net negative resistance at the device terminals creates circuit-matching problems.

Interestingly, the two popular TED materials have important differences with regard to these active-layer regions. Indium phosphide, because of its higher peak-to-valley ratio, has more negative resistance associated with the active n-layer region but its comparatively poor low-field mobility (4000 $cm^2/V \cdot sec$) also results in a longer dead-space region. The full benefits of indium phosphide, therefore, can only be realized by removing this dead-space effect (Colliver et al., 1974). Improved output powers and efficiencies can be realized by creating a high-field current-limiting cathode contact that essentially injects hot electrons into the device n layer. This injecting contact thus removes the parasitic loss region associated with the normal ohmic n^+-cathode contact. Conversely, gallium arsenide has a lower negative resistance owing to the n-layer active zone, but its higher low-field mobility (8000 $cm^2/V \cdot sec$) gives a shorter dead-space region and therefore lower parasitic loss. Unfortunately, this latter fact means that there is little to gain from an injecting cathode contact on gallium arsenide.

The ability of computer simulation to provide a better understanding of the internal behavior of devices is particularly valuable in the study of TED operating modes at millimeter-wave frequencies. For the case of gallium arsenide, the simulations verify the existence of harmonic operation in W-band oscillators as seen in earlier practical work (Eddison and Brookbanks, 1982). Most of the gallium arsenide oscillators commercially available for frequencies above 75 GHz are n^+-n-n^+ structures with n-layer lengths of 1.7 to 2.7 μm at doping levels close to 1.10^{16} cm^3. However, both computer simulation and practical characterization reveals that these devices exhibit optimum fundamental operation in the 30–50-GHz frequency range. It is therefore hardly surprising that above 75 GHz these devices generate useful power only by harmonic extraction–enhancement. The main properties of such harmonic generators, as shown by the simulation, can be summarized by referring to Fig. 5 and 6.

Figure 5 shows that a high-voltage component at the fundamental frequency f_0 is vital if significant power is to be extracted at the harmonic frequency $2f_0$ (or possibly $3f_0$). It is important to note that the optimum harmonic power output corresponds to very low levels of fundamental output, because the essential requirement is for high rf fields to be across the device at the fundamental frequency, i.e., reactive loading at f_0. This feature can be regarded as more evidence of the need for high fields to increase the electron rethermalization rate and hence the upper frequency limit of the device. Figure 6 displays the harmonic output behavior alone, and it is

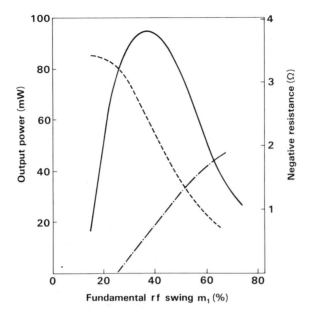

FIG. 5   Fundamental frequency output power, harmonic output power, and negative resistance as a function of the fundamental rf voltage swing across a harmonic GaAs TEO: ——, P_1; ---, R_1; —·—, P_2. (From Friscourt *et al.*, 1983. © 1983 IEEE.)

clear that optimum output occurs at low harmonic rf voltages with very low values of negative resistance. This collapse of negative resistance in the device places constraints on the oscillator matching circuit, and matching low-impedance devices in practical oscillator circuits to extract maximum output power can be very difficult.

The simulation work shows that the harmonic gallium arsenide device is a low-power, low-efficiency device. Under realistic operating-temperature conditions, the predictions suggest an optimum performance of 35 mW at 0.9% efficiency at frequencies approaching 100 GHz. This performance level is only slightly above the best gallium arsenide results reported. To improve on this power and efficiency level, it is necessary to produce a fundamental frequency device; for gallium arsenide this would require unrealistic material growth needs (1 μm active length, 2.10^{16} cm^3 doping density) and low active-layer operating temperatures. Indeed, no power at all has yet been generated from such short active-length gallium arsenide devices. Conversely, the simulations show that the fundamental frequency device using indium phosphide is the best approach to higher millimeter-wave powers. Figure 7 summarizes the predictions for indium phosphide up to 220 GHz, and a significant decrease in fundamental frequency power is

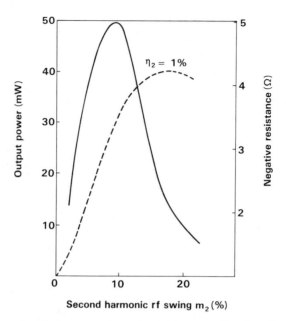

FIG. 6 Harmonic output power and negative resistance as a function of second harmonic rf voltage swing across a harmonic GaAs TEO: ——, R_2; --, P_2. (From Friscourt et al., 1983. © 1983 IEEE.)

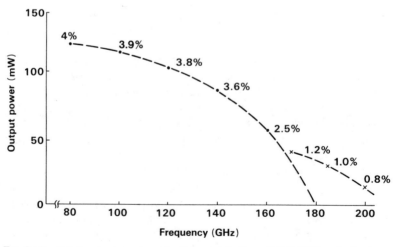

FIG. 7 Predicted output power and efficiency of n^+–n–n^+ InP TEOs: ●, fundamental mode; ×, harmonic mode; $T = 220°C$; $T_{amb} = 20°C$. (Courtesy University of Lille.)

seen as 150 GHz is approached. At higher frequencies harmonic operation could be used to achieve a predicted 10-mW output power at 220 GHz. The simulations of n^+-n-n^+ millimeter-band TEDs clearly show the theoretical advantages of indium phosphide over gallium arsenide. In the remainder of this section, practical oscillator devices will be described and their measured behavior will be related to these predictions.

C. MATERIAL GROWTH

The TED semiconductor material is normally grown by vapor-phase epitaxy on highly conducting n^+ substrates. The active (n) region is grown with a carrier density of 5×10^{15} to 1.5×10^{16} atoms/cm and lengths (L) of 1.0 to 2.5 μm, these values being dependent upon the required operating frequency, efficiency, and impedance level. To achieve adequate control of epitaxial growth and to identify those material parameters that critically affect device rf performance, it is essential to establish good material characterization facilities. Perhaps the most important facility is the carrier concentration profiler, which provides the information shown in Fig. 8. The

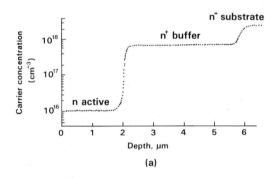

(a)

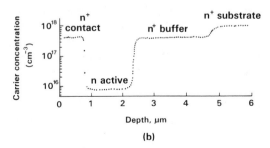

(b)

FIG. 8   Profiles of typical millimeter-wave TEO devices: (a), $n-n^+$ flat layer; (b), n^+-n-n^+ layer. [Courtesy Plessey Research (Caswell) Limited.]

equipment used in this case was an electrochemical impurity profiler manufactured by Polaron; the main advantage of this profiler is that it allows accurate monitoring of doping level variations, layer thickness, and interface sharpness for all the epitaxial regions as well as for the substrate.

When the basic growth facilities have been established, the choice of material profile for the particular TED application has to be made. The two obvious material profiles that can be used for the TEO device are the three-layer n^+-n-n^+ and two-layer $n-n^+$ structures, as shown in Fig. 8. Of these two profiles the three-layer structure produces the least efficient devices owing to the dead-space problems mentioned previously. Conversely, the use of injecting contacts to the n layer of an $n-n^+$ structure can, under the correct conditions, produce high power and efficiency results from indium phosphide. In the case of gallium arsenide the shorter dead-space region and lower peak-to-valley velocity ratio suggest only small benefits from the two-layer structure.

With regard to the choice of material profile, consideration has to be given to the secondary performance features of the two options. First, although the three-layer device is inherently less efficient, the use of ohmic contacts to the n^+ regions provides improved reliability and thermal stability. Against this, the injecting contact two-layer structure demonstrates considerable advantages in power and efficiency (Crowley et al., 1980). Unfortunately, the Schottky-like injecting contact has a strong current–temperature dependence (dI/dT), a feature that could lead to difficulties in producing a stable device. To understand the reasons for this effect, the operation of the injecting contact has to be considered in relation to the material velocity–field characteristic of Fig. 9. Ideally, electrons must be injected into the n layer with the satellite valley velocity, and this can be accomplished only by creating a high-field region at the cathode. Clearly, from the velocity–field curve, there is a relationship between the magnitude of this high field and the injected electron velocity. By controlling the carrier velocity the device current density is also controlled, and the ratio of this injected current density J to the threshold current density J_c is known as the contact injection ratio (J/J_c). Now it can be shown that there is a particularly strong relationship between the injection ratio and the device dc–rf conversion efficiency (Brookbanks and Buck, 1981). If a Schottky-like injecting contact is used, changes in device temperature cause variations in contact barrier height and therefore in the magnitude of the contact region's high field. From the velocity–field relationship, the temperature-induced changes to the high-field region directly affect the injected electron velocity. Hence wide variations in conversion efficiency and output power will result from the effects of temperature on the contact injection ratio (see Fig. 10).

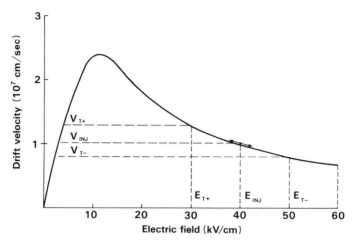

FIG. 9 Electron velocity–field characteristics showing dependence of injection character-istics on InP device temperature: E_{inj}, nominal injecting field; E_{T+}, injecting field for higher operating temperature; E_{T-}, injecting field for lower operating temperature.

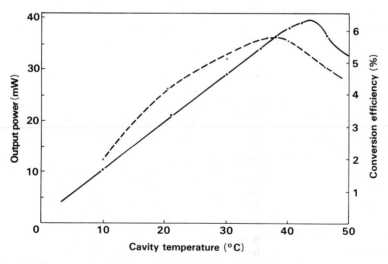

FIG. 10 Output power and conversion efficiency dependence on temperature of low-injection-ratio 90-GHz $n-n^+$ InP TEO: --, efficiency; —·—, output power; $F = 90$ GHz; $V = -4.2$ V.

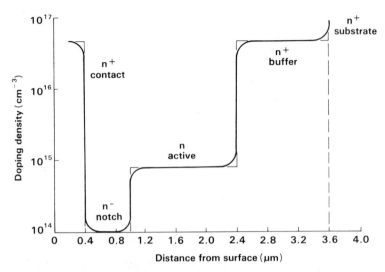

FIG. 11   Schematic diagram of an InP cathode-notch TED layer profile.

Although there are great advantages to the $n-n^+$ structures, there is also a
need to reduce the unwelcome temperature effects. One possible means of
achieving the necessary high cathode fields is by a cathode notch in the
material profiles, as depicted in Fig. 11. Here the cathode field is inherent in
the material structure, and stable ohmic contacts can be used. However, it
is not yet clear whether this grown-material cathode would provide the
necessary high-efficiency contact action.

D.   DEVICE DESIGN AND THERMAL ANALYSIS

All TEDs present the designer with two major problems at millimeter-
wave frequencies. First, the devices have low available negative resistances,
which means that parasitic losses such as semiconductor resistance, skin
effect loss, and contact resistance must be minimized. This low negative
resistance also makes circuit matching difficult, because circuit losses
increase at millimeter wavelengths. Second, all the TEDs have dc–rf conversion efficiencies of only a few percent, and consequently nearly all of the
device input power (e.g., 98 %) has to be removed as waste heat. Hence it is
clear that great care must be taken to minimize both the semiconductor
parasitic resistances and the thermal impedances of the final device geometry.

A common solution to these problems is to fabricate the devices as integral
heat sink (IHS) structures, as shown schematically in Fig. 12. Here the
ability to place a plated metal heat sink within a few micrometers of the
device's heat-generating active region gives clear thermal advantages. The

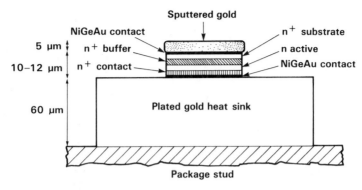

FIG. 12   Schematic diagram of InP integral heat sink device structure.

heat is thus allowed to spread evenly from the semiconductor into the heat sink without the hot-spot nonuniformities often experienced when the semiconductor is simply bonded onto a package heat sink. Another important advantage of the IHS device process is that the plated heat sink's structural rigidity allows thinning of the semiconductor to 10 to 12 μm in thickness. In this way most of the semiconductor substrate material and its associated parasitic resistances are removed.

When the basic device structure has been chosen, together with the semiconductor material and its profile (e.g., $n-n^+$ or n^+-n-n^+), the design exercise is a complex iterative process in which the device's thermal impedance is usually the limiting parameter. Essentially, a device diameter must be chosen so that sufficient input power can be achieved to generate the desired rf output power without overheating the device. Because some of the most important semiconductor parameters involved are temperature sensitive, thermal analysis has to be carried out using accurate computer modeling techniques. From such an analysis thermal impedance and input power can be found as a function of device diameter, as shown in Fig. 13 for the case of a 90-GHz n^+-n-n^+ indium phosphide device. To first order, the thermal impedance is inversely proportional to the device radius, whereas device input power is proportional to device area. When these factors are combined, as in Fig. 14, the required design data of output power, efficiency, and operating temperature as a function of device diameter become available.

Examination of these results shows that increasing the device diameter increases the operating temperature to such an extent that the output power becomes thermally limited and the conversion efficiency consequently suffers. This thermal limitation is brought about by the high-threshold field of indium phosphide and the consequent high bias voltage. Fortunately, this material offers two possible solutions for the gold IHS device. First,

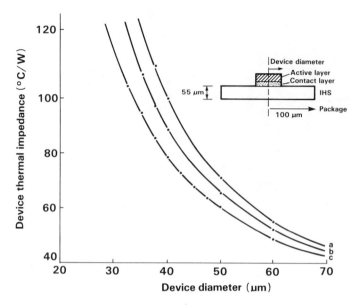

FIG. 13   Calculated thermal impedance and dc input power of an InP n^+-n-n^+ input IHS TED as a function of device diameter. Values for input power are: curve a, 5 W; curve b, 3 W; curve c, 1 W.

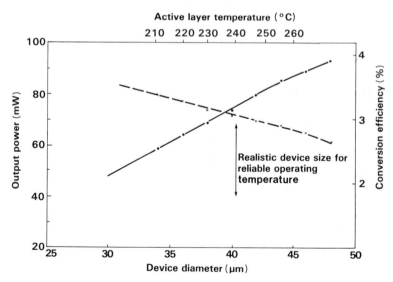

FIG. 14   Calculated output power (●), conversion efficiency (×), and active-layer temperature of an InP n^+-n-n^+ TED as a function of device diameter. The ambient temperature is assumed to be 30°C.

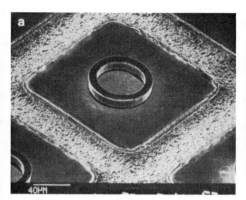

FIG. 15   (a) Annular and (b) cross structure InP devices.

the use of an injecting contact n–n^+ device will reduce device input power, and hence temperature, through the cathode contact's current-limiting action. Second, if an n^+–n–n^+ profile has to be used, then novel device structures could be used to improve the IHS device's thermal properties. In particular, manufacturing the devices as annular or cruciform geometries (see Fig. 15) rather than as conventional solid disks can give a 25–30% reduction in the spreading resistance contribution to the overall thermal impedance. A further benefit of these novel geometries is that their widths can be chosen to be less than twice the skin depth of the substrate region to minimize skin-effect losses and guarantee current spreading into the active regions.

Although the foregoing design and thermal analysis arguments have been developed with reference to indium phosphide, the conclusions drawn also hold for gallium arsenide. In general terms it can be shown that the thermal disadvantages caused by indium phosphide's high threshold field and high current density are offset by gallium arsenide's poor conversion efficiency, longer active lengths, and higher bias voltages, all of which are inherent in the harmonic operating mode. From these considerations it is clear that any further improvements in the power capability of TE devices can be realized only through either improved conversion efficiences or better thermal environments. The former option can be achieved only by the indium phosphide n–n^+ injecting cathode contact device, but the use of type IIA diamond heat sinks would offer clear advantages to all the TE devices.

The possible benefits of a diamond heat sink technology can be estimated by modeling the electrical and thermal behavior of typical 94-GHz devices. Table II presents the results of such an analysis for conventional gold integral heat sink (IHS) devices and their diamond heat sink equivalents, with the IHS

TABLE II

THE EFFECTS OF DIAMOND HEAT SINKING ON TYPICAL 94-GHz TEO DEVICES[a]

Device Heat Sink	IHS	Diamond (1)	Diamond (2)
Gallium arsenide (105 μm ϕ)			
Nominal bias voltage (V)	4.6	4.6	5.5
Thermal impedance (°C/W)	37.5	20	21.5
Input power (W)	5.5	5.72	7.0
Active-layer operating temperature (°C)	266	164	200
Output power (mW)	30	52	66.5
dc–rf conversion efficiency (%)	0.55	0.91	0.95
Indium phosphide (40 μm ϕ)			
Nominal bias voltage (V)	6.0	6.0	7.5
Thermal impedance (°C/W)	98	47	49
Input power (W)	2.0	2.2	2.7
Active-layer operating temperature (°C)	246	153	183
Output power (mW)	40	56	72
dc–rf conversion efficiency (%)	2.0	2.6	2.65

[a] A 50°C ambient temperature has been assumed for all devices to account for typical laboratory system operating temperatures. See text for descriptions of the diamond heat sink devices.

results being based on the known performance of existing devices. The two possible operating conditions considered for the equivalent diamond heat sink devices are the following.

(1)   The device is operated at a bias voltage identical to the value used in the IHS device.

(2)   The same device is operated at a higher bias level to take advantage of the lower active-layer temperatures inherent in using diamond heat sinks. The bias voltage chosen is an estimate of the value that gives maximum output power, i.e., the electronic power limit at which the power pushing characteristic of the device (dP/dV) becomes zero.

The data in Table II indicate that gallium aresenide gains most from the reduced thermal impedance of the diamond heat sink. This is to be expected because the reduced temperature stability of electronic carriers in gallium arsenide gives it a higher-power temperature coefficient ($dP/dT = -0.026$ dB/°C) compared with that of indium phosphide ($dP/dT = -0.012$ dB/°C). It might therefore be concluded that as long as conversion efficiency is not a critical factor, gallium arsenide would be an attractive alternative to indium phosphide for diamond heat sink TEOs. However, the matching of gallium arsenide devices might make it difficult to extract the higher available power.

This is particularly true if devices of increased area and hence lower rf impedance are considered. Conversely, the device–circuit matching required by the indium phosphide TEO device is generally straightforward, probably because of the higher bias voltages and lower currents needed for this fundamental frequency device. Therefore, the possibilities of using a larger area of indium phosphide on diamond can be investigated with some confidence. Once again, thermal and electrical modeling reveals the advantages of diamond, and Fig. 16 shows the estimated output power at 94 GHz as a function of the diameter of the indium phosphide device. It can be seen that powers greater than 100 mW are suggested at only moderate active-layer operating temperatures. Use of standard waveguide resonant-cap circuits would make it possible to extract powers approaching 100 mW, whereas lower impedance circuits would probably be necessary to exceed 100 mW power levels. Circuits are available for matching very low impedance fundamental frequency IMPATT devices at around 94 GHz; therefore it is likely that few difficulties would be encountered in matching the large-area indium phosphide devices. For even greater power levels than these, a similar analysis carried out for n–n^+ indium phosphide injecting-contact devices shows predicted power levels at 94 GHz of 150 mW (Fank, 1983).

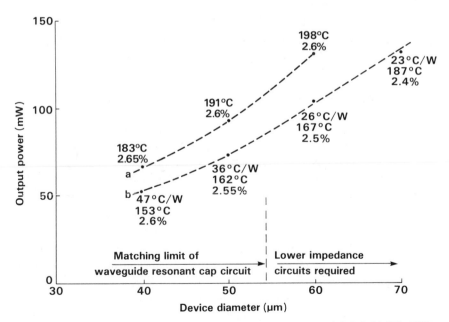

FIG. 16   Predicted output power capabilities of a diamond heat sink InP 94-GHz TED. Bias voltage is 7.5 V for curve a and 6.0 V for curve b. The ambient temperature is assumed to be 50°C.

There is little doubt that the use of type IIA diamond heat sinks would give considerable benefits to millimeter-wave TEO devices. However, the complexities involved in bonding III–V semiconductors directly to diamond could be severe. If moderate improvements of output power are required to 50–70-mW levels, then gallium arsenide would be an attractive solution, particularly because it is less prone to damage from mechanical stress than indium phosphide. Conversely, if 100-mW power levels are a real need, then indium phosphide offers the only solution; such power could be generated at reasonable operating temperatures using topologies of existing fundamental frequency circuits. Although these predictions have been made purely in terms of the output power benefits of lower operating temperatures, there are of course reliability benefits in reducing operating temperatures. Figure 17 presents the results of extensive endurance testing carried out on gallium arsenide TEDs; the data show that a mean time to failure (MTTF) of better than 5×10^4 h is possible at active-layer operating temperatures below 280°C. A point to note from these results is the superior reliability of the TEDs when compared with the silicon IMPATT. Preliminary endurance testing on indium phosphide has revealed no reliability problems, and MTTFs of better than 2×10^4 h are predicted at normal operating temperatures (240–260°C).

The purpose of this section is to illustrate the complexities and demands placed on the device manufacturer in designing a millimeter-wave transferred-

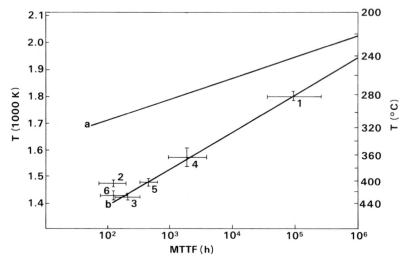

FIG. 17   Reliability of two-terminal devices (developed at Plessey Research, Ltd.): curve a, TiPdAu–GaAs IMPATT, $\theta = 2.5$ eV; curve b, AuGeIn–GaAs TEO (IHS cw K–M band).

electron device. The thermal limitations of the TED have to be emphasized, and the possible advantages of type IIA diamond heat sinks are an important pointer to future development. During the design process the crucial effect of device operating temperatures on the expected TED reliability is highlighted as a vital consideration.

E. DEVICE MANUFACTURING TECHNOLOGY

Efficient operation at millimeter-wave frequencies demands device structures capable of removing heat efficiently while minimizing parasitic losses. The fabrication of these diodes therefore requires optimization of the device geometry and metal contact technology in striving for good thermal design and device reliability.

The most common fabrication process for the TED at these frequencies is the integral heat sink (IHS) process, represented schematically in Fig. 18. In this process a high-quality contact metallization is first put down onto the epitaxial material surface. The type of metal system employed varies from manufacturer to manufacturer and also depends on the TED material profile. For the n^+-n-n^+ profile, ohmic contacts with bases of evaporated InGeAu or GeAuNi are sintered to give the required low contact resistivity. In the case of the $n-n^+$ profile, this epiface metallization stage determines the cathode contact properties, and great care is taken in the sintering of the metal system (InGeAu or GeAuNi) to tailor the behavior of the contact injection ratio. Following epiface metallization, a gold heat sink is electroplated onto this layer to an overall depth of 50 to 60 μm. The gold-plated

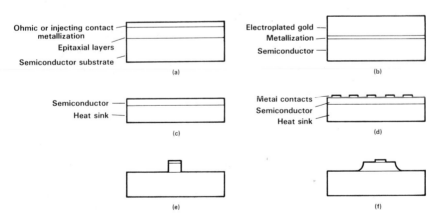

FIG. 18 Simplified schematic description of TED processing techniques: (a), epiface metallization; (b), heat sink fabrication; (c), semiconductor thinning; (d), substrate contact definition; mesa formation showing (e) photoetched structure and (f) chemical etched structure.

supported slice is now strong enough to permit removal of most of the substrate material.

The skin depth in the substrate acts as a constraint on the allowed device dimension if skin-effect losses and nonuniform current injection are not to degrade device performance. Therefore, the material thickness has to be closely controlled. This is achieved by employing chemical polishing or bubble etching techniques to thin the semiconductor slice from the back surface. Following this stage the substrate contact is defined using conventional metallizing and photoengraving procedures. The individual mesa diodes are then isolated. At lower frequencies, where the device diameter is relatively large, simple chemical etching techniques can be employed because the skirt produced by this method presents no significant problem with respect to area control. However, at higher frequencies, where devices are getting very small, photoetching techniques are used to produce vertical-sided devices whose area can be controlled more carefully than is possible with conventional chemical etching. The vertical nature of the etched surface can also be used to improve the thermal properties and reduce skin-effect losses. This can be done by increasing the periphery of the device by producing ring and cross structures.

The integral heat sink is now cut into cubes, each supporting one diode chip. These are then bonded individually into conventional copper-based microwave packages. The quality of the bond is important for high-power operation and techniques such as thermocompression, alloy soldering or ultrasonic bonding can be used successfully. The encapsulated diodes, whether for amplifiers or for oscillators, are thus ready for rf assessment.

F.   OSCILLATOR PRIMARY CHARACTERISTICS

The internal physics of the transferred-electron effect, practical device design, and fabrication technologies have been discussed in detail. We can now compare the behavior of the real device with the foregoing theoretical approach. Two important material growth parameters that must be established before optimum oscillator performance can be achieved are the dependence of optimum operating frequency on the active-layer length l and the correspondence of dc–rf conversion efficiency with the active-layer product $n \times l$. Figures 19 and 20 present these characteristics for both gallium arsenide and indium phosphide ohmic contacted n^+-n-n^+ devices in the millimeter-wave frequency range. The measured frequency-versus-length behavior shows that, as theoretically predicted, the higher effective electron velocity in indium phosphide leads to longer active-layer lengths for a given fundamental operating frequency. This information and the efficiency versus "$n \times l$ product" behavior can be used to design devices for TED operation from 30 to 140 GHz.

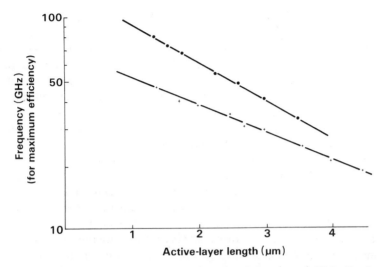

FIG. 19   Measured frequency against active-layer length for n^+–n–n^+ TEOs (developed at Plessey Research, Ltd.): ●, InP fundamental frequency; +, GaAs fundamental frequency.

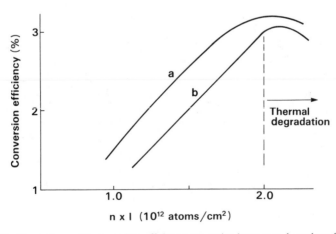

FIG. 20   Dependence of conversion efficiency on active-layer $n \times l$ product for fundamental frequency n^+–n–n^+ InP (curve a, 75 GHz) and GaAs (curve b, 40 GHz) TEDs. The active-layer temperature is 250°C.

An investigation into the operating modes of TEOs throughout this frequency range has established that fundamental frequency operation is possible up to at least 120 GHz from indium phosphide but only up to 60–70 GHz from gallium arsenide (Eddison and Brookbanks, 1982). At frequencies higher than these, both materials operate in a harmonic extraction–enhancement mode, with gallium arsenide ceasing to generate significant power above 110 GHz, whereas indium phosphide is already capable of 10-mW power levels at 120 GHz and 1 mW at 140 GHz. Thus the theoretically predicted higher operating frequency potential of indium phosphide is now practically verified.

The output power and efficiency performances of the n^+–n–n^+ indium phosphide devices have been encouraging. Operating in a fundamental frequency mode, these devices have produced the high power and efficiency results presented in Table III.

A comparison of performances of current cw devices of both indium phosphide and gallium arsenide n^+–n–n^+ TEOs is also presented in Fig. 21. These results show that the theoretically expected advantages of indium phosphide in terms of output power and efficiency are borne out by the performance of the real device. Additionally, as material optimization is carried out for operation in the 90–140-GHz range, it is expected that the output power and efficiency capabilities predicted in Fig. 21 will be approached.

The work described herein on n^+–n–n^+ cw devices reveals a moderate improvement in power and efficiency from indium phosphide over gallium arsenide. Similarly, the n–n^+ current-limiting cathode-contacted device

TABLE III

rf PERFORMANCE OF MILLIMETER-WAVE InP
TRANSFERRED-ELECTRON OSCILLATORS[a]

Highest powers			Highest efficiencies		
Frequency (GHz)	Power (mW)	Efficiency (%)	Frequency (GHz)	Power (mW)	Efficiency (%)
50	210	4	57	140	4.75
70	160	3	75	70	3.25
—	—	—	80	60	3.25
80	120	3	94	40	2.2
90	50	2	98	30	2.0
—	—	—	106	19	1.3

[a] From Eddison and Davies (1982a).

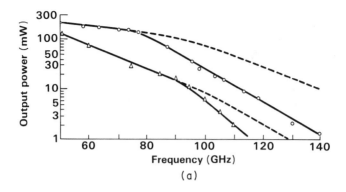

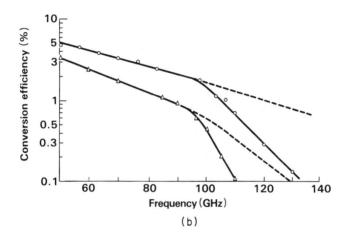

FIG. 21   (a) Output power and (b) conversion efficiency capabilities of n^+-n-n^+ TEOs: ——, measured values; — —, predicted values; $\triangle$, GaAs; $\bigcirc$, InP. (From Eddison and Davies, 1982b. © 1982 IEEE.)

exhibits the impressive cw power and efficiency potential of indium phosphide, as shown in Fig. 22 and Table IV.

From these practical results it appears that the current-limiting contact on $n-n^+$ has, as previously suggested, overcome the dead-space losses inherent to the n^+-n-n^+ structure. Unfortunately, there is some temperature instability associated with the Schottky-barrierlike cathode contact that can lead to practical system limitations.

Pulsed operation of two-layer indium phosphide produces some outstanding peak powers with a best result to date at 60 GHz of conversion efficiencies as high as 12% for peak power levels of 1 W (Eddison et al., 1981).

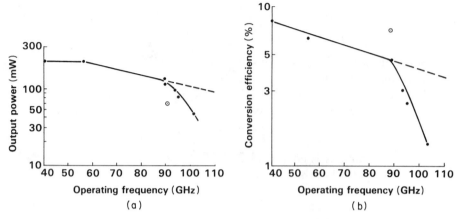

FIG. 22   (a) Output power and (b) conversion efficiency capabilities of n–n^+ InP TEOs: ●, Varian results; ○, Plessey results. (Courtesy Varian and Plessey.)

At higher frequencies than these the pulsed operation of smaller devices can give peak powers of 240 mW at 90 GHz with an efficiency of 4.3 % (Varian) or, conversely, efficiencies of 9 % at 80 GHz for a 70-mW peak power (Plessey). Although still at an early stage of development, the pulsed operation of indium phosphide n–n^+ structures suggests useful potential for oscillator systems.

TABLE IV

rf PERFORMANCE OF FIXED-FREQUENCY
cw InP n–n^+ GUNN OSCILLATORS

Device manufacturer	Frequency (GHz)	Power (mW)	Efficiency (%)
Varian	89.5	125	3.27
Varian	89.6	107	3.48
Varian	93.1	91	2.96
Varian	90.1	100	2.80
Varian	93.15	79	2.81
Varian	94.48	71	2.48
Varian	94.8	68	2.50
Varian	94.9	63	2.40
Varian	100.5	44	1.52
Varian	89.7	35	4.70
Plessey	90.0	50	7.20

G. Oscillator Secondary Characteristics

1. FM Noise Performance

Some of the results quoted in the previous section represent the highest powers and efficiencies ever realized from the transferred-electron device at millimeter wavelengths. However, in themselves they are not sufficient to ensure the successful application of these devices, because often the second-order parameters of the oscillator are the system designer's prime concern. The TED is primarily a local-oscillator device and for this reason its most important parameter is final oscillator noise. This is particularly true of the fm noise output, which in many cases is the critical factor in overall system performance.

A detailed discussion of FM-noise measurement techniques is not given here because this topic is covered in Chapter 7 as phase-noise measurement. The results of the oscillator FM-noise measurements are presented in this chapter as noise-to-carrier ratios N/C versus carrier offset modulation frequency f_m. The frequency deviation (Δf) can be calculated as follows:

$$N/C = 10 \log_{10}(f_m/\sqrt{2}\,\Delta f), \tag{3}$$

where single-sideband and 1-Hz bandwidths are assumed.

Before the FM noise performance of gallium arsenide and indium phosphide TEOs is discussed, it is pertinent to consider the operating modes of these devices. In the millimeter-wave frequency range, gallium arsenide devices operate in a harmonic extraction–enhancement mode that results in apparently high Q_e oscillator load-pulling behavior. Thus it is difficult to determine the Q of the gallium arsenide device's local fundamental frequency environment. Conversely, the indium phosphide devices operate in the more efficient fundamental frequency mode in which the determination of the oscillator Q_e is relatively straightforward.

The interpretation of noise measurement is made difficult if there are uncertainties in oscillator Q. This problem is clearly illustrated when the concept of noise measure M is used to relate the behavior of different device–circuit combinations. Noise measure is a "figure of merit," according to Kurokawa (1968), that attempts to remove circuit-related effects from a given measurement and thereby to permit comparison of device noise on an equitable basis. The values of M in decibels are simply computed from the measured noise frequency deviation Δf as follows:

$$M = 10 \log_{10}[(P_0/kTB)(Q_L \, \Delta f/f_0)^2], \tag{4}$$

where P_0 is the oscillator output power in watts, k the Boltzmann constant, T the device operating temperature in kelvins, B the measurement bandwidth in hertz, Q_L the oscillator-loaded Q, Δf the FM noise frequency deviation in hertz, and f_0 the oscillator frequency in hertz.

From this relationship the oscillator's loaded Q value Q_L is clearly an important factor. Although Q_L cannot be readily measured, the oscillator external Q (Q_e) can be determined using standard load-pulling or injection-locking techniques. Under normal oscillator-matching conditions, at which the device is producing close to its optimum output power, the two Q values are approximately equal (i.e., $Q_e = Q_L$). Usually, therefore, M is defined in terms of Q_e and the noise measure is used as a valuable tool for comparing different active devices. This is not the case for harmonic operation, in which the uncertainties in Q_e and its unknown relationship to Q_L can lead to large errors in determining M. Above 70 GHz the comparison of indium phosphide and gallium arsenide noise behavior is made complex by the difference in operating modes.

In the lower millimeter-wave region these difficulties do not arise and M can be used with confidence. Figure 23 shows the noise measure behaviors against carrier offset frequency (f_m) of indium phosphide and gallium arsenide $n^+–n–n^+$ TEOs. At close-to-carrier frequencies both devices have similar strong flicker noise ($\propto 1/f_m$) components. As higher offset frequencies are approached a thermal noise region is reached (for GaAs, $f_m > 500$ KHz; for InP, $f_m > 1$ MHz), and here indium phosphide emerges as the quieter device. The lower thermal noise of indium phosphide agrees with the lower diffusion coefficient of electrons in this material compared with gallium arsenide.

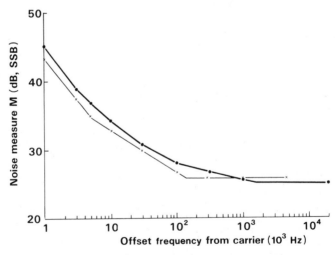

FIG. 23   Comparison of noise measure M for GaAs ($\times$) and InP ($\bullet$) K-band TEOs fundamental modes. $F_0 = 35 \pm 1$ GHz. (Developed at Plessey Research, Ltd.)

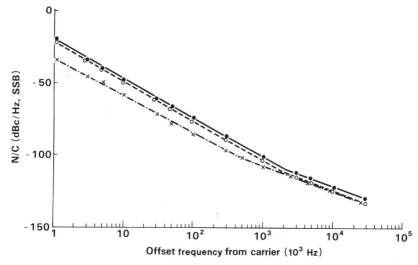

FIG. 24 FM noise performance of InP and GaAs M-band TEOs (developed at Plessey Research, Ltd.): ●, InP, 60 mW, 2.5% Q_e = 125; ○, GaAs, 29 mW, 1% Q_e = 690; ×, GaAs, 12 mW, 0.2% Q_e = 1390. (From Eddison and Davies, 1982a. © 1982 IERE.)

Noise measurements at 80 and 94 GHz using delay-line discriminator test equipment (Simmons and Smith, 1982) indicate that indium phosphide and gallium arsenide devices can exhibit similar FM noise outputs. This fact is seen in the two upper curves of Fig. 24, which reveal almost identical noise-to-carrier ratios for the two materials despite the fact that the indium phosphide device generates more power at twice the efficiency of the apparently higher-Q gallium arsenide device.

Improved noise behavior can be realized at the expense of lower power and efficiency (e.g., the lower curve in Fig. 24), but this would be a severe drawback in some systems. Although the figures quoted here are for the three-layer (n^+-n-n^+) device structure, work on $n-n^+$ devices reveals similar noise results.

In summary, the measurements taken to date show little difference between gallium arsenide and indium phosphide. This fact is readily seen at frequencies below 60 GHz, where the noise measure concept can be used with confidence. At higher frequencies the comparison is more difficult, but the evidence suggests that indium phosphide already offers a competitive noise performance allied to an improved power and efficiency capability.

2. AM Noise Performance

The very low AM noise outputs of today's solid-state oscillator devices places severe demands on the noise measurement equipment in terms of its

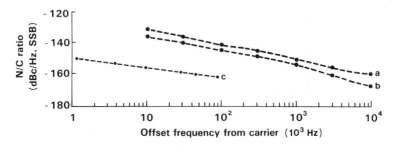

FIG. 25   AM noise performance of InP and GaAs K- and M-band TEOs (developed at Plessey Research, Ltd.): curve a, 25-mW GaAs TED; curve b, 45-mW InP TED; curve c, 150-mW GaAs and InP TEDs, 40 GHz.

sensitivity, dynamic range, and detection threshold. Even at the lower microwave frequencies, quite sophisticated direct detection techniques have to be employed (Ondria, 1968) to achieve meaningful oscillator measurements. This situation is made more difficult at millimeter wavelengths, and there is a resultant lack of published results for AM noise at frequencies above 30 GHz. Those measurements that have been carried out show oscillator AM noise levels close to the system sensitivity, and therefore the following results have to be regarded as measures of the worst possible oscillator noise output.

At L-band frequencies (40–60 GHz) AM noise measurements can be attempted using a sophisticated carrier-suppression technique that provides good system sensitivity (Dean, 1978). Characterization of indium phosphide and gallium arsenide oscillators with this equipment gives identical AM noise outputs at very low levels from both materials (Fig. 25). Similarly, at M band (60–100 GHz) the measured AM noise levels are low, but here indium phosphide appears to be marginally quieter than gallium arsenide.

The very low noise levels displayed in Fig. 25 together with the extreme difficulty in detecting the oscillator noise sidebands prove that AM noise is unlikely to be a problem for any TEO devices. These devices can therefore be regarded as near-ideal local oscillator units, particularly when used in balanced mixer configurations. Under these conditions the effects of the TED noise output is reduced still further by the AM rejection properties of the balanced mixers.

3.   Voltage and Temperature Stability

An understanding of oscillator voltage and temperature stability can be gained by considering the behavior of practical TEOs. A typical gallium arsenide device's variations in output power and frequency with bias voltage (dP/dV and dF/dV) are displayed in Fig. 26. As the bias voltage (and thus

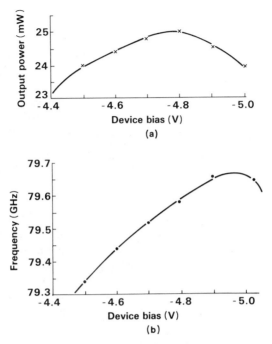

FIG. 26   Voltage behavior of a typical n^+–n–n^+ GaAs millimeter-wave TEO: (a), output power; (b), frequency.

the input power) is raised, the oscillator's output power increases until a thermally limited maximum rf output power is reached. Normally, the frequency–voltage law also has a maximum closely associated with the power peak. In contrast the three-layer indium phosphide device exhibits a more linear behavior as typified by the practical oscillator results shown in Fig. 27. Unlike gallium arsenide, this device's output power increases mono-tonically with bias voltage, whereas its oscillation frequency reduces linearly with bias voltage. These features are a consequence of indium phosphide's more temperature-stable velocity–field characteristics, which results in the device output power and conversion efficiency being relatively insensitive to active-layer temperature and hence to input power. This more stable velocity–field characteristic is seen in the device performance with tempera-ture, which shows power and frequency variations (dP/dT and dF/dT) of −0.013 dB/°C and −8 MHz/°C, respectively. These figures compare with conventional M-band gallium arsenide results of −0.03 dB/°C and from −6 to −8 MHz/°C.

The voltage and temperature characteristics of both n^+–n–n^+ TED materials detailed here are compatible with existing system requirements.

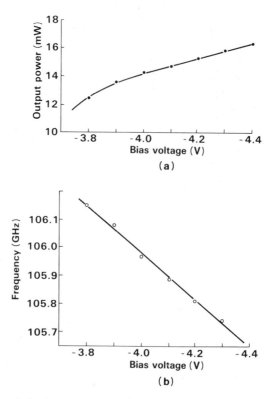

FIG. 27   Voltage behavior of a typical n^+–n–n^+ InP millimeter-wave TEO: (a), output power, 1.6 dB/V; (b), frequency, −900 MHz/V. (From Eddison and Davies, 1982b. © 1982 IEEE.)

The different voltage dependencies seen from indium phosphide and gallium arsenide point to different applications for these materials. Because there are no difficulties associated with frequency and power maxima, linear electronic bias voltage control can be implemented simply with indium phosphide TEOs. For example, frequency temperature compensation to track a known transmitter temperature drift becomes possible using basic electronic bias voltage control techniques. On the other hand, selecting the correct bias voltage for a gallium arsenide device, usually at the power maximum, will provide improved immunity to power supply ripple of both the oscillator frequency and power output. This could be an important advantage if power supply upconversion noise is a system constraint.

Although three-layer indium phosphide devices show well-controlled voltage and temperature stability, the more efficient two-layer n–n^+ device is not quite so stable. As has been explained, the n–n^+ device relies on an injecting cathode contact to achieve high output power and conversion efficiencies. The cathode current-limiting contact is Schottky-like in behavior, and as such it exhibits a positive dependence of bias current on operating temperature. Figure 28 illustrates the type of terminal current voltage characteristics that this contact imposes on the complete two-layer device. In the cw mode the bias current does not display the classic "drop back" above threshold but rather increases with bias voltage. This effect is caused by diode self-heating, which lowers the cathode contact's effective barrier height and thereby reduces its current-limiting action. Similarly, any changes in ambient temperature conditions lead to variations in the device current–voltage characteristics. Thus the device voltage and temperature stabilities are a function of the cathode contact properties and not of the bulk semiconductor behavior.

For these reasons successful production of a stable cw two-layer indium phosphide device is not straightforward. During fabrication the cathode contact must be carefully tailored to exhibit the correct injection ratio. Too high an injection ratio gives poor output power and efficiency characteristics with a low current–temperature stability. Conversely, lowering the injection ratio can produce optimum power and efficiency but at the expense of a high positive current–temperature coefficient (dI/dT). Correct choice of the cathode current-limiting conditions can give good rf performance with acceptable voltage and temperature stability. However, it is unlikely

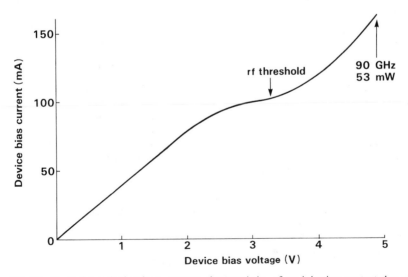

FIG. 28   Typical terminal voltage–current characteristics of an injecting contacted n–n^+ InP TEO.

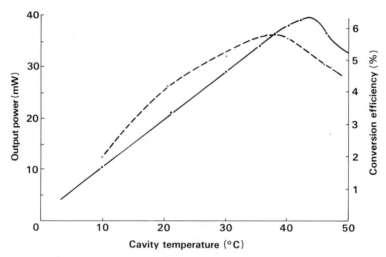

FIG. 29   Output power (——) and conversion efficiency (– –) dependence on temperature of low-injection-ratio n–n^+ InP TEO. $F = 90$ GHz, $V = -4.2$ V.

that the linear behavior of the three-layer device will be achieved from the n–n^+ device structure.

Because the cathode contact is such a dominating feature, it is not possible to quote typical stability figures of frequency and power with voltage for this device structure. Indeed, the parameter values for a given device can be dependent on operating temperature and circuit-matching conditions. Hence the power–temperature law of Fig. 29 can only be regarded as a qualitative representation.

A good deal of research is being performed to examine the question of hot-electron injection in indium phosphide. Topics presently under investigation including notch and spike material profiles as well as alternative contacting techniques. Undoubtedly, as this work progresses more detailed information on the stability parameters of the high-efficiency indium phosphide TEO will emerge.

H.   OSCILLATOR APPLICATIONS

A great deal of attention has been paid to the basic operating principles and characteristics of the transferred-electron oscillator. It is now possible to illustrate how circuit designers have used the TEO to satisfy a wide range of system requirements. In general terms all the TEO devices are rugged, reliable elements that tolerate a wide range of circuit-matching conditions. They are also relatively inexpensive and only simple dc power supplies are necessary for successful operation. The TEO has thus become a popular option for many millimeter-wave system applications, and examples are given here to illustrate the versatility of the device.

One of the disadvantages of the TEO is that even with indium phosphide it is inherently inefficient, and single devices can only be used for low-power applications. Conversely, the devices exhibit high available negative resistance, and consequently circuit matching is straightforward. Hence combining more than one TED has become an attractive way to obtain low-noise, high-power signal sources. An example of the potential of this approach is a waveguide circuit in which eight gallium arsenide devices are combined to produce a 1 W output at 45 GHz (Young El-Ma and Cheng Sun, 1980). A two-diode waveguide wafer module is the basic building block and four of these units can be cascaded with a claimed combining efficiency of 90%.

A waveguide modular approach can also be utilized with n–n^+ indium phosphide devices at 90 GHz. Combining up to four such devices can produce cw power levels of 0.25 W with conversion efficiencies greater than 1% (Sowers et al., 1982). Table V presents a summary of this work, and the photograph in Fig. 30 emphasizes how straightforward power-combining indium phosphide devices can be.

Frequency agility is increasingly becoming a requirement for oscillator systems, and mechanical and electronic tuning-range parameters are therefore of great interest. Here the high rf impedances and wide negative resistance bandwidths of all the transferred-electron devices are real advantages. Mechanical tuning bandwidths as high as 20% can be achieved from both indium phosphide and gallium arsenide devices using reduced-height waveguide circuits at frequencies up to 60 GHz. These wide bandwidths point to successful test equipment uses, but for many systems electronic frequency control is a greater need. Varactor tuning of millimeter-wave oscillators is a recent area of device and circuit research. Conventional waveguide circuits

TABLE V

rf Performance of Two- and Four-Diode Combining Circuits[a]

Diode	F (GHz)	P_0 (mW)	η (%)	η_{comb} (%)
EE198-397, EE198-401	89.55	170	2.89	93
EE198-397, EE198-401	90.25	150	2.7	82
EE198-207, EE198-213	91.3	97	1.6	106
EE271B-8, EE271B-26, EE172B-31, EE271B-24	98.6	260	1.6	92.85
EE268-22, EE268-26, EE268-6, EE268-15	90.8	230	1.4	106.5

[a] Courtesy Varian.

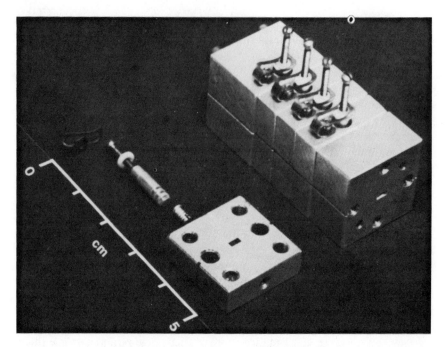

FIG. 30   A 90-GHz InP TEO power combiner. (From Sowers *et al.*, 1982. © 1982 IEEE.)

such as the one shown schematically in Fig. 31 are commonly used for narrow-band tuning applications below 70 GHz. The varactor diode is post-mounted approximately $\lambda g/2$ behind and in parallel with the TED. This approach can realize up to 5 % bandwidths at 62 GHz with output powers of 32 mW and $2\frac{1}{2}$ % bandwidths at 60 GHz with output powers as high as 75 mW from indium phosphide.

Circuit analysis reveals that the ultimate tuning range of the parallel configuration is limited by the "parasitics" associated with the TED and varactor mounting structures. These limitations are not exhibited by the series configuration shown in Fig. 32, which has realized 6 % bandwidth at 42 GHz with output powers of 60 mW from gallium arsenide devices (Ondria, 1981). This impressive performance shows the potential of the series connection technique and there is little doubt that this approach will be seen in future systems.

In some applications there is a need for signal sources capable of better long-term stability and noise performance than is possible from a free-running oscillator. Such requirements can be met by phase locking the oscillator signal to a high-stability, low-frequency oscillator (e.g., a crystal-controlled reference). Although the circuit realization of phase-locked oscillators is complex, excellent noise performance gains such as those of

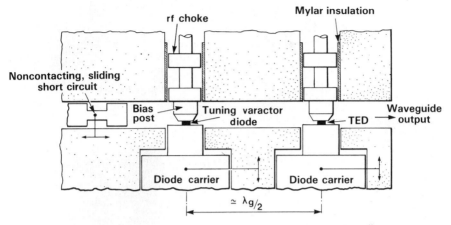

FIG. 31 Schematic diagram of a parallel-tuned waveguide oscillator circuit.

Fig. 33 can be achieved. The system illustrated utilizes gallium arsenide devices; however, indium phosphide could also be used. Indeed, the more linear bias voltage behavior of indium phosphide may ease the realization.

The linear bias voltage behavior of indium phosphide offers an alternative approach for some applications. For instance, frequency–temperature compensation becomes possible using simple electronic bias voltage control techniques. Just such an approach is used for the development of an indium phosphide millimeter-wave high-stability oscillator. Shown in Fig. 34 is the

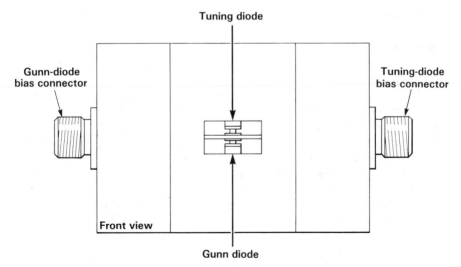

FIG. 32 Schematic diagram of a series-tuned waveguide oscillator circuit. (Courtesy Alpha Ind.)

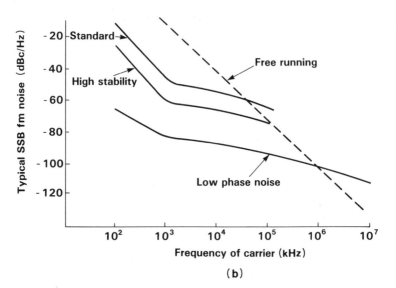

FIG. 33   (a) A phase-locked GaAs millimeter-wave source and (b) its FM noise performance capabilities. (Courtesy Hughes Aircraft.)

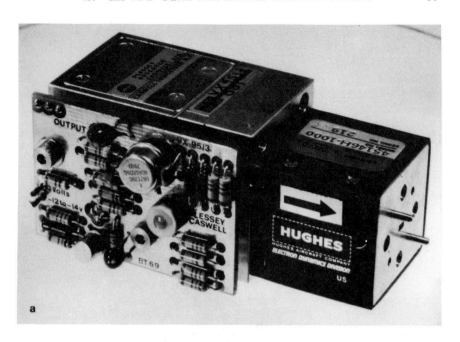

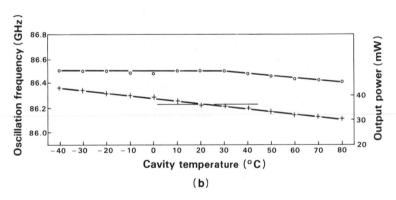

FIG. 34   (a) Temperature-compensated, cavity-stabilized InP oscillators and (b) their compensated frequency and power performance. The bias voltage correction is -14.58 mV/°C; $dP/dT = -0.013$ dB/°C; $\bigcirc$, frequency variation, -100 MHz; $+$, power variation, -1.6 dB; $dF/dT = -833$ kHz/°C; $dp/dT = -0.013$ dB/°C. (Courtesy Hughes Aircraft.)

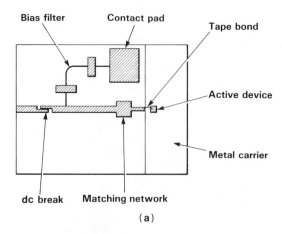

FIG. 35   (a) Circuit layout and (b) a completed InP microstrip oscillator circuit. (From Smith and Simmons, 1983. © 1983 IEE.) The circuit length is 6 mm and the circuit width is 5 mm.

compensated frequency and power performance of the oscillator over a temperature range of -40 to $80°C$. Here a bias voltage temperature correction of -14.58 mV/°C results in an overall output power variation of -0.013 dB/°C and a frequency variation of -830 kHz/°C for a $120°C$ temperature range. This characteristic is clearly superior to any currently available gallium arsenide M-band TEO performance.

All of the oscillator applications use metal waveguide circuit structures. These components are inherently large, heavy, and expensive to manufacture. They are also vulnerable to harsh environmental conditions, particularly shock and vibration. There is a need for minature, rugged oscillators that are inexpensive and easy to manufacture. Also, such devices must be capable of integration into the architectures of complete millimeter systems. Bearing in mind the number of system components now fabricated in the microstrip medium, there is a great potential for a low-noise stable oscillator in microstrip. Just such an oscillator can be demonstrated using an indium phosphide TEO (Smith and Simmons, 1983). This microstrip oscillator produces a 40-mW output power with a 1.4% conversion efficiency at 81 GHz. A schematic diagram of the microstrip circuit together with a photograph of the unit are displayed in Fig. 35 to illustrate its size in relation to a conventional waveguide structure.

Preliminary noise measurements on the microstrip oscillator reveal only a small FM noise degradation compared with devices from the same layer operated in waveguide. It is therefore concluded that the fundamental-mode indium phosphide TEO will provide the basis of fully integrated microstrip millimeter-wave systems. Conversely, the gallium arsenide TEO is not an easy device to integrate above 70 GHz owing to its harmonic operating mode, which demands careful circuit matching at both the fundamental and harmonic frequencies.

III. Amplifier Devices

A. INTRODUCTION TO REFLECTION AMPLIFIER OPERATION

Transferred-electron amplifier (TEAs) can be used to provide rf power amplification at any frequency where TE oscillators are employed. A typical layout for such an amplifier is illustrated in Fig. 36. This reflection amplifier includes a circulator to couple the input signal into the TED device and to separate the input from the amplified output signal. The purpose of the broadband matching network is to control the amplifier's frequency response and to help stabilize the active device. To avoid spurious oscillation the amplifier circuit must present the TE device's negative resistance with a higher positive matching resistance. Hence the circuit matching for an

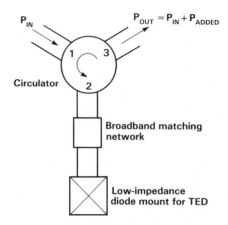

FIG. 36   Schematic diagram of a circulator-coupled reflection amplifier.

amplifier is a more difficult task than is generally the case for an oscillator. In spite of the circuit complexities, the TEA offers advantages in terms of dynamic range, output power ruggedness, and resistance to spike burnout.

Although good TEA performances are attainable below K band (i.e., < 20 GHz), little interest is now shown in the TEA at these frequencies because the modern gallium-arsenide metal–semiconductor field-effect transistor (MESFET) and silicon bipolar transistors possess far superior noise performance at similar levels of output power. Submicron-gate-length FET devices are now becoming available at frequencies in the 20–35 GHz range, but they cannot yet provide power levels of more than a few milliwatts. Therefore in the millimeter-wavelength frequency range the TEA has emerged as the only solid-state solution for power amplification.

Although the TEA cannot provide the multiwatt output powers of some vacuum tube devices, it can produce medium power levels of several hundred milliwatts. This, coupled with its lower noise performance, better reliability, and less complex power-supply needs, makes the TEA an ideal transmitter-driver amplifier. For low-noise receiver applications the TEA does not exhibit the noise figure performances of the low-noise mixer system, but its power handling abilities make it a solution to many system requirements.

The question of choice of material for TEAs at millimeter wavelengths has a strong parallel with the choice of materials for oscillator applications. The active TEA devices are usually very similar to the TEO device structures in terms of material profile and subsequent fabrication technologies. Hence, the electron-transfer considerations of Section II.A are also relevant to the amplifier device operating principles. Thus it can be shown that for the higher output powers, efficiencies, gains, and operating frequencies, indium phosphide is a better choice of TEA material than gallium arsenide. This choice is further supported when the critical noise figure–noise measure parameters are considered, as detailed in the following section.

B.   Amplifier Noise Performance

As for the case of TEO devices, it is convenient to characterize two-terminal reflection amplifiers by their intrinsic noise measure M. The noise measure is related to the important noise figure F by

$$M = (F - 1)(1 - 1/G), \tag{5}$$

where G is the amplifier power gain. Once again, the measure M has the advantage that it is invariant if the device is embedded in a lossless circuit.

The choice of the semiconductor material for low-noise amplifier (LNA) applications is largely dependent on the material's noise measure properties. Now it is known that the transferred-electron device's thermal noise output is strongly related to the diffusion coefficient–mobility ratio $[D(E)/\mu(E)]$ of the semiconductor material. Indeed, the work of Thim (1971) shows that under uniform field (E_0) conditions the TEA device noise measure tends to the asymptotic limit for low $n \times l$ values of

$$M \to [qD(E_0)/kT \, | \mu\varepsilon_0 \, |]. \tag{6}$$

From these basic material parameters it can readily be established that indium phosphide should exhibit lower thermal noise with an amplifier noise measure approximately 6 dB lower than the equivalent gallium arsenide device. Hence indium phosphide emerges as the optimum material for reflection amplifier uses.

Having selected the material, the conditions for low-noise-figure operation must be determined. Amplifier characterization reveals that the noise measure of a TEA is a strong function of $n \times l$ product; this can be clearly seen in Fig. 37. Here the noise measure dependence of indium phosphide n^+–n–n^+ devices is displayed over a wide range of $n \times l$ products. An interesting feature of this investigation is that there is no noise degradation over an operating frequency range of 10 to 40 GHz. This shows the excellent high-frequency behavior of indium phosphide, confirmed by some of the noise figures realized at 60 and 94 GHz.

Material profile has also emerged as a crucial determinant of TEA noise measure. Figure 38 shows three possible material profiles for amplifier devices. The first of these is the n^+–n–n^+ flat profile, which has the noise figure behavior detailed previously. The other two profiles are the n^--notch and p-notch structures, which exhibit superior noise performances. Computer simulations of the TEA reveal that this improvement in noise measure is due to the flattening of the field profile within the device active layer. In

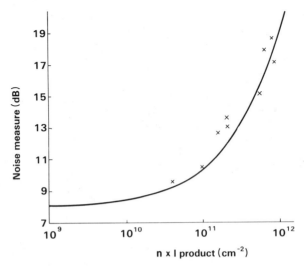

FIG. 37   The dependence of amplifier noise measure on device $n \times l$ product: ——, J-band results; $\times$, K-band results.

fact, optimum noise measure occurs under uniform field conditions where the field value is approximately twice the threshold field (Sitch and Robson, 1976). A comparison of the computed field profiles for a notch device with that of a flat profile device (Fig. 39) shows large differences in electric field behavior for the two cases. By careful choice of the notch doping level and length, a uniform field and hence an optimum low-noise device can be produced.

C.   DEVICE DESIGN AND MANUFACTURE

Unlike the TEO the design of an amplifier device is not generally dominated by the thermal limitations discussed earlier (Section II.D). Rather, amplifier design involves a trade-off between the overall noise figure requirements and the need to achieve a given gain and added power performance. These conflicting needs generally necessitate the use of a multistage amplifier solution with each stage containing a specially tailored design. Essentially, the active device's $n \times l$ product is selected for each stage, bearing in mind the relationships between $n \times l$ and several important amplifier parameters.

It is known that devices with low carrier concentration (subcritically doped) exhibit a low noise measure. However, these devices, which are not capable of generating spontaneous rf output power in the absence of an input signal, have a low power-added capability and a poor gain–band-width product ($G^{1/2}B$). Thus the subcritically doped device is a sensible

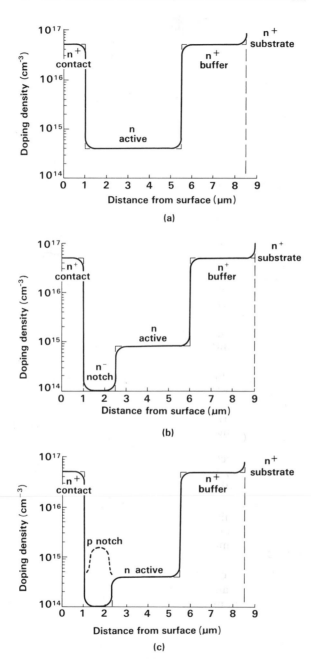

FIG. 38 Schematic diagrams of K-band TEA device material profiles: (a), flat layer profile; (b), cathode n-notch profile; (c), cathode p-notch profiles.

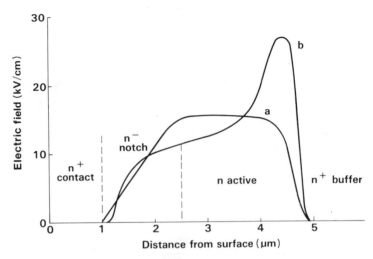

FIG. 39   Calculated electric field profiles for flat layer and notch layer TEA devices. Curve a, 1.5 μm, 0.5 × 10^{15} notch; 3.0 μm, 1 × 10^{15} active. Curve b, 4.5 μm, 1 × 10^{15} active.

input-stage amplifier element. Conversely, the devices with higher carrier concentration (supercritically doped) are ideally suited to higher power amplification where theoretically 100% of the device's available power as an oscillator can be added in the amplification mode (Hines, 1970). In practice it is difficult to achieve greater than 70% addition with any degree of stability and freedom from spurious oscillations. The supercritically doped device possesses noise measures approaching the thermal region values of the equivalent oscillator structure.

To design the individual amplifier devices these arguments have to be quantified with some accuracy. This can only be done by using a computer simulation technique to model the TEA structure (Sitch and Robson, 1976). The results of such a simulation are shown in Fig. 40 and 41, carried out for a 5-μm-thick indium phosphide device with doping densities of 1, 3, and 9 × 10^{14}. The noise measure is seen to decrease as the doping density (i.e., $n \times l$) is decreased, but it has the asymptotic behavior toward a minimum noise measure seen in the experimental results. These theoretical results also show the usable bandwidth increasing with the device $n \times l$ product. Calculations of the available negative chip resistance as a function of $n \times l$ product illustrate the trade-offs that have to be made in the TEA design. Although the low $n \times l$ device has good noise performance, it has a reduced operating bandwidth and low negative resistance values, which give the device a reduced gain bandwidth product. The low negative resistance also demands that the device and circuit parasitic resistances must be minimized.

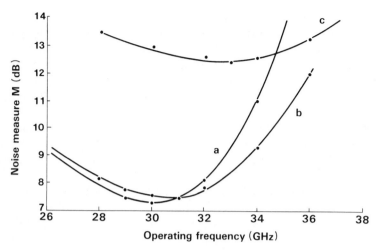

FIG. 40   Computer simulation results showing the effect of the $n \times l$ product on noise measure for a K-band n^+-n-n^+ TEA. The $n \times l$ product is 5×10^{10} for curve a, 15×10^{10} for curve b, and 45×10^{10} for curve c; $l = 5 \ \mu m$.

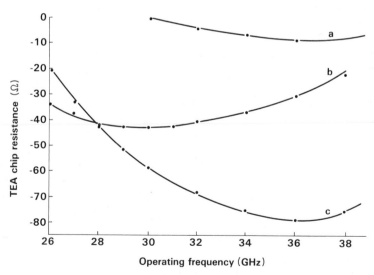

FIG. 41   Computer simulation results showing the effect of the $n \times l$ product on available negative resistance for a K-band TEA. The $n \times l$ product is 5×10^{10} for curve a, 15×10^{10} for curve b, and 45×10^{10} for curve c.

Conversely, the higher $n \times l$ devices have wide operating bandwidths and large negative resistance values at the expense of a degraded noise performance. These devices have better gain bandwidth products, but their high negative resistances can make it difficult to stabilize them against oscillation.

From the foregoing arguments it can be seen that computer simulations are an important tool in the TEA design process. The theoretical models are used in an iterative manner to select device $n \times l$ product and diameter to satisfy the amplifier noise figure and gain bandwidth specifications with a reasonable stability margin. In most of the amplifier stages the device input power density is low and the resultant active-layer operating temperatures are generally below 200°C. However, this is not necessarily the case for the high-power stages, where the $n \times l$ products and input power densities approach the values of an oscillator device. Therefore the same thermal design approach as detailed earlier for the oscillator (Section II.D) has to be used for the TEA power stages, with the recognition that circuit stabilization considerations will limit the power-added levels to approximately 70% of the power capability of the device as an oscillator. When thermal analysis is being carried out, account has to be taken of the ultimate amplifier reliability. Because the TEA device has the same structure and contacting technology as the oscillator device, the same reliability as a function of temperature can be assumed (e.g., Fig. 17). With this information, predictions can be made of the likely MTTF of the TEA, and it is clear that the low operating temperatures of the low-power devices are not likely to present any lifetime problems. The only lifetime determinant of any high-power device is the operating temperature. In the worst case, this device has the same operating temperatures and therefore the same reliability as an oscillator unit. However, for most medium-power amplifiers and all low-power amplifiers operating temperatures of the active device are lower than in any TEO with a correspondingly longer lifetime expectancy.

The TEA simulations and device design have been described in relation to the $n^+ - n^- n^+$ structure; however, the general approach and conclusions apply to the notch profile device. For these notch devices the computer simulations show the expected improvements in noise performance. Further, a given noise measure can be realized by a notch device with a higher $n \times l$ product than would be the case for a uniformly doped device. Thus higher gain bandwidth products can be achieved for a particular noise performance using notch devices. For the high power stages the current-limiting action of the notch cathode can also produce power and efficiency advantages with consequent thermal benefits. Therefore, the notch device offers many advantages to the amplifier designer. However, realization of the necessary material doping profiles can be a problem; in contrast, the $n^+ - n - n^+$ structure is relatively easy to grow.

After the amplifier devices have been designed, their fabrication closely follows that of the transferred-electron oscillator device, because the two classes of TED are very similar in form. The material profiles, contacting technologies, and geometries of the TEA and TEO devices are almost identical and therefore the processing details previously given (Section II.E) apply to the amplifier device.

In addition to the thermal and electrical problems associated with the active device design and fabrication, the TEA also demands complex circuit-matching techniques. Generally the amplifier designer's objective is a broad operating bandwidth with a tight gain-ripple specification that places stringent demands on the circulator and matching-circuit behavior. Unlike the oscillator circuit, which is usually a narrowband design problem, the amplifier circuit calls for wideband modeling and design procedures. In fact, not only does the circuit design have to ensure good in-band performance, but also the need to stabilize against out-of-band spurious oscillations dictates controlled circuit behavior well outside the operating frequency range of the amplifier. There are normally four distinct features of amplifier circuit design that must be addressed.

(1) The microwave packaging techniques used with the TEA must be such that there are no resonances presented at or near the desired frequencies of operation.

(2) A low-impedance matching network is required at the effective device terminals to present the correct circuit conductance. This is normally adjusted to give the required gain–bandwidth response and to prevent spontaneous oscillation.

(3) A broadband matching network is required to transform the effective circuit impedance to a level compatible with the circulator input impedance. The design of this section will affect the gain–bandwidth and gain–ripple responses.

(4) The ferrite circulator response must be broadband. In general there must be no highly nonuniform characteristics over a bandwidth typically twice that of the amplifier bandwidth. This factor usually limits the useful bandwidth of negative resistance reflection amplifiers employing Y-junction circulators.

D. PRACTICAL REFLECTION AMPLIFIER PERFORMANCE

1. *Primary Characteristics*

There has not yet been as high a level of interest in amplifier devices as in oscillators and therefore a comprehensive review of practical millimeter-wave amplifier capabilities cannot be given. However, from the computer

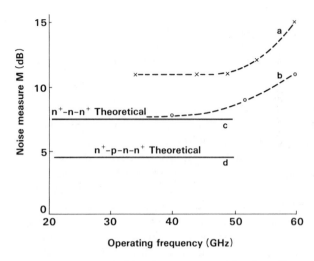

FIG. 42   Predicted and measured TEA noise measures as a function of frequency: curve a, n^+-n-n^+ measured (Varian and Plessey); curve b, $n^+-p-n-n^+$ measured (Varian); curve c, n^+-n-n^+ theoretical; curve d, $n^+-p-n-n^+$ theoretical.

simulations and the known oscillator device performances, sensible predictions can be made as to the noise figure and power-added behavior over the 30–100 GHz frequency range. Figure 42 shows the noise figure predictions made by Sitch and Robson (1976) for indium phosphide at frequencies up to 50 GHz. For an assumed 10-dB gain these simulations suggest 7.5 dB for the n^+-n-n^+ device and 5.5 dB for the notch device. When the practical results achieved to date at K and M band are plotted, it is clear that the real device noise figures are fairly constant over a wide frequency range albeit somewhat higher than the predicted level. Work is being carried out to improve these 10–15-dB noise measures by careful tailoring of the notch profile devices. The best result realized so far is a noise measure of 7.8 dB at 40 GHz (Fank and Crowley, 1982) from an n^--notch device, and confident predictions of 6-dB noise measures from a p-notch structure at 60 GHz are being made (Fank, 1983).

It is probably easier to estimate the power-added abilities of the indium phosphide TEA, because there are already many published oscillator results for both the ohmic contact n^+-n-n^+ structure and the injecting contact $n-n^+$ structure. Figure 43 presents these estimates with the best amplifier results so far realized. The medium powers of 100 to 200 mW presently seen are well below the ultimate performances of indium phosphide, but the other amplifier demands of bandwidth, stability, and gain flatness will always restrict the output powers below the state-of-the-art levels.

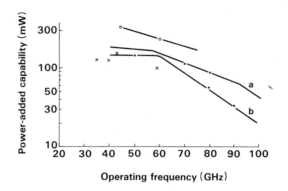

FIG. 43 Predicted and measured power-added capabilities of single InP TEA IHS devices: ○, target for n^+-p-n-n^+ devices; ×, results for practical n^+-n-n^+ devices; curve a, n-n^+; curve b, n^+-n-n^+.

2. Secondary Characteristics

Just as for TEO devices the applicability of an amplifier often depends on its secondary performance parameters. A given noise figure and output power level is of no great use unless the amplifier has the desired gain bandwidth product, gain flatness, linearity, and temperature stability. Many of these characteristics are dependent on the interactions between the device and the circuit, therefore the performance of specific amplifiers will be used to illustrate the general TEA behavior.

The first amplifier to be considered is a three-stage K-band unit designed as a transmitter tube drive amplifier, for which a photograph and schematic diagram are shown in Fig. 44. At these low millimeter frequencies the long active lengths of the indium phosphide devices give rise to thermal limitations, and therefore gallium arsenide devices have to be used for the high-power output stages. Hence at these frequencies there is a mixture of device types with indium phosphide in the low-noise, low-power stages and gallium arsenide in the later higher-noise, high-power stages. This unit exhibits a 24-dB gain over a 3-GHz bandwidth, with a typical noise figure of 14 dB and a saturated output power capability in excess of 120 mW. The active devices are mounted in waveguide cavities of very reduced height, and are matched to the full-height output port via a broadband multisection Tchebyschev transformer. Figure 45 shows a typical gain–frequency plot for this Plessey amplifier. The transfer characteristics for this amplifier (Fig. 46) display good linearity and well-behaved compression characteristics. Similar work on a five-stage unit gives 0.5 W cw power levels with small signal gains of 45 dB over a 6-GHz bandwidth in mid K band. The excellent power outputs

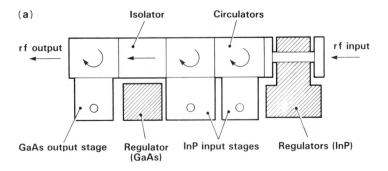

Fig. 44 (a) Schematic diagram of a three-stage K-band InP/GaAs TEA and (b) a typical unit developed at Plessey Research Ltd.

of this unit demonstrate that there is a genuine lower-noise, solid-state alternative to some traveling-wave tube (TWT) amplifiers.

At higher frequencies indium phosphide becomes the optimum choice for all the TEA stages, as illustrated by the performance of the 50–60-GHz unit shown schematically in Fig. 47. The three-stage unit uses flat profile devices to furnish the highest power in the fewest number of stages. Even though optimum low-noise devices are not utilized, the good noise performance of Fig. 47 can be realized. The gain of the complete amplifier is shown for both low- and high-level inputs in Fig. 48, and the amplifier is linear up to power

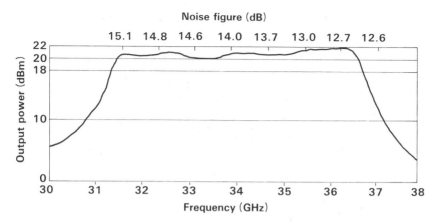

FIG. 45 Typical gain and noise figure performances for a Plessey three-stage *K*-band amplifier. [Courtesy Plessey Research (Caswell) Limited.]

levels 10 dB below the saturated output power, as illustrated in the transfer characteristics of Fig. 48. The TEA exhibits these stable performances over wide temperature ranges, as the power output versus temperature behavior of Fig. 49 shows. These typical temperature stabilities ($dP/dT = -0.025$ to -0.03 dB/°C) are an inherent feature of the ohmic contacted transferred-electron device and straightforward thermal compensation techniques could

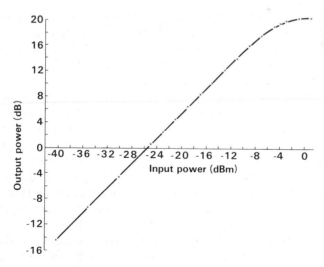

FIG. 46 Measured transfer characteristics of a Plessey three-stage *K*-band amplifier. [Courtesy Plessey Research (Caswell) Limited.]

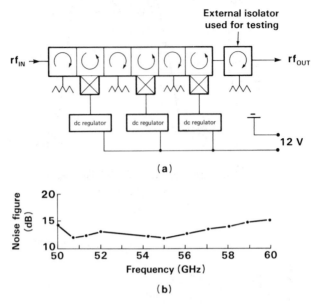

(a)

(b)

FIG. 47   (a) Schematic diagram of an *L*-band wideband InP amplifier and (b) its measured noise performance. (From Sowers *et al.*, 1982. © 1982 IEEE.)

easily be used to provide full operating temperature ranges from −40 to 70°C.

An indication of the high-frequency potential of the indium phosphide reflection amplifier is given by the 94–96-GHz two-stage unit of Fig. 50. This prototype amplifier has a small signal gain of 12 dB and is the precursor of further developments toward a low noise figure (10 dB) and a medium-power (100 mW) performance level.

Throughout the development of all of these TEAs, the Y-junction circulator has been a crucial component. This ferrite circulator component must have a wideband response with no nonuniform characteristics over a bandwidth of approximately twice that of the amplifier bandwidth. This very stringent performance specification is required to avoid the onset of any out-of-band spurious oscillations while also providing good in-band behavior. Indeed, the effects of nonideal circulator performance on reflection amplifier performance has been the subject of many theoretical and experimental investigations [see, e.g., Sard (1959), Okean (1966), or Bates and Khan (1980)]. Experience shows that circulators with low insertion losses (0.3 dB) together with return loss and isolation figures better than 15 dB must be realized over wide bandwidths if acceptable millimeter-wave TEA performances are to be obtained.

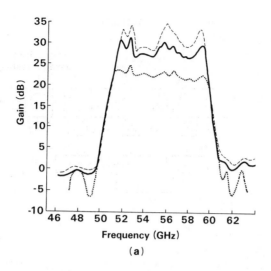

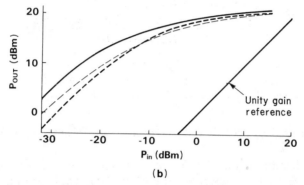

FIG. 48   (a) Power gain at various input drive levels for an L-band amplifier: $T = 35°C$; $\cdots$, $P_{in} = -5$ dBm; ———, $P_{in} = -15$ dBm; ---, $P_{in} = -25$ dBm. (b) Transfer characteristics of an L-band amplifier: $T = 30°C$; ———, 54 GHz; ———, 58 GHz; ———, 56 GHz. (From Sowers et al., 1982. © 1982 IEEE.)

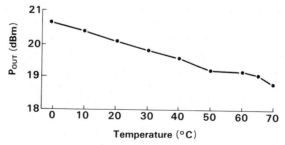

FIG. 49   Amplifier output power as a function of temperature: $P_{in} = 3$ dBm; $F = 56$ GHz. (Courtesy Varian.)

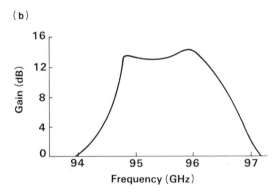

FIG. 50   (a) A 95-GHz InP amplifier and (b) its measured small-signal gain characteristics. (Courtesy Varian.)

IV. Conclusion and Future Prospects

The practical oscillator and amplifier performances described herein confirm many of the theoretical predictions related to indium phosphide and gallium arsenide transferred-electron devices. Higher operating frequency limits, better output powers and conversion efficiencies, as well as more temperature-stable characteristics are now proven features of indium phosphide. The lower noise characteristics of indium phosphide have also been proven for amplifier devices, but so far oscillator noise measurements display a close parity between the two materials. Overall it now seems that above 40–50 GHz indium phosphide offers the best performance parameters for oscillator and amplifier applications, whereas in the lower frequency ranges gallium arsenide is still the best choice for all but the lowest-noise TEA stages. Although at the time of writing gallium arsenide has the considerable advantage of being readily available commercially, this situation will shortly change. Plessy has now released InP devices commercially in pilot production quantities. Manufacturing capabilities are also now being established at Varian for indium phosphide devices, oscillators, and amplifiers. As oscillator devices, the aim specifications for the two-layer structures are 175 mW at 56 GHz and 75 mW at 94 GHz (Fank, 1983), whereas the three-layer structures are producing 140 mW at 57 GHz and 40 mW at 90 GHz with reasonable manufacturing yields. In the case of amplifier devices the performance figures depend critically on circuit behavior, but the systems described are intended for commercial applications.

With regard to the future there is still much developmental work being undertaken, particularly on indium phosphide. Optimum design oscillators with a stable current-limiting cathode, diamond heat sink, and a thin IHS process are being proposed for power outputs of 300 mW at 56 GHz and 150 mW at 94 GHz (Fank, 1983). Similarly, p-notch amplifier devices are under development aimed at 6-dB noise measures at 60 GHz; amplifier chains with a 10-dB noise figure at a 10-GHz bandwidth (100 mW) are also envisaged at frequencies up to 100 GHz. The performance improvements likely to accrue from gallium arsenide will be more modest owing to its greater maturity. Improved low-impedance circuit-matching techniques are likely to realize 50-mW powers from harmonic devices at 90 GHz, but these devices will not achieve the efficiencies of the alternative indium phosphide units.

Regardless of material used there is still a great deal of subsystem development needed. Increasingly, systems will demand oscillator and amplifier components with better electronic control of frequency and power as well as higher power output levels. Although these are important features, the most critical parameters for the future of millimeter-wave systems are overall system cost, size, and ruggedness. It could readily be argued from these considerations that the future success of millimeter-wave applications is dependent upon the integration of complete systems into planar transmission media such as microstrip or dielectric waveguide. Bearing in mind the high-quality integrated mixer, limiters, and modulator systems already available, it seems clear that the first of the power generating and amplifying devices that can be integrated into such systems will have a real edge in future equipments. It is suggested that the microstrip oscillator performance of indium phosphide TEOs could be a significant factor in this development.

ACKNOWLEDGMENTS

It is a pleasure to acknowledge my fellow workers at Plessey Research (Caswell) Ltd., who have provided high-quality semiconductor material (P. Giles and staff), consistent device processing (I. Davies, A. M. Howard, and staff), detailed rf assessment and device design (D. M. Brookbanks, K. F. Flanagan, C. R. Green, F. A. Myers, D. C. Smith, T. J. Simmons, and R. L. Tebbenham), as well as Mrs. H. Barbour and K. Jenkins for the typing and artwork. I would also particularly like to thank F. B. Fank of Varian for providing comprehensive indium phosphide oscillator and amplifier results. Information has also been freely given by Tom Cantle of Auriema (Hughes), P. A. Rolland (University of Lille), and J. Ondria (M. E. D. L., formerly Alpha Industries).

REFERENCES

Bates, B. D., and Khan, P. J. (1980). *IEEE Microwave Symp. Dig.*, pp. 174–176.
Brookbanks, D. M., and Buck, B. J. (1981). *Proc. 8th Biennial Cornell Elec. Eng. Conf.*, pp. 405–414.
Colliver, D. J., Irving, L. D., Patterson, J. E., and Rees, H. D. (1974). *Electron. Lett.* **10**, 221–222.
Crowley, J. D., Sowers, J. J., Janis, B. A., and Fank, F. B. (1980). *Electron. Lett.* **16**, 705–706.
Dean, M. (1978). "Development of Q-band FM Noise Measurement Equipment," personal communication.
Eddison, I. G., and Brookbanks, D. M. (1982). *Electron. Lett.* **18**, 308–309.
Eddison, I. G., and Davies, I. (1982a). *Radio Electron. Eng.* **52**, 529–533.
Eddison, I. G., and Davies, I. (1982b). *IEEE Microwave Symp. Dig.*, Paper Y6.
Eddison, I. G., Davies, I., Howard, A. M., and Brookbanks, D. M. (1981). *Electron. Lett.* **17**, 948–949.
Fank, F. B. (1983). "Current Status and Future Objectives in InP Devices," personal communication.
Fank, F. B. and Crowley, J. D. (1982). *Microwave J.* **25**(9), 143–147.
Fawcett, W., and Hill, G. (1975). *Electron. Lett.* **11**, 80–81.
Friscourt, M. R., Rolland, P. A., Cappy, A., Constant, E., and Salmer, G. (1983). *IEEE Trans. Electron Devices* **ED-30**, 223–229.
Hines, M. E. (1970). *IEEE Trans. Electron Devices* **ED-17**, 1–8.
Jones, D., and Rees, H. D. (1972). *Electron. Lett.* **8**, 363–364.
Jones, D., and Rees, H. D. (1975). *Electron. Lett.* **11**, 13–14.
Kroemer, H. (1978). *Solid-State Electron,* **21**, 61–7.
Kurokawa, K. (1968). *IEEE Trans. Microwave Theory Tech.* **MTT-16**, 214–240.

Okean, H. C. (1966). *IEEE Trans. Microwave Theory Tech.* **MTT-14**, 323–336.
Ondria, J. (1968). *IEEE Trans. Microwave Theory Tech.* **MTT-16**, 767–781.
Ondria, J. (1981). *Proc. 11th Eur. Microwave Conf.*, Amsterdam, pp. 888–893.
Sard, E. W. (1959). *IRE Trans. Microwave Theory Tech.* **MTT-11**, 288–293.
Simmons, T. J., and Smith, D. C. (1982). *Electron. Lett.* **16**, 308–309.
Sitch, J. E., and Robson, P. N. (1976). *IEEE Trans. Electron Devices* **ED-23**, 1086–1094.
Smith, D. C., and Simmons, T. J. (1983). *Electron. Lett.* **19**, 222–223.
Smith, D. C., and Tebbenham, R. L. (1979). *Proc. 9th Eur. Microwave Conf.*, pp. 538–542.
Sowers, J. J., Crowley, J. D., and Fank, F. B. (1982). *IEEE MTT-S Int. Microwave Symp. Dig.*, Dallas, Texas, pp. 503–505.
Thim, H. (1971). *Electron. Lett.* **7**, 106–107.
Young El-Ma and Cheng Sun (1980). *IEEE Trans. Microwave Theory Tech.* **MTT-28**, 1460–1463.

CHAPTER 2

Nonradiative Dielectric Waveguide

Tsukasa Yoneyama *

Research Institute of Electrical Communication
Tohoku University
Sendai, Japan

I. Introduction

In recent years there have been significant developments in the application of such millimeter-wave systems as line-of-sight communications, radar, radiometry, and remote sensing. These advances have been supported in large part by the spectacular progress achieved in the field of device and component technology. Outstanding advantages of using millimeter waves are operational capability under cloudy, hazy, and foggy weather conditions and physically small-aperture antennas. Atmospheric absorption peaks caused by oxygen molecules and water vapor are even beneficial to terrestrial and satellite communication systems that are required to prevent signals from

---

* Present address: Department of Electronics and Information Engineering, Ryukyu University, Nishiharacho, Okinawa, 903-01, Japan.

overshooting their designated ranges and sometimes to gain a degree of covertness. To meet the ever increasing need for millimeter-wave systems, integrated-circuit approaches are being increasingly employed, although quasi-optical techniques also show promise for use at the higher end of the millimeter-wave spectrum.

At the lower end of the millimeter-wave spectrum, from 30 to 140 GHz, the use of microstrips, slot lines, and fin lines is still common, but conduction losses along these printed transmission lines tend to increase drastically with higher operating frequencies. Therefore, less lossy guiding media are being studied with the expectation of realizing high-performance millimeter-wave integrated circuits. Dielectric waveguides (Knox, 1976; Knox and Toulios, 1970) have been introduced as a candidate for fulfilling such a low loss requirement, because excellent dielectric materials with loss tangents as small as 10^{-4} are currently available in the millimeter-wave frequency range. Indeed, the insulated image guide has been used for the development of a V-band communication transmitter and receiver (Kietzer et al., 1976).

In principle, the reduction of transmission loss is always possible if the dielectric waveguides are made thin enough in terms of the wavelength so that most of the transmitted power is carried in the surrounding lossless air region. However, the loose confinement of fields within the dielectric medium is by no means a practical solution to the realization of low-loss dielectric waveguides, because it is accompanied by intolerable radiation levels at curved sections and discontinuities. The simultaneous reduction of transmission and radiation losses has been an ultimate but never attained goal of dielectric waveguide technology. Using low-loss, high-dielectric material to concentrate more fields within the dielectric strip has been found to be effective in suppressing radiation (Knox, 1976). The trapped-image guide utilizes a metal trough to reflect radiation waves back into the image guide and has succeeded in considerably reducing loss at 90° bends (Itoh and Adelsek, 1980).

The nonradiative dielectric waveguide (NRD guide) (Yoneyama and Nishida, 1981) is a newly proposed dielectric waveguide that can almost completely suppress radiation at curved sections and discontinuities without spoiling the inherent low-loss nature of dielectric waveguides. The NRD guide resembles the H guide both in structure and in field configuration, except that the separation of the metal plates is less than half a wavelength. Because of this reduced metal plate separation and an electric field configuration that is predominantly parallel to the metal plates, parasitic radiation, if any, is suppressed below cutoff. This is not possible with the conventional H guide because, to attain small transmission loss rather than radiation suppression, its metal plate separation is greater than the wavelength.

Although the NRD guide is similar to the H guide in structure, the idea for it came from the metal waveguide evanescent-mode filter, which basically consists of dielectric chips and strips housed in a below-cutoff metal waveguide. Because cutoff is eliminated in the presence of the dielectric medium, each component inserted in the waveguide serves as a resonator or a connecting guide in the filter structure. In the NRD guide, the metal waveguide is replaced with a below-cutoff parallel-plate waveguide. An additional advantage of the NRD guide is that there are no metal walls perpendicular to the electric field, thus eliminating the major source of conduction loss.

Fortunately, the NRD guide is amenable to rigorous analysis for obtaining various required design equations. Field expressions, operational diagrams, transmission loss curves, bandwidths of the single-mode operation, dispersion curves, and coupling coefficients of the NRD guide are derived and presented as a function of various structural and operational parameters, based on theoretical analysis.

Prototype passive components of the NRD guide, such as transitions between a metal waveguide and an NRD guide, matched terminations, variable shorts, ring resonators, bandpass filters, and directional couplers, were fabricated and measured to demonstrate their applicability to millimeter-wave integrated circuits (Yoneyama and Nishida, 1983). NRD-guide bends were also studied experimentally (Yoneyama et al., 1982). No radiation at all occurred at curved sections, as expected, but reflection rather than radiation caused serious trouble at very sharp bends with radii of curvature nearly identical to the guide wavelength. A method of reducing the reflection has been proposed and tested successfully. Also, a qualitative theory has been developed to interpret measured results of NRD-guide bends.

II.   Field Expressions of NRD Guide

Figure 1 shows a typical configuration of NRD guide, which consists of straight and/or curved strips and sometimes specially shaped dielectric objects sandwiched between parallel metal plates separated by a distance of less than half a wavelength. It is essential to note that the electric field of the operating mode is predominantly parallel to the metal plates. As is well known, if two parallel metal plates are separated by a distance less than half a wavelength, electromagnetic waves with the electric field parallel to the plates cannot propagate between the plates because of their below-cutoff nature. However, if dielectric strips with a proper dielectric constant are placed between the metal plates, the cutoff is eliminated and electromagnetic waves can propagate freely along the dielectric strips, whether straight or curved. Because radiated waves, if any, decay rapidly outside the strip, bends and junctions can be incorporated into complicated integrated circuits.

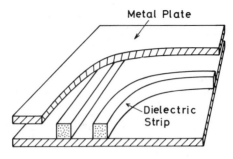

FIG. 1   Schematic representation of nonradiative dielectric waveguide.

To understand the wave-guiding mechanism in the NRD guide, field analysis of a straight guide with a cross-sectional view as depicted in Fig. 2 will be made first. The metal plate separation is a, the width of the strip is b, and the relative dielectric constant of the strip material is ε_r, respectively. The x and y rectangular-coordinate axes are taken to be normal and parallel to the metal plates, respectively, and the z axis is in the direction of transmission.

Modes of the NRD guide are hybrid in nature, having both electric and magnetic components in the longitudinal direction. They are usually referred to as longitudinal-section magnetic (LSM) and longitudinal-section electric (LSE) modes (Collin, 1960). In the LSM mode the magnetic field is parallel to the air–dielectric interfaces, whereas in the LSE mode the electric field is parallel. Specifically, the operating mode of the NRD guide is the lowest LSM mode, or the LSM_{10} mode according to this scheme of mode classification.

Because there is no H_y component in LSM modes, field analysis starts with solving

$$(\nabla^2 + k^2)E_y = 0, \tag{1a}$$

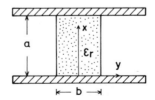

FIG. 2   Cross-sectional view of an NRD guide: a, plate separation; b, the width of the strip; ε_r, the relative dielectric constant of the strip material; x and y, rectangular coordinate axes normal and parallel to the metal plates. The z axis is in the direction of transmission.

and other field components are then determined by the following relations:

$$(k^2 + \partial^2/\partial y^2)E_x = \partial^2 E_y/\partial y\, \partial x, \tag{1b}$$

$$(k^2 + \partial^2/\partial y^2)E_z = \partial^2 E_y/\partial y\, \partial z, \tag{1c}$$

$$(k^2 + \partial^2/\partial y^2)H_x = -j\omega\varepsilon\, \partial E_y/\partial z, \tag{1d}$$

$$(k^2 + \partial^2/\partial y^2)H_z = j\omega\varepsilon\, \partial E_y/\partial x, \tag{1e}$$

where

$$k^2 = \begin{cases} \varepsilon_r k_0^2, & |y| \le b/2, \\ k_0^2, & |y| \ge b/2, \end{cases} \tag{2a}$$

$$\varepsilon = \begin{cases} \varepsilon_r \varepsilon_0, & |y| \le b/2, \\ \varepsilon_0, & |y| \ge b/2. \end{cases} \tag{2b}$$

In these equations, ε_0 and k_0 represent the permittivity and wave number of the surrounding medium (air in most cases), respectively.

The LSM even modes, which are characterized by the electric wall boundary condition at $y = 0$, are of primary importance because they include the operating mode of the NRD guide as a constituent. Solving Eq. (1a) for E_y under the requirements that E_y vanishes on the metal plates and εE_y is continuous across the air–dielectric interfaces and then substituting the resultant equation into Eqs. (1b–e) leads to field expressions for the dielectric region ($|y| \le b/2$),

$$E_x = -\frac{Aq}{h^2 \cos(qb/2)}\left(\frac{m\pi}{a}\right)\cos\left(\frac{m\pi}{a}x\right)\sin(qy)\exp(-j\beta z), \tag{3a}$$

$$E_y = \frac{A}{\cos(qb/2)}\sin\left(\frac{m\pi}{a}x\right)\cos(qy)\exp(-j\beta z), \tag{3b}$$

$$E_z = j\frac{Aq\beta}{h^2 \cos(qb/2)}\sin\left(\frac{m\pi}{a}x\right)\sin(qy)\exp(-j\beta z), \tag{3c}$$

$$H_x = -\frac{\omega\varepsilon_r\varepsilon_0 A\beta}{h^2 \cos(qb/2)}\sin\left(\frac{m\pi}{a}x\right)\cos(qy)\exp(-j\beta z), \tag{3d}$$

$$H_z = j\frac{\omega\varepsilon_r\varepsilon_0 A}{h^2 \cos(qb/2)}\left(\frac{m\pi}{a}\right)\cos\left(\frac{m\pi}{a}x\right)\cos(qy)\exp(-j\beta z); \tag{3e}$$

and for the air region ($|y| \geq b/2$),

$$E_x = \mp \frac{\varepsilon_r A P}{h^2} \left(\frac{m\pi}{a}\right) \cos\left(\frac{m\pi}{a} x\right) \exp\left[-p\left(|y| - \frac{b}{2}\right) - j\beta z\right], \qquad (4a)$$

$$E_y = \varepsilon_r A \sin\left(\frac{m\pi}{a} x\right) \exp\left[-p\left(|y| - \frac{b}{2}\right) - j\beta z\right], \qquad (4b)$$

$$E_z = \pm j \frac{\varepsilon_r A P \beta}{h^2} \sin\left(\frac{m\pi}{a} x\right) \exp\left[-p\left(|y| - \frac{b}{2}\right) - j\beta z\right], \qquad (4c)$$

$$H_x = -\frac{\omega \varepsilon_r \varepsilon_0 A \beta}{h^2} \sin\left(\frac{m\pi}{a} x\right) \exp\left[-p\left(|y| - \frac{b}{2}\right) - j\beta z\right], \qquad (4d)$$

$$H_z = j \frac{\omega \varepsilon_r \varepsilon_0 A}{h^2} \left(\frac{m\pi}{a}\right) \cos\left(\frac{m\pi}{a} x\right) \exp\left[-p\left(|y| - \frac{b}{2}\right) - j\beta z\right]; \qquad (4e)$$

where

$$\beta^2 = h^2 - (m\pi/a)^2, \qquad m = 1, 2, \ldots, \qquad (5a)$$

$$h^2 = \varepsilon_r k_0^2 - q^2 = k_0^2 + p^2. \qquad (5b)$$

The upper and lower signs in Eq. (4) apply to the air regions defined by $y \geq b/2$ and $y \leq -b/2$, respectively. The field equations given here guarantee continuities not only of εE_y but also of H_x and H_z across the air–dielectric interfaces. Letting the remaining field components E_x and E_z be continuous across the same interfaces gives

$$q \tan(qb/2) = \varepsilon_r p \qquad (6a)$$

with the auxiliary relation

$$q^2 + p^2 = (\varepsilon_r - 1)k_0^2. \qquad (6b)$$

Equation (6a) is exactly the same as the characteristic equation for the even TM modes of the dielectric-slab waveguide and yields a system of eigenvalues p_n and q_n with indices of zero and even integers.

Field analysis of the LSM odd modes can be carried out in the same manner and results in the characteristic equation

$$q \cot(qb/2) = -\varepsilon_r p \qquad (7)$$

with the same auxiliary relation as Eq. (6b). Although their derivation is straightforward, full field expressions for the odd modes are not needed in the following discussion and hence are not cited here. The characteristic equation (7) is identical to that for the odd TM slab modes, and its solutions constitute another system of eigenvalues p_n and q_n with indices of odd integers. Thus a complete system of eigenvalues p_n and q_n with indices extending over zero and all positive integers can be obtained by mixing eigenvalues for the even and odd modes alternatively. Referring to Eqs. (5a) and (5b), the corresponding phase constant β_{mn} is given by

$$\beta_{mn} = \sqrt{\varepsilon_r k_0^2 - (m\pi/a)^2 - q_n^2} = \sqrt{k_0^2 - (m\pi/a)^2 + p_n^2}. \qquad (8)$$

A mode specified by a pair of mode numbers m and n is called the LSM_{mn} mode.

It is interesting to note here that the expression for phase constants of the NRD-guide modes can also be derived by the effective dielectric constant (EDC) method (Knox and Toulios, 1970; McLevige et al., 1975; Itoh, 1976). To demonstrate this, the original NRD guide in Fig. 3a is decomposed into two-dimensional equivalent structures, as shown in Figs. 3b and 3c, respectively. First, the slab waveguide in Fig. 3b is solved for the TM modes to calculate the effective dielectric constant as follows:

$$\varepsilon_{re} = \varepsilon_r - q_n^2/k_0^2 = 1 + p_n^2/k_0^2, \qquad n = 0, 1, 2, \ldots, \qquad (9)$$

where p_n and q_n are solutions of Eq. (6a) or Eq. (7) under the auxiliary condition given by Eq. (6b). Then the dielectric-filled parallel-plate waveguide, shown in Fig. 3c, is analyzed for the TE modes having the electric field parallel to the metal plates to derive the following expression for the phase constant:

$$\beta_{mn} = \sqrt{\varepsilon_{re} k_0^2 - (m\pi/a)^2}, \qquad m = 1, 2, \ldots. \qquad (10)$$

Substituting Eq. (9) into Eq. (10) results in Eq. (8). Thus the EDC method applied in the proper sequence is proved to be valid for the analysis of the NRD-guide transmission characteristics as well.

Another interpretation of the wave-guiding mechanism in the NRD guide is that TM surface waves propagate down a zigzag path, bouncing back and forth between the metal plates just as plane waves do in rectangular metal waveguides. This observation suggests that a close similarity exists between

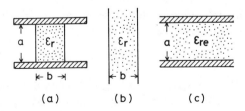

FIG. 3   Cross-sectional views of (a) NRD guide, (b) its equivalent dielectric slab waveguide, and (c) parallel-plate metal waveguide filled with dielectric medium having an effective dielectric constant ε_{re}.

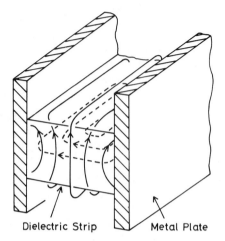

Dielectric Strip   Metal Plate

Fig. 4   Schematic diagram of the field lines of the dominant LSM_{10} mode in NRD guide:
——, electric field; – – – magnetic field. The fields are hybrid in nature having both electric and magnetic longitudinal components but no magnetic component perpendicular to the air–dielectric interfaces. (From Yoneyama *et al.*, 1982. © 1982 IEEE.)

the field configurations in the NRD guide and the metal waveguide. A rough sketch of field lines of the dominant LSM_{10} mode is shown for reference in Fig. 4. They are not very much different from those of the rectangular metal waveguide except for the presence of the longitudinal electric-field component. This field resemblance of both waveguide structures seems to support the expectation that any *E*-plane metal waveguide components including bends, junctions, couplers, and even filters can be transferred into the realm of NRD-guide technology without much difficulty. Thus the advantage of the NRD guide for use in millimeter-wave integrated circuits is obvious.

III.   Operational Diagram of NRD Guide

The requirement for radiation suppression in the NRD guide is fulfilled if the separation of the metal plates is less than half a wavelength, i.e.,

$$a < \tfrac{1}{2}\lambda_0, \tag{11}$$

where $\lambda_0 = 2\pi/k_0$ is the free-space wavelength. The requirement that all the modes except the dominant one be below cutoff is expressed, referring to Eq. (8), as

$$\lambda_{s0}/2 < a < \lambda_{s0}, \lambda_{s1}/2, \tag{12}$$

where $\lambda_{sn} = 2\pi/h_n$ is the guide wavelength of the nth slab mode, h_n being the corresponding propagation constant as given by Eq. (5b).

The operational diagrams for polystyrene ($\varepsilon_r = 2.56$) and fused quartz ($\varepsilon_r = 3.8$) NRD guides are prepared by calculating Eqs. (11) and (12), and they are presented in the a/λ_0 and $(b/\lambda_0)\sqrt{\varepsilon_r - 1}$ coordinates as shown in Fig. 5. Specifically, the diagram is drawn in such a way that each mode is above cutoff in the upper region bounded by the corresponding critical or cutoff curve ($m\lambda_{sn} = 2a$) and below cutoff in the lower region, respectively. Radiated waves are suppressed throughout the region on the left side of the vertical line $a/\lambda_0 = 0.5$, as implied by Eq. (11). Thus single-mode operation is assured in the region enclosed by the curves $\lambda_{s0} = 2a$ and $\lambda_{s1} = 2a$ and a straight section of $\lambda_0 = 2a$, as is inferred from the foregoing description. In designing the NRD guide, the transverse dimensions a and b of the dielectric strip have to be determined so that the corresponding design point falls somewhere on the single-mode area of the operational diagram. But further consideration is required for optimizing the performance of the NRD guide from various viewpoints, such as transmission loss, bandwidth of the

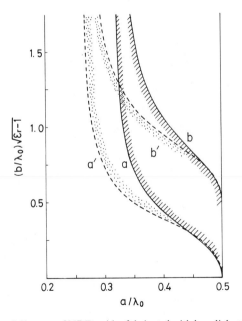

FIG. 5 Operational diagram of NRD guides fabricated with low dielectric materials: curves a and a', $\lambda_{S0} = 2a$; curves b and b', $\lambda_{S1} = a$; ——, polystyrene ($\varepsilon_r = 2.56$); ---, fused quartz ($\varepsilon_r = 3.8$). Single-mode operation is guaranteed in the hatched and dotted regions for the respective dielectric materials.

single-mode operation, and radiation-suppression capability, as will be described later.

It should be noted that the critical curve $\lambda_{s0} = a$ associated with the LSM_{20} mode does not yet appear in the operational diagram. This is always the case for dielectric material whose dielectric constant is smaller than a threshold value of 6.8. For higher dielectric material, say, alumina ($\varepsilon_r = 9.5$), however, the LSM_{20} mode influences the diagram significantly and reduces the area of the single-mode operating region, as demonstrated in Fig. 6. This is not desirable from a practical point of view because the bandwidth of the single-mode operation is consequently reduced. Elimination of the LSM_{20}-mode critical curve from the diagram is possible if the dielectric strip is thinned and insulated from the metal plates by means of low dielectric overlays (Yoneyama *et al.*, 1983a). This version of the modified NRD guide is called the insulated NRD guide and is a future goal of NRD-guide research.

It is important to recognize that, although higher modes and radiated waves are evanescent in nature, field intensities associated with them cannot be ignored in close proximity to circuit discontinuities. To estimate the field

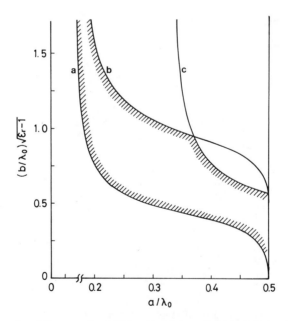

FIG. 6   Operational diagram of an NRD guide fabricated with high dielectric material such as alumina ($\varepsilon_r = 9.5$): curve a, $\lambda_{s0} = 2a$; curve b, $\lambda_{s1} = 2a$; curve c, $\lambda_{s0} = a$. The extra LSM_{20} mode with the critical curve c acts to reduce the area of the single-mode operating region.

behavior at discontinuities, the decay constants of evanescent waves must be known. The decay constant of the LSM_{11} higher mode is given by

$$\alpha_{11} = \sqrt{(\pi/a)^2 - h_1^2}, \tag{13}$$

whereas that of radiated waves is

$$\alpha_0 = \sqrt{(\pi/a)^2 - k_0^2}. \tag{14}$$

Calculated results of these equations are presented in Fig. 7 in the form of equi-decay-constant curves on the operational diagram for Teflon strips ($\varepsilon_r = 2.04$). Equi-transmission-loss curves, also included in Fig. 7, will be discussed in Section IV. Each equi-decay-constant curve is a composite of two components, one for the LSM_{11} mode in the upper region with the ordinate larger than 0.5 and the other for radiated waves in the remaining lower region. Actually, the upper parts of the equi-decay-constant curves are parallel to the LSM_{11}-mode critical curve, whereas the lower parts are parallel to the vertical axis for radiation suppression.

The separation of the metal plates should be as large as possible to reduce the conduction loss in circuits. Because a higher mode attenuation of about

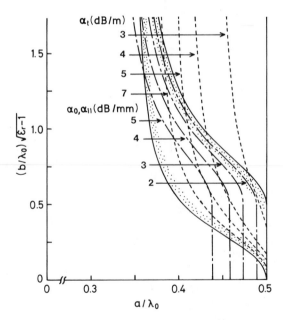

FIG. 7   Transmission loss curves (– – –) of the dominant LSM_{10} and mode decay-constant curves (—— - ——) of the below-cutoff LSM_{11} mode and radiated waves superposed on the operational diagram. Teflon ($\varepsilon_r = 2.08$) is assumed for strip material.

5 dB/mm is adequate in practice, the plate separation of $a/\lambda_0 = 0.45$ would be recommended as a proper choice, referring to Fig. 7.

IV.   Loss Characteristics of NRD Guide

Although transmission loss is not the sole criterion for selecting any specific waveguide for use in millimeter-wave integrated circuits, it is still an important factor. The transmission-loss constant α_t of the NRD guide is represented as

$$\alpha_t = \alpha_c + \alpha_d, \tag{15}$$

where α_c and α_d are the conduction and dielectric loss constants, respectively. They can be calculated by the well-known traditional perturbation approaches as follows:

$$\alpha_c = W_c/2P_t, \tag{16a}$$

$$\alpha_d = W_d/2P_t, \tag{16b}$$

where W_c and W_d are powers dissipated per unit longitudinal length owing to the conduction and dielectric losses, respectively, and P_t is the total power carried by the dominant LSM_{10} mode of the NRD guide. In this case these quantities are calculated using the field expressions given by Eqs. (3) and (4), as follows:

$$W_c = 2R_s \int_{-\infty}^{\infty} |H_z(x = 0)|^2 \, dy$$

$$= R_s \frac{\omega^2 \varepsilon_r^2 \varepsilon_0^2 |A|^2}{h_0^4 \cos(q_0 b/2)} \left(\frac{\pi}{a}\right)^2 \left[b + \left(1 + \frac{q_0^2}{\varepsilon_r p_0^2}\right) \frac{\sin(q_0 b)}{q_0}\right], \tag{17a}$$

$$W_d = \omega \varepsilon_r \varepsilon_0 \tan \delta \int_{-b/2}^{b/2} \int_0^a (|E_x|^2 + |E_y|^2 + |E_z|^2) \, dx \, dy$$

$$= \frac{\omega \varepsilon_r \varepsilon_0 \tan \delta |A|^2 a}{4h_0^2 \cos^2(q_0 b/2)} \left[\varepsilon_r k_0^2 b + (\varepsilon_r k_0^2 - 2q_0^2) \frac{\sin(q_0 b)}{q_0}\right], \tag{17b}$$

$$P_t = -\int_{-\infty}^{\infty} \int_0^a E_y H_x^* \, dx \, dy$$

$$= \frac{\omega \varepsilon_r \varepsilon_0 |A|^2 \beta_{10} a}{4h_0^2 \cos^2(q_0 b/2)} \left[b + \frac{(\varepsilon_r - 1)k_0^2}{p_0^2} \frac{\sin(q_0 b)}{q_0}\right]. \tag{17c}$$

In these equations the asterisk denotes the conjugate complex, tan δ is the loss tangent of the dielectric material, and R_s is the surface resistance of the metal plates given by

$$R_s = \sqrt{\omega\mu_0/2\sigma}, \qquad (18)$$

where σ is the conductivity of the metal plates and μ_0 the free-space permeability. Substituting Eq. (17) into Eq. (16) yields

$$\alpha_c = 2R_s \frac{\omega\varepsilon_r\varepsilon_0}{h_0^2\beta_{10}a}\left(\frac{\pi}{a}\right)^2 \frac{q_0 b + (1 + q_0^2/\varepsilon_r p_0^2)\sin(q_0 b)}{q_0 b + (\varepsilon_r - 1)k_0^2/p_0^2 \sin(q_0 b)}, \qquad (19a)$$

$$\alpha_d = \frac{\varepsilon_r k_0^2 \tan\delta}{2\beta_{10}} \frac{q_0 b + (1 - 2q_0^2/\varepsilon_r k_0^2)\sin(q_0 b)}{q_0 b + (\varepsilon_r - 1)k_0^2/p_0^2 \sin(q_0 b)}. \qquad (19b)$$

Numerical calculation of the attenuation constant is made at 50 GHz by assuming Teflon ($\varepsilon_r = 2.04$, tan $\delta = 1.5 \times 10^{-4}$), polystyrene ($\varepsilon_r = 2.56$, tan $\delta = 9 \times 10^{-4}$) and alumina ($\varepsilon_r = 9.5$, tan $\delta = 10^{-4}$) for the dielectric strip material and copper ($\sigma = 5.8 \times 10^7$ S/m) for the metal plates. Assumed values of material constants are taken from both measured and manufacturers' data. The metal-plate separation is fixed at 2.7 mm for the three cases, but the strip width is varied (3.5 mm for Teflon, 2.4 mm for polystyrene, and 0.93 mm for alumina) so that the cutoff frequencies of the respective guides are nearly identical. The results are presented in Fig. 8.

Conduction loss is unexpectedly small in spite of the small separation of the metal plates. The reason for this suppressed conduction loss is that the electric field of the operating mode is predominantly parallel to the metal plates, just as in the case of the H_{01} circular metal waveguide. The dielectric loss of each waveguide is almost the same as the infinite dielectric-medium loss, indicating that most of the electromagnetic energy is confined within the dielectric strip. For the polystyrene guide, the dielectric loss is dominant because its loss tangent is the greatest among the three dielectric materials discussed here. Teflon and alumina are excellent dielectric materials at millimeter-wave frequencies, hence dielectric losses are reduced to levels less than or equal to those of conduction losses. The total transmission loss of a suitably designed NRD guide is expected to be one order of magnitude less than that of the microstrip line at 50 GHz.

Further information on loss characteristics of the NRD guide can be obtained by drawing equi-transmission-loss curves of the dominant LSM_{10} mode on the operational diagram. Such curves are calculated for Teflon strips at 50 GHz and added to Fig. 7. In general, the transmission loss decreases toward the upper right in the diagram. This implies that with an increase in metal-plate separation and dielectric-strip width, the transmission loss becomes smaller. The larger width of the dielectric strip serves

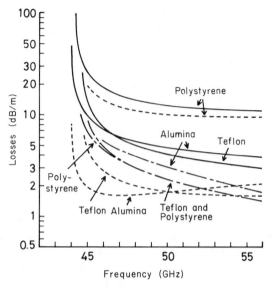

FIG. 8   Theoretical transmission loss curves of NRD guides: ——, total loss; ---, dielectric loss; ———-—, conduction loss. Teflon ($\varepsilon_r = 2.04$, $\tan \delta = 1.5 \times 10^{-4}$), polystyrene ($\varepsilon_r = 2.56$, $\tan \delta = 9 \times 10^{-4}$), and alumina ($\varepsilon_r = 9.5$, $\tan \delta = 10^{-4}$) are assumed for strip materials.

to lower the cutoff frequency of the NRD guide. This reduces the conduction loss considerably, just as in the H_C · circular metal waveguide in which an oversized guide is used to lower the cutoff frequency. Thus total transmission loss decreases because of the reduction in conduction loss for a dielectric strip with a large width.

The Q factor is another important design parameter that specifies the loss characteristics of integrated-circuit components. In general, the unloaded Q factor (Q_0) of the resonator is defined as

$$Q_0 = \omega \, \frac{\text{stored energy in resonator}}{\text{dissipated power in resonator}}. \tag{20}$$

When applied to a straight section of dielectric strip in the NRD guide, this equation reduces to

$$Q_0 = \omega[(W_e + W_m)/2\alpha_t P_t] \tag{21}$$

so long as the field deformation at both ends of the dielectric strip is negligible. This is true for a dielectric strip that is long compared to the guide wavelength, but the end effect cannot be ignored for a short dielectric strip or a chip.

In Eq. (21), the total transmission loss α_t and transmitted power P_t of the dominant mode have already been calculated. The other quantities W_e and W_m are the stored electric and magnetic energy per unit length of the strip, respectively, which prove to be equal in magnitude. Thus, by using the field expressions (3) and (4), the sum of W_e and W_m is obtained as follows:

$$W_e + W_m = 2W_m = \mu_0 \int_{-\infty}^{\infty} \int_0^a [|H_x|^2 + |H_z|^2] \, dx \, dy$$

$$= \frac{\varepsilon_r^2 \varepsilon_0 k_0^2 a |A|^2}{4 h_0^2 \cos^2(q_0 b/2)} \left[b + \left(1 + \frac{q_0^2}{\varepsilon_r p_0^2} \right) \frac{\sin(q_0 b)}{q_0} \right]. \quad (22)$$

Substituting Eqs. (15), (17c), and (22) into Eq. (21) yields

$$Q_0 = \frac{\varepsilon_r}{2\alpha_t} \frac{k_0^2}{\beta_{10}} \left[1 - \frac{[(\varepsilon_r - 1)/\varepsilon_r] q_0^2 [\sin(q_0 b)/q_0 b]}{p_0^2 + (\varepsilon_r - 1) k_0^2 [\sin(q_0 b)/q_0 b]} \right]. \quad (23)$$

Theoretical values of the unloaded Q factor around 50 GHz are presented for Teflon, polystyrene, and alumina materials in Fig. 9. The transverse dimensions of dielectric strips are assumed to be $b = 3.5$ mm for Teflon, 2.4 mm for polystyrene, and 0.93 mm for alumina, and $a = 2.7$ mm for all three cases. Material constants of dielectrics are the same as those assumed for the loss calculation in Fig. 8.

The Q factor of the alumina guide is the largest of the three because of the large dielectric constant and small loss tangent. A Q factor higher than 3000 can be attained for the Teflon guide. In spite of its large loss tangent, the polystyrene guide has a practical Q factor in excess of 1000. These high values of the unloaded Q factor are quite favorable for the construction of small insertion-loss NRD-guide filters.

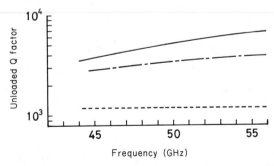

FIG. 9   Theoretical unloaded Q factors of straight dielectric strips in NRD guides: Teflon (————, $\varepsilon_r = 2.04$, $\tan \delta = 1.5 \times 10^{-4}$), polystyrene (–––, $\varepsilon_r = 2.56$, $\tan \delta = 9 \times 10^{-4}$), and alumina (————, $\varepsilon_r = 9.5$, $\tan \delta = 10^{-4}$) are assumed for dielectric materials.

V.   Other Properties of NRD Guide

A.   BANDWIDTH OF SINGLE-MODE OPERATION

The upper and lower ends f_u and f_l of the frequency range for the single-mode operation can be determined mathematically by solving the following equations:

$$a = \min(\pi/h_1, 2\pi/h_0, \lambda_0/2) \qquad \text{for} \quad f_u, \qquad (24a)$$

$$a = \pi/h_0 \qquad \qquad \qquad \text{for} \quad f_l. \qquad (24b)$$

Equation (24a) yields the lowest cutoff frequency among those for the LSM_{11} mode, LSM_{20} mode, and radiated waves, whereas Eq. (24b) gives the cutoff frequency of the dominant LSM_{10} mode. The relative bandwidth in percentage can be calculated by the following definition:

$$BW = 2(f_u - f_l)/(f_u + f_l) \times 100\%. \qquad (25)$$

Numerical data are presented versus the ratio $(b/a)\sqrt{\varepsilon_r - 1}$ for polystyrene, fused quartz, and alumina strips in Fig. 10. Smooth curves of the relative bandwidth are obtained for low-dielectric materials such as polystyrene and fused quartz, but for the alumina strip the curve is crooked at two points

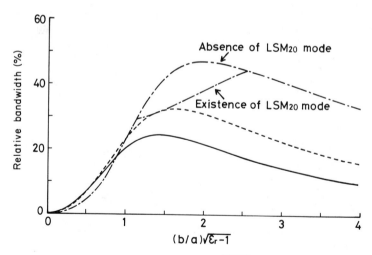

FIG. 10   Relative bandwidth as a function of $(b/a)\sqrt{\varepsilon_r - 1}$ for polystyrene (——, $\varepsilon_r = 2.56$), fused quartz (– – –, $\varepsilon_r = 3.8$), and alumina (—·—, $\varepsilon_r = 9.5$) strip materials. Bandwidth of the alumina NRD guide is limited by the presence of the LSM_{20} mode for certain values of $(b/a)\sqrt{\varepsilon_r = 1}$.

where the LSM_{11} and LSM_{20} modes change their roles as the second higher mode in the NRD guide. Because the LSM_{20} mode is odd with respect to the midplane between the metal plates of the guide, it may not be excited if the guide is carefully kept symmetrical in structure. In such a case the bandwidth is upward limited by the cutoff frequency of only the LSM_{11} mode, even for the high dielectric strip, and hence increases further as indicated in Fig. 10. It is also seen that the maximum bandwidth is attained at a certain value of $(b/a)\sqrt{\varepsilon_r - 1}$ for a given dielectric material. The maximum bandwidth and the required aspect ratio b/a can be determined as a function of the dielectric constant ε_r of the strip. Calculated results are shown in Fig. 11. The maximum bandwidth of the NRD guide increases until the dielectric constant of the strip reaches the threshold value of 6.8, at which point the LSM_{20} mode begins to influence guide characteristics, and then maintains a nearly constant value of 45% if the LSM_{20} mode is considered as the second higher mode. But if only the LSM_{11} mode is taken into account, the maximum bandwidth still increases as the dielectric constant increases beyond the threshold value, as plotted in Fig. 11.

It should be mentioned that the bandwidth achievable with the NRD guide is enough for practical purposes, because it covers a 20-GHz frequency band around 50 GHz. To realize such a wide-band NRD guide, the aspect ratio b/a has to be optimized properly, according to Fig. 11.

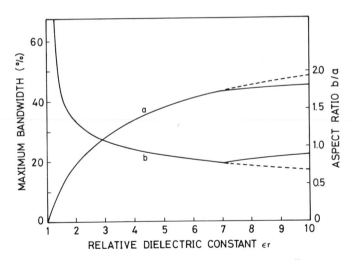

FIG. 11   Achievable maximum bandwidth (curve a) and corresponding aspect ratio b/a (curve b) as a function of relative dielectric constant ε_r of strip material: ——, existence of LSM_{20} mode; – – –, absence of LSM_{20} mode. The extra LSM_{20} mode limits the maximum bandwidth when the relative dielectric constant is larger than a threshold value of 6.8.

B. DISPERSION CHARACTERISTICS

The propagation constant of the dominant LSM_{10} mode of the NRD guide can be obtained by letting $m = 1$ and $n = 0$ in Eq. (8), which gives

$$\beta_{10} = \sqrt{\varepsilon_r k_0^2 - q_0^2 - (\pi/a)^2} = \sqrt{k_0^2 + P_0^2 - (\pi/a)^2}. \qquad (26)$$

Figure 12 represents calculated results of dispersion characteristics for alumina strips having ratios $(b/a)\sqrt{\varepsilon_r - 1} = 1.0$, 1.5, and 2.5. The dispersion curve of the alumina-filled metal waveguide is also presented for reference. The ordinate is the ratio of the wavelength λ in infinitely extending alumina to the guide wavelength λ_g, and the abscissa is the ratio of frequency f to the cutoff frequency f_c of the dominant mode. Dispersion curves of NRD guides are found to be very nearly identical to that of the metal waveguide, especially when the aspect ratio is large. Considering this, the NRD guide can be regarded as a dielectric-filled metal waveguide with respect to dispersion characteristics. Although the dispersion characteristics are very similar, the bandwidths of single-mode operation are different for the two waveguides, as previously discussed, and the NRD guide is rather dispersive in nature within the frequency range of single-mode operation.

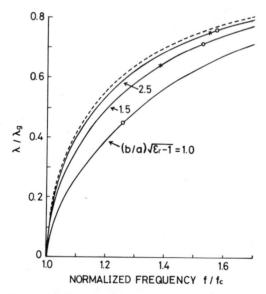

FIG. 12   Dispersion curves of alumina NRD guides (——) compared with the curve of an alumina-filled metal waveguide (---). The ordinate is the ratio of the wavelength in alumina to the guide wavelength, and the abscissa is frequency normalized with the cutoff frequency. Bandwidth is limited by the generation of the LSM_{11} ($\bigcirc$) or LSM_{20} ($\times$) mode as indicated on the dispersion curves.

C. COUPLING CHARACTERISTICS

The directional coupler is a basic component in millimeter-wave integrated circuits and is useful for the construction of various functional devices such as mixers (Paul and Chang, 1978), filters (Aylward and Williams, 1978), and power dividers (Paul and Yen, 1981). Coupling characteristics of simple parallel dielectric strips in the NRD guide are considered as a basis for such applications. Figure 13 represents a cross-sectional view of the NRD-guide coupling structure having a coupling spacing d, and also a coordinate system to be used for analysis. The coupling mechanism between a pair of parallel waveguides is usually interpreted as resulting from the superposition of even and odd modes propagating longitudinally with slightly different propagation constants β_e and β_o, respectively. Thus the coupling coefficient between the waveguides is given by

$$C = \tfrac{1}{2}(\beta_e - \beta_o), \tag{27}$$

and the coupling length required for the complete power transfer from one guide to the other is calculated by

$$L_c = \tfrac{1}{2}\pi/C = \pi/(\beta_e - \beta_o). \tag{28}$$

To obtain the even and odd mode propagation constants β_e and β_o, characteristic equations must be derived. Assuming LSM-type fields, Eq. (1a) is solved to obtain

$$E_y = A \exp(-py) \sin[(m\pi/a)x], \qquad y > b + \tfrac{1}{2}d, \tag{29a}$$

$$= [B \cos(qy) + C \sin(qy)] \sin[(m\pi/a)x], \qquad b + \tfrac{1}{2}d > y > \tfrac{1}{2}d, \tag{29b}$$

$$= [D \cosh(py) + E \sinh(py)] \sin[(m\pi/a)x], \qquad \tfrac{1}{2}d > y > 0, \tag{29c}$$

where factors A through E are unknowns to be determined and a common factor $\exp(-j\beta z)$ is implied. The electric wall is assumed at $y = 0$ for even modes, whereas the magnetic wall is assumed for odd modes. Therefore, $E = 0$ holds for even modes and $D = 0$ for odd modes. By substituting

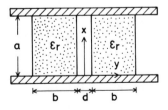

FIG. 13   Cross-sectional view of coupled dielectric strips in NRD guide: d, coupling spacing. (From Yoneyama et al., 1983b. © 1983 IEEE.)

Eq. (28) into Eq. (1b), another field component E_x, which is needed for determining the characteristic equations, is derived as follows:

$$E_x = -A(p/h^2)(m\pi/a)^2 \exp(-py) \cos[(m\pi/a)x], \qquad y > b + \tfrac{1}{2}d; \qquad (30a)$$

$$= (q/h^2)(m\pi/a)[-B \sin(qy) + C \cos(qy)] \cos[(m\pi/a)x],$$
$$b + \tfrac{1}{2}d > y > \tfrac{1}{2}d; \quad (30b)$$

$$= (p/h^2)(m\pi/a)[D \sinh(py) + E \cosh(py)] \cos[(m\pi/a)x],$$
$$\tfrac{1}{2}d > y > 0. \quad (30c)$$

Requiring continuity of E_x and εE_y across the air–dielectric interfaces yields the following expressions for the desired characteristic equations:

$$\varepsilon_r p \tanh(\tfrac{1}{2}pd) = q \frac{q \tan(qb) - \varepsilon_r p}{\varepsilon_r p \tan(qb) + q} \qquad \text{even mode,} \qquad (31a)$$

$$\varepsilon_r p \coth(\tfrac{1}{2}pd) = q \frac{q \tan(qb) - \varepsilon_r p}{\varepsilon_r p \tan(qb) + q} \qquad \text{odd mode,} \qquad (31b)$$

where

$$p^2 + q^2 = (\varepsilon_r - 1)k_0^2. \qquad (32)$$

Once Eqs. (31a) and (31b) are solved, it is easy to calculate the corresponding propagation constant β_e and β_o according to Eq. (8).

Figure 14 shows numerical data for the coupling coefficient as a function of the coupling spacing d for polystyrene, fused quartz, and alumina strips. Ordinate scales on the right side represent the corresponding normalized coupling length required for the complete power transfer, as given by Eq. (28). The coupling coefficient is found to be large when the value of $(b/a)\sqrt{\varepsilon_r - 1}$ is small. Since a small value of $(b/a)\sqrt{\varepsilon_r - 1}$ means a low dielectric constant and narrow width of the strip, fields of one waveguide have a small transverse decay constant and can reach the other guide with sufficient intensity; hence the coupling is enhanced.

Another feature of the coupling coefficient is the exponential dependence on the coupling spacing as long as the spacing is not small compared with the wavelength. This can be understood if Marcuse's approximate theory for the coupling coefficient (Marcuse, 1971) is applied to the present coupling structure. His theory gives

$$C = K \exp(-p_0 d), \qquad (33)$$

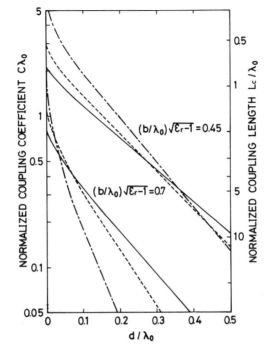

FIG. 14 Normalized coupling coefficient and normalized coupling length versus normalized coupling spacing: polystyrene (——, $\varepsilon_r = 2.56$), fused quartz (- - -, $\varepsilon_r = 3.8$), and alumina (—·—, $\varepsilon_r = 9.5$) are assumed for strip materials. $a/\lambda_0 = 0.45$. (From Yoneyama and Nishida, 1981. © 1981 IEEE.)

where

$$K = \frac{\varepsilon_r p_0^2 q_0^2}{\beta_{10}\{\frac{1}{2}[q_0^2 + (\varepsilon_r p_0)^2]p_0 b + \varepsilon_r(p_0^2 + q_0^2)\}}, \tag{34a}$$

$$\beta_{10} = \sqrt{\beta_0^2 - (\pi/a)^2}. \tag{34b}$$

Here, β_{10} is the propagation constant of the dominant NRD-guide mode, whereas β_0 is that of the TM slab mode. The exponential dependence of the coupling coefficient on the guide spacing is obvious because of the presence of an exponential factor in Eq. (33). Because Eq. (33) is simple in form, it can easily be applied to nonparallel coupling structures as well (Yoneyama et al., 1983b). Disagreement between the exact and approximate curves occurs only for very small values of the coupling spacing.

It should be noted that Eq. (33) is the same as the expression for the coupling coefficient of parallel TM slab waveguides (Marcuse, 1971), except that the propagation constant β_{10} of the NRD guide is replaced with the propagation constant β_0 of the TM slab mode. The propagation constant β_{10} is smaller than that of the slab waveguide because of the presence of the term $(\pi/a)^2$ under the square root in Eq. (34b). This means that the coupling coefficient of the NRD guide is considerably larger than that of the slab waveguide and hence probably larger than those of other dielectric waveguides. The large coupling coefficient of the NRD guide provides a practical advantage in that couplers for use in millimeter-wave integrated circuits can be reduced in size.

VI. Measurements of Prototype NRD-Guide Circuit Components

Because NRD-guide characteristics have previously been discussed in theoretical terms, this section will be devoted to a description of prototype circuit components fabricated with polystyrene dielectric and their measurements at 50 GHz (Yoneyama and Nishida, 1983). Actually, Teflon is much preferable to polystyrene from a standpoint of low loss characteristics, as is manifested by comparing their measured loss tangents 1.5×10^{-4} and 9×10^{-4}, respectively, at 50 GHz. In spite of that, polystyrene is recommended for strip material, especially at the prototype stage of development, because of its good machinability, moderate price, and suitable dielectric constant ($\varepsilon_r = 2.56$). A height of 2.7 mm and a width of 2.4 mm were chosen as the cross-sectional dimensions of the strip. Brass was employed for the top and bottom plates. Figure 15 is a photograph of several circuit components and the transition horn used for measurements.

FIG. 15   Various fabricated components for NRD guide designed at 50 GHz. Left to right: 90° bend, 180° bend, bandpass filter, ring resonator, directional coupler, transition horn.

A. Transitions

For carrying out measurements effectively, transitions of high-return losses are needed to transfer power between the NRD guide and the metal waveguide. Figure 16 shows details of the fabricated transition. The horn is flared in the *E* plane and tapered in the *H* plane from the broad dimension of the metal waveguide to the metal-plate separation of the NRD guide. The dielectric strip was provided with an *E* plane taper of about 3 cm in length and inserted 5 cm into the transition. Because the width of the dielectric strip was designed to be 2.4 mm, i.e., the same as the narrow dimension of the metal waveguide, the strip was tightly held within the horn at its mouth and throat. Teflon strips, however, require an additional *E* plane taper to match the size of the metal waveguide; its width has to be somewhat larger than 2.4 mm because of the smaller relative dielectric constant. This is one of the reasons why polystyrene is preferable to Teflon in prototype fabrication.

To realize the matched termination needed when measuring transition return losses, thin resistive sheets were attached to the *H* plane surfaces of the strip at some distance from the horn. Several sizes of horns were prepared, and their return losses were measured as a function of frequency. One of the measured results is presented in Fig. 17 for a horn whose length and aperture width were both 14.5 mm. The VSWR does not exceed 1.15 over the frequency range of measurements even though it includes the contribution owing to the residual reflection of the termination. To extract the reflection coefficient of the transition alone, a curve-fitting technique was applied to find an

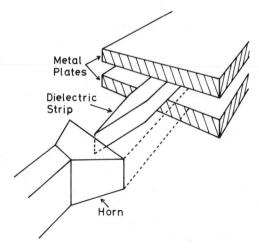

Metal
Plates

Dielectric
Strip

Horn

FIG. 16 Detail of a transition horn for 50-GHz NRD guide. The horn is flared in the *E* plane and tapered in the *H* plane to match the metal-plate separation of the NRD guide.

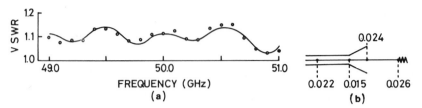

FIG. 17   (a) Measured reflection characteristics (○) of the transition. The transition horn
is 14.5 mm in both horn length and aperture width. The curve (——) was estimated assuming
reflection at four typical discontinuities. (b) Magnitudes of the assume reflection coefficients.
(From Yoneyama and Nishida, 1983.)

estimated curve for the observed data in the sense of least square terms.
The solid curve in Fig. 17 represents such a curve, which was calculated
based on assumed reflection at four points, namely, at the tip of the tapered
strip, at the throat and mouth of the horn, and at the tip of the matched
termination. Magnitudes of the reflection coefficients assumed at these
points are presented in Fig. 17b.

By adding all but one value for the termination, the possible maximum
reflection coefficient of only the transition is estimated to be 0.061, and the
corresponding return loss is estimated to be 24.3 dB. Even for a more compact transition horn having a length of 4.5 mm and an aperture width of
6.5 mm, a return loss as high as 18.5 dB has been obtained. In general, the
values of return losses of NRD guide transitions are very high compared with
those of dielectric-rod waveguide transitions. Because radiation does not
occur at the transition at all, transmitted power can be transferred between
a metal waveguide and an NRD guide without any appreciable power loss.
The NRD-guide transition is quite similar in performance to its fin-line
counterpart (Meier, 1974).

B.   MATCHED TERMINATIONS

Matched terminations for NRD-guide integrated-circuit applications were
constructed with resistive film. The resistive film used here, about 0.17 mm
in thickness, is a multilayer structure of Mylar–SiO_2–NiCr–SiO_2–Mylar
with adhesive coating on one side, and the NiCr core itself is only 0.1 μm or
less in thickness, depending on its resistivity. A narrow rectangle with dimensions of 2.7 mm and 15 mm was cut from the film to cover the H plane
surfaces of the strip. It was provided with a taper at the front end, 5 mm in
length, as shown by the insert in Fig. 18. Usually, the resistive film is attached
to the E planes of a waveguide, but there is no choice other than the H plane
surfaces in the NRD guide. When large absorption was required, sheets of
film were piled in layers. Standing-wave ratios of fabricated terminations were

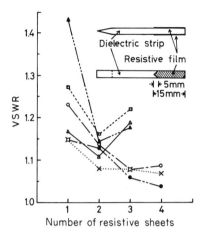

FIG. 18   Standing-wave ratios of fabricated matched termination as a function of the number of layered resistive sheets and specific resistivity: — - - - —, 500 $\Omega/\square$; —, 400 $\Omega/\square$; ——, 300 $\Omega/\square$; ···, 200 $\Omega/\square$; ————, 100 $\Omega/\square$; — — — —, 50 $\Omega/\square$. (From Yoneyama and Nishida, 1983.)

measured as a function of the number of layers and film resistivity. The results are summarized in Fig. 18.

The obtained minimum VSWR is somewhat less than 1.05 for the four-layer 50-Ω film termination, being comparable to or better than those of commercial metal-waveguide terminations. In this respect, it should be noted that the performance of the termination depends on the width of the strip. The narrower the dielectric strip, the greater the absorption; however, the reflection at the tip of the termination increases as well.

C. TRUNCATED ENDS AND VARIABLE SHORTS

Because no radiation occurs at any discontinuities in the NRD guide, complete reflection of waves is expected to take place at the truncated end of the dielectric strip. To confirm this prediction, standing-wave patterns along truncated dielectric strips were measured. Figure 19 is an example of measured standing-wave patterns. The dielectric strip extends over the negative region of z and is truncated at $z = 0$. A VSWR in excess of 30 dB was observed, showing that incident waves are completely reflected at the truncation.

Note that the field maximum of the VSWR pattern does not occur just at the truncated end ($z = 0$), but is shifted toward the generator. Thus the truncated end exhibits an inductive reactance. This aspect of the truncation must be taken into account in designing various circuit components such as dielectric resonators. In the case of Fig. 19, the normalized reactance of the

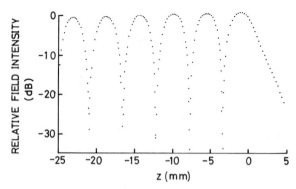

FIG. 19 A measured standing-wave pattern along a dielectric strip extending over the negative region of z and truncated at $z = 0$: $a = 2.7$ mm, $b = 2.4$ mm, $\varepsilon_r = 2.56, f = 50$ GHz.

truncated end is 1.15. The field intensity decreases beyond the end of the dielectric strip with a decay constant of about 4.6 dB/mm.

Complete reflection can be realized with another scheme, as shown in Fig. 20. In this arrangement a rectangular metal wafer is inserted into a slit cut in the midplane of the dielectric strip parallel to the metal plates. Because the field is below cutoff in the presence of the metal wafer, all incident waves are reflected back at the front edge of the wafer so long as it is not narrow enough for waves to overshoot. The metal wafer can move in the longitudinal direction in the slit, hence a variable short can be realized. Measurements of standing-wave patterns were made at 35 GHz using a Teflon strip with transverse dimensions of $a = 4.0$ mm and $b = 3.5$ mm and a metal wafer with dimensions of 2.0 mm × 20.0 mm and a thickness of 0.2 mm. The standing-wave pattern shown in Fig. 21 reveals that complete reflection actually takes place at the variable short ($z = 0$). This structure is very convenient for use in circuit components as a matching element.

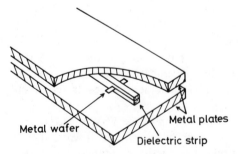

FIG. 20 Variable short for NRD guide. A movable metal wafer makes the waveguide section below cutoff behind its front edge; hence, complete reflection takes place if the wafer is long enough compared with the attenuation length.

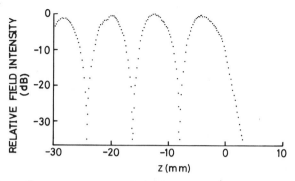

FIG. 21 A standing-wave pattern produced by the NRD guide's variable short: $a = 4.0$ mm, $b = 3.5$ mm, $\varepsilon_r = 2.04$, $f = 35$ GHz. The front edge of the metal wafer is set at $z = 0$.

D. RING RESONATORS

The ring resonator, shown in Fig. 22, is important in various filter structures, especially in channel-dropping filters (Itanami and Shindo, 1978). The resonator of this type exhibits very sharp band-rejection characteristics. The condition for perfect band rejection is given by Matthaei *et al.* (1964) as follows:

$$1 - K^2 = \exp(-2\alpha_t l), \tag{35}$$

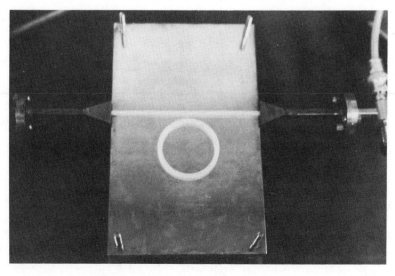

FIG. 22 A fabricated polystyrene ring resonator.

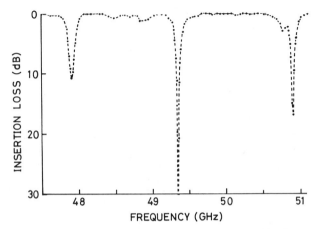

FIG. 23   Measured band-rejection characteristics of an NRD-guide ring resonator. The mean radius of the ring is 10.5 mm, and the coupling spacing is 2.0 mm. (From Yoneyama and Nishida, 1983.)

where K is the voltage-coupling factor between the main guide and the ring and α_t and l are the attenuation constant and mean path length of the ring, respectively. The attenuation constant consists of the transmission and radiation losses. Because there is no radiation loss in the NRD guide, α_t is smaller than that of other dielectric waveguides. In addition, the coupling coefficient is inherently large in the NRD guide, as stated in Section V.C; hence Eq. (35) holds if the coupling spacing between the two guides is properly adjusted. In the image-guide-type ring resonator, however, α_t is large because of radiation loss, and the coupling coefficient is not large; therefore Eq. (35) does not hold in this case, even if the ring is touched with the main arm to enhance the coupling.

Figure 23 shows measured band-rejection characteristics of a fabricated ring resonator with a mean radius of curvature of 10.5 mm and a coupling spacing of 2.0 mm. A band rejection of about 30 dB has been observed. This value should be compared with the band rejection of 15 dB that has been achieved with an image-guide ring resonator (Solbach, 1976).

E.   BANDPASS FILTERS

Because the metal plate separation is smaller than half a wavelength, a filter structure similar to a metal-waveguide evanescent-mode filter can be constructed with the NRD guide. To confirm this idea, a two-pole bandpass filter of maximally flat response with a 3-dB bandwidth of 1 GHz was fabricated at a center frequency of 50 GHz. The filter consisted of a pair of dielectric chips inserted between input and output strips, as shown in Fig. 24.

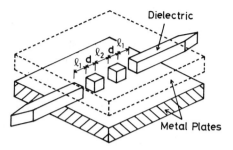

FIG. 24   Schematic representation of a two-pole bandpass filter. (From Yoneyama and Nishida, 1983.)

The required coupling coefficient and Q factor of chips were 0.014 and 70, respectively (Zverev, 1967). The actual chip length and gap spacings were experimentally determined to be $d = 2.5$ mm, $l_1 = 2.5$ mm, and $l_2 = 3.5$ mm, respectively, with d, l_1, and l_2 as defined in Fig. 24.

The measured frequency response of the fabricated filter is presented in Fig. 25 together with the theoretical curve. The center frequency was lowered slightly to 49.6 GHz because of the lack of a precise finish in the chip resonators. The insertion loss was about 1 dB at the center frequency. This insertion loss was possibly caused by the relatively large loss tangent of polystyrene

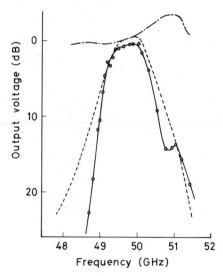

FIG. 25   Calculated (−−−) and measured (—○—) response of a maximally flat two-pole bandpass filter: —·—, reference; $d = 2.5$ mm; $l_1 = 2.5$ mm; $l_2 = 3.5$ mm. (From Yoneyama and Nishida, 1983.)

and probably in part by poor adjustment of the filter elements. Considerable reduction of the insertion loss can be expected if a high-quality dielectric such as Teflon is used instead of polystyrene. Indeed, unloaded Q factors of Teflon chip resonators have been experimentally found to be more than three times as great as those of polystyrene chips.

Although the principle of operation is identical for both the metal-waveguide evanescent-mode filter and the NRD guide filter, the design theory of the former is only an approximation for the latter because the field behavior in the gap regions is different. An elaborate analysis is required to develop a design theory for NRD-guide filters.

F.   DIRECTIONAL COUPLERS

To confirm the applicability of the NRD-guide couplers to millimeter-wave integrated circuits, the frequency dependence of the coupling performance was measured at 50 GHz. Measurements were made for symmetric and nonsymmetric couplers, as shown in Fig. 26. Two types of symmetric couplers were fabricated, one consisting of 180° bends with a curvature

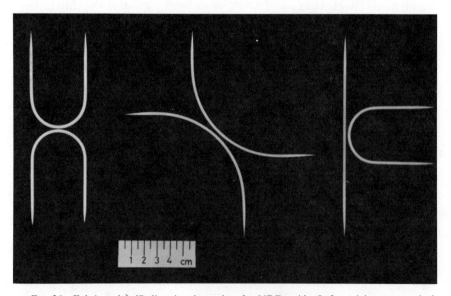

FIG. 26   Fabricated 3-dB directional couplers for NRD guide. Left to right: symmetrical coupler consisting of 180° bends with a curvature radius of 20 mm and a coupling spacing of 1.3 mm; symmetrical coupler consisting of 90° bends with a curvature radius of 65 mm and a coupling spacing of 1.9 mm; nonsymmetrical coupler consisting of a straight guide and a 180° bend with a curvature radius of 20 mm and a coupling spacing of 1.7 mm. (From Yoneyama et al., 1983b. © 1983 IEEE.)

radius of 20 mm and the other consisting of 90° bends with a curvature radius of 65 mm. The nonsymmetric coupler was constructed by combining a straight guide with a 180° bend having a curvature radius of 20 mm.

Because of their importance in many applications, quadrature hybrids were examined first. The coupling spacings needed for half-power transfer are 1.3 and 1.9 mm for the symmetric couplers with curvature radii of 20 and 65 mm, respectively, and 1.7 mm for the nonsymmetric coupler with a curvature radius of 20 mm. Theoretical and measured data are shown in Fig. 27, in which $|S_{21}|$ and $|S_{31}|$ represent the scattering coefficients at the direct and coupled ports, respectively. In theoretical calculations, Eq. (33) is modified to take into account the field deformations at the curved sections of the couplers (Yoneyama *et al.*, 1983b). Agreement between theory and measurements is satisfactory, and well-balanced outputs are obtained over a frequency range of about 1 GHz.

Another example is the in-phase power divider constructed as shown in Fig. 28. Power is fed to port 1 and extracted from ports 3 and 5. The curvature radius of the 180° bends that were used was 20 mm. Because the structure is so complicated, the coupling theory cannot accurately predict the performance

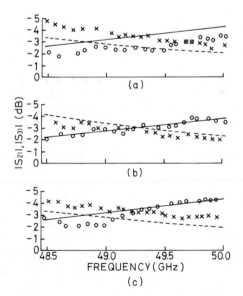

FIG. 27 Calculated and measured coupling characteristics of NRD-guide 3-dB directional couplers consisting of (a) identical 180° bends with a curvature radius of 20 mm, (b) identical 90° bends with curvature radius of 65 mm, and (c) a straight guide and a 180° bend with curvature radius of 20 mm. For direct-port scattering coefficients $|S_{21}|$: ---, theoretical; ×, measured. For coupled-port scattering coefficients $|S_{31}|$: ——, theoretical; ○, measured. (From Yoneyama *et al.*, 1983b. © 1983 IEEE.)

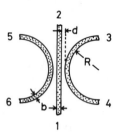

FIG. 28   Schematic representation of a power divider consisting of a straight guide and symmetrically arranged 180° bends. Power is fed into port 1 and detected at ports 3 and 5. (From Yoneyama *et al.*, 1983b. © 1983 IEEE.)

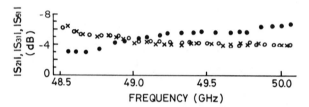

FIG. 29   Measured frequency characteristics of a power divider: ●, $|S_{21}|$; ×, $|S_{31}|$; ○, $|S_{51}|$. The curvature radius of 180° bends is 20 mm, and the coupling spacing is 1.4 mm. (From Yoneyama *et al.*, 1983b. © 1983 IEEE.)

of the power divider. Therefore, the construction had to be done experimentally, and the best result was achieved for a coupling spacing of 1.4 mm. The measured data are shown in Fig. 29. Well-balanced outputs of 4 dB below the input could be obtained at ports 3 and 5 over a frequency range of about 1 GHz, and the remaining power was detected at direct port 2.

VII.   Properties of Bends in NRD Guide

Bends are essential for constructing various functional devices for millimeter-wave integrated circuits. Use of the NRD guide is very attractive because undesirable radiation is completely suppressed at curved sections. In this connection, it may be interesting to examine properties of bends in the NRD guide. A cutaway view of the NRD-guide 90° bend is shown in Fig. 30 together with the cylindrical coordinates (r, θ) to be used for analysis. Figure 31 is a photograph of fabricated bends. Bending losses were measured by the substitution method, using a straight strip equal in length to the bend as a standard of comparison to remove the effect of transmission losses. Figure 32 shows the results of bending-loss measurements for curvature radius

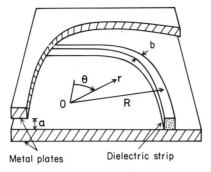

FIG. 30 Cutaway view of an NRD-guide bend with a mean radius of curvature R. Cylindrical coordinates (r, θ) are used for the analysis. (From Yoneyama et al., 1982. © 1982 IEEE.)

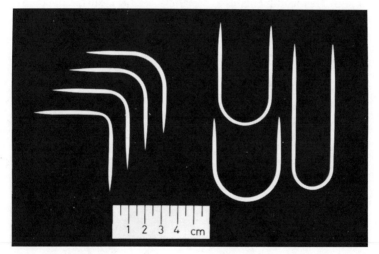

FIG. 31 Fabricated 90° and 180° bends used in measurements. Dielectric strips of some sharp bends are narrowed at curved sections to reduce reflection.

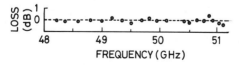

FIG. 32 Bending losses of an NRD-guide 90° bend with a curvature radius R of 20 mm and a strip width b of 2.4 mm: ---, reference level; ○, measured. (From Yoneyama et al., 1982. © 1982 IEEE.)

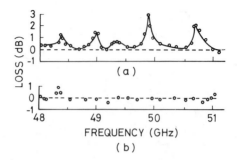

FIG. 33   Bending losses of an NRD-guide 90° bend with a curvature radius R of 16 mm:
—, reference level; ○, measured. (a) Bending losses caused by reflection are observed for a bend
with a regular strip width b of 2.4 mm; (b) losses are eliminated by reducing the strip width b
to 2.2 mm. (From Yoneyama et al., 1982. © 1982 IEEE.)

$R = 20$ mm. The bending losses are negligible. This is surprising considering
the large amount of radiation that might be caused at a bend of such a small
curvature radius in the image guide. The bending losses for the curvature
radius $R = 16$ mm are shown in Fig. 33a. Several peaks appeared in the
loss curve. These peaks may be attributed to reflection rather than to radiation. Among several ways tried for eliminating the reflection, narrowing
the strip at the bend was found to be most effective. In fact, the reflection
peaks disappeared as shown in Fig. 33b with a reduction in the width of the
strip by 0.2 mm along the curved section. This is in contrast to image-guide
bends, in which the bending losses increase because of the increased radiation
if the strip is narrowed at the bend. This technique of reducing the reflection
can also be applied to sharper bends. In Fig. 34, bending losses are shown for

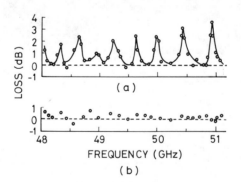

FIG. 34   Bending losses of an NRD-guide 180° bend with a curvature radius R of 12 mm:
—, reference level; ○, measured. (a) Loss curve for a bend with a regular strip width b of 2.4 mm;
(b) peaks are eliminated by reducing the strip width b to 2.0 mm. (From Yoneyama et al., 1982.
© 1982 IEEE.)

a 180° bend whose radius of curvature is as small as 12 mm and whose strip width is 2.0 mm, revealing that narrowing the strip is useful in this case as well.

The reason for the suppression of reflection at the bend with the reduced width can be understood qualitatively by referring to the effective dielectric-constant profile. The effective dielectric constant of the curved NRD guide is given as follows (Yoneyama *et al.* 1982):

$$\varepsilon_e = [\varepsilon/\varepsilon_0 - (\lambda_0/2a)^2] \exp(2w/R), \tag{36}$$

where

$$w = R \ln(r/R) \tag{37}$$

and R and r are the mean radius of bending and the radial coordinate as shown in Fig. 30, respectively. Equation (36) is plotted in Fig. 35 for a typical case. Because the effective dielectric constant is always negative outside the core region, electromagnetic tunneling (Snyder and Love, 1975) never occurs, and hence there is no radiation at all at bends. Therefore, it may be said again that the bending losses in the NRD guide are not due to radiation but are due to reflection caused by the field mismatch between the straight and curved guides.

The transverse field distribution at the bend is subject to two effects acting against each other. The raised edge in the core region of the effective dielectric-constant profile forces the field to shift outward and at the same time to lift the outer slope upward. On the contrary, the surrounding negative

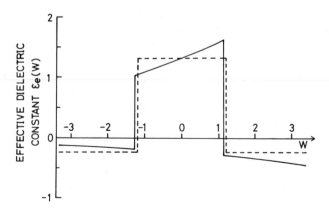

FIG. 35   Effective dielectric constant profile of a typical NRD-guide bend: ---, $R = \infty$; ——, $R = 11$ mm; $a = 2.7$ mm; $b = 2.4$ mm; $\varepsilon_r = 2.56$; $f = 50$ GHz. The negative dielectric constant outside the core region acts to suppress both radiation and reflection by decreasing the field mismatch between a straight and a curved section. (From Yoneyama *et al.*, 1982. © 1982 IEEE.)

dielectric constant has a net effect of shifting the field inward. The field distribution settles itself at a position where these two counteracting effects balance. Therefore, it can be understood that there is an optimum width of the dielectric strip that minimizes the field mismatch between the straight and curved guides for a given curvature radius of a bend. The widths of the curved dielectric strips adopted in experiments can certainly be such optimum ones. The minimum practical curvature radius of the polystyrene NRD guide bend achieved with this method is found to be about 10 mm at 50 GHz. Further information about NRD-guide bends is given in the literature (Yoneyama et al., 1982).

VIII. Conclusion

In this chapter the newly proposed nonradiative dielectric waveguide (NRD guide) has been reviewed from theoretical and experimental points of view. The NRD guide is recommended for use in millimeter-wave integrated circuits because it almost completely suppresses radiation at curved sections and at discontinuities of circuits without spoiling the low-loss nature inherent in dielectric waveguides. The radiation-suppression capability of the NRD guide is owing to the below-cutoff nature of the parallel-plate metal waveguide whose plate separation is smaller than half a wavelength. The theoretical transmission loss of the Teflon NRD guide is found to be as small as 4 dB/m at 50 GHz, being one order of magnitude less than that of the microstrip line. Various waveguide characteristics have also been calculated and presented as a function of structural and operational parameters.

Prototype circuit components of practical importance, such as filters and directional couplers, were fabricated with polystyrene and successfully measured at 50 GHz. Complete reflection was observed at truncated ends and at variable shorts of the NRD guide, clearly demonstrating the non-radiative nature of this waveguide. Furthermore, measurements of bending losses showed that, as expected, there is no radiation at all at most bends, but that reflection occurs at very sharp bends because of the field mismatch between straight and curved sections of the strip. A qualitative theory predicts that an optimum strip width for reflection minimization exists for a given bending radius.

Low-dielectric materials such as Teflon and polystyrene are preferable for the fabrication of the NRD guide because of convenient sizes achievable at millimeter-wave frequencies, but high-dielectric alumina must be used when more miniaturized circuits are required. In such a case, however, the generation of an extra higher mode causes a deterioration in the performance of the waveguide. To overcome this difficulty, the insulated NRD guide has been proposed (Yoneyama et al., 1983a). The insulated NRD guide, though not described here in detail, is expected to be very attractive for the construction of millimeter-wave integrated circuits in the future, because advances are being made in precise ceramic processing technology.

The operational principle of the NRD guide can be generalized to any dielectric object that is symmetrical with respect to the midplane between the metal plates of the guide. Such dielectric objects include dielectric resonators of various shapes, ferrite disks for nonreciprocal devices, and semiconductor diodes for detectors, mixers, and oscillators. By sandwiching not only passive but also active and nonreciprocal components in a below-cutoff parallel-plate waveguide, a totally integrated system of millimeter-wave circuits may be realized in a single housing without the aid of external connecting waveguides. Presently, the emphasis of NRD guide research is being shifted from passive components toward active, nonlinear, and nonreciprocal components.

ACKNOWLEDGMENTS

The author would like to thank Professor S. Nishida of the Research Institute of Electrical Communication, Tohoku University, for his support and encouragement.

REFERENCES

Aylward, M. A., and Williams, N. (1978). *8th Eur. Microwave Conf.* Paris, France, pp. 319–323.
Collin, R. E. (1960). "Field Theory of Guided Waves." McGraw-Hill, New York.
Itanami, T., and Shindo, S. (1978). *IEEE Trans. Microwave Theory Tech.* **MTT-26**, 759–764.
Itoh, T. (1976). *IEEE Trans. Microwave Theory Tech.* **MTT-24**, 821–827.
Itoh, T., and Adelseck, B. (1980). *IEEE Trans. Microwave Theory Tech.* **MTT-28**, 1433–1436.
Kietzer, J. E., Kaurs, A. R., and Levin, B. J. (1976). *IEEE Trans. Microwave Theory Tech.* **MTT-24**, 797–803.
Knox, R. M. (1976). *IEEE Trans. Microwave Theory Tech.* **MTT-24**, 806–814.
Knox, R. M., and Toulios, P. P. (1970). *Proc. Symp. Submillimeter Waves*, Polytechnic Institute of Brooklyn, New York.
Marcuse, D. (1971). *Bell Syst. Tech. J.* **50**, 1791–1816.
Matthaei, G. L., Yound, L., and Jones, E. M. T. (1964). "Microwave Filters, Impedance Matching Networks and Coupling Structures." McGraw-Hill, New York.
McLevige, W. V., Itoh, T., and Mittra, R. (1975). *IEEE Trans. Microwave Theory Tech.* **MTT-23**, 788–794.
Meier, P. J. (1974). *IEEE Trans. Microwave Theory Tech.* **MTT-22**, 1209–1216.
Paul, J. A., and Chang, Y. W. (1978). *IEEE Trans. Microwave Theory Tech.* **MTT-26**, 751–754.
Paul, J. A., and Yen, P. C. H. (1981). *IEEE Trans. Microwave Theory Tech.* **MTT-29**, 948–953.
Snyder, A. W., and Love, J. D. (1975). *IEEE Trans. Microwave Theory Tech.* **MTT-23**, 134–141.
Solbach, K. (1976). *IEEE Trans. Microwave Theory Tech.* **MTT-24**, 879–881.
Yoneyama, T., and Nishida, S. (1981). *IEEE Trans. Microwave Theory Tech.* **MTT-29**, 1188–1192.
Yoneyama, T., and Nishida, S. (1983). *Int. J. Inf. Millimeter Waves* **4**, 439–449.

Yoneyama, T., Yamaguchi, M., and Nishida, S. (1982). *IEEE Trans. Microwave Theory Tech.* **MTT-30**, 2146–2150.

Yoneyama, T., Fujita, S., and Nishida, S. (1983a). *IEEE/MTT-S Int. Microwave Symp. Dig.*, Boston, Massachusetts, pp. 302–304.

Yoneyama, T., Tozawa, N., and Nishida, S. (1983b). *IEEE Trans. Microwave Theory Tech.* **MTT-31**, 648–654.

Zverev, A. I. (1967). "Handbook of Filter Synthesis." Wiley, New York.

CHAPTER 3

Groove Guide for Short Millimetric Waveguide Systems

Yat Man Choi* and Douglás J. Harris

Department of Physics, Electronics, and Electrical Engineering
University of Wales Institute of Science and Technology
Cardiff, United Kingdom

* Present Address: Department of Electronic Engineering, Hong Kong Polytechnic, Kowloon, Hong Kong.

I. Introduction

Groove guide is potentially attractive as a low-loss waveguide for frequencies above 100 GHz, when the loss characteristics, dimensional tolerances, and power-handling capacity of more conventional guides such as rectangular waveguide and microstrip become progressively more prohibitive. [See Griemsmann (1964), Nakahara and Kurauchi (1964, 1965), Tischer (1963b), and Yee and Audeh (1965).] The basic guide consists of two identical parallel conducting walls containing grooves that face each other, as shown in Fig. 1. The direction of propagation is along the groove.

This guide, which supports a transverse electric mode, offers the possibility of a waveguide system having large dimensions compared with the operating wavelength. It can operate as a low-loss, low-dispersion, single-mode, and high-power transmission waveguide, the loss of which is about an order of magnitude less than that of dominant TE_{10}-mode rectangular waveguide. Because operation is well removed from cutoff, the guide dispersion is very low, and it therefore has a very high potential bandwidth. With a proper choice of groove dimensions, the guide can be made to transmit only the first-order mode. Higher-order modes leak transversely from the central

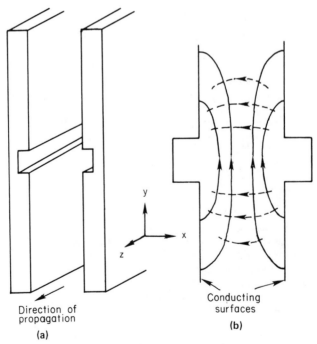

FIG. 1   (a) Cross section and (b) field distribution for single-groove guide: ——, E; – – –, H.

region, and they can be removed by suitably placed absorbent material at the top and bottom of the transverse extremities of the guide cross section, or allowed to radiate if the guide edges are left open.

The guide is simple to construct and no dimensions are critical to a small fraction of a wavelength. Sections of groove guide can be connected together simply with very low additional losses. The low tolerance on these dimensions makes the guide easy to manufacture even for very short wavelengths.

In view of the relatively large cross-sectional dimensions, groove guide is not likely to be attractive for use below 100 GHz except for high-power or long-distance transmission. At frequencies above 100 GHz, however, it becomes progressively more attractive, and a complete system in groove guide including generation, transmission, manipulation, and detection could be ideal. This guide has been given detailed consideration by the research group at the University of Wales Institute of Science and Technology. A comparison with alternative guide types (Harris, 1980) shows that, unlike most of the other possibilities, there is no major obstacle to higher-frequency operation. This chapter surveys previous work and gives an account of the theoretical and experimental work at UWIST on the guide itself, on double-groove guide, and on components that have been developed. These include IMPATT sources, detectors, transitions to rectangular guide, bends, couplers, standing-wave detectors, and attenuators. It is also shown that coupling between guide lengths can be carried out with low dimensional tolerances.

II. Theory of Single-Groove Guide

A. ANALYTICAL APPROACHES

Groove guide has been previously analyzed by a conformal mapping technique (Bava and Perona, 1966; Tischer, 1963a, b; Yee and Audeh, 1965) and by a transverse resonance technique (Nakahara and Kurauchi; 1964, 1965; Griemsmann, 1964; Schiek and Schlosser, 1964; Ruddy, 1965; Snurnikova, 1970, 1971) to obtain an approximate but adequate solution for guides with rectangular-cross-section grooves. The latter technique has now been used and extended (Choi, 1982; Lee, 1978) and the results of the analysis and computations are given in this chapter.

Theoretically, the fields in the guide can be expressed by an infinite series of orthogonal functions in a rectangular coordinate system. Although they are too complex to solve rigorously, an approximate method considering some predominant terms of the series can be used for the analysis of characteristics such as the cutoff wavelength, the field distribution, and the attenuation constant.

The modes are divided into groups, namely, the TE_o^s, TE_e^s, TM_o^s, TM_e^s, TE_o^f, TE_e^f, TM_o^f, and TM_e^f modes. The subscripts "o" and "e" denote the odd- and even-order modes, and the superscripts "s" and "f" denote a mode group with an effective short circuit ($E_x = 0$) and open circuit ($E_x \neq 0$) on the symmetrical x–z plane, respectively. The TE_o^s-mode group has the unusual and valuable properties outlined in Section I. Attention is concentrated on this mode group.

The guide can be divided into three regions: the grooved region, designated by A, and two evanescent side regions, designated by B. The boundary plane between them is $y = \pm a$. By virtue of symmetry, only one quarter of the cross section need be considered. In each region the propagation mode is expanded in terms of normal modes in a rectangular waveguide, with the propagation direction along the z axis. The characteristic equation may be obtained by considering the continuity conditions of the fields at the boundary planes, at $y = \pm a$.

B. Basic Expressions of the TE_o^s Modes

In this case, the TE modes are to be obtained that can exist in the guide when the x–z plane is effectively replaced by a conducting plane. Introducing the boundary conditions of Fig. 2, we obtain the following for the field components along the z-direction:

Region A, $0 \leq y \leq a$,

$$H_{zA} = \sum_p A_p k_c^2 \sin(k_{xA} x) \cos(k_{yA} y); \tag{1}$$

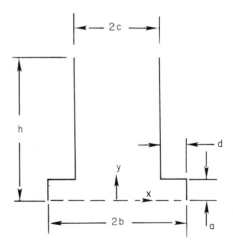

FIG. 2   Boundary conditions and dimensions for TE_o^s modes ($E_x = 0$ for $y = 0$).

Region B, $a \le y \le h$,

$$H_{zB} = \sum_r B_r k_c^2 \sin(k_{xB} x) \exp[-k_{yB}(y - a)]; \tag{2}$$

where A_p, B_r are the amplitude coefficients of field components in region A and region B, respectively; $k_c = (k_o^2 - \beta^2)^{1/2}$, the cutoff wave number; $k_{xA} = (p\pi/2b)$, the wave number in the x direction in region A (p, odd numbers); $k_{xB} = (r\pi/2c)$, the wave number in the x direction in region B (r, odd numbers); k_{yA} is the wave number in the y direction in region A; jk_{yB} is the wave number in the y direction in region B; $k_o = (2\pi/\lambda_o) = \omega\sqrt{(\mu_o \varepsilon_o)}$, the wave number in free space; $\beta = (2\pi/\lambda_g)$, the propagation constant in the z direction; ω is the angular frequency; ε_o, μ_o are the permittivity and permeability of air, respectively; λ_o, λ_g are the wavelength in the air and in the guide, respectively.

The other field components in the transverse directions can be obtained from the Maxwell curl equations. The field equations of the TE_o^s modes are shown here.

Region A:

$$H_{zA} = \sum_p A_p k_c^2 \sin(k_{xA} x) \cos(k_{yA} y),$$

$$E_{xA} = j \sum_p A_p \omega\mu_o k_{yA} \sin(k_{xA} x) \sin(k_{yA} y),$$

$$H_{yA} = j \sum_p A_p \beta k_{yA} \sin(k_{xA} x) \sin(k_{yA} y),$$

$$E_{yA} = j \sum_p A_p \omega\mu_o k_{xA} \cos(k_{xA} x) \cos(k_{yA} y),$$

$$H_{xA} = -j \sum_p A_p \beta k_{xA} \cos(k_{xA} x) \cos(k_{yA} y).$$

Region B:

$$H_{zB} = \sum_r B_r k_c^2 \sin(k_{xB} x) \exp[-k_{yB}(y - a)],$$

$$E_{xB} = j \sum_r B_r \omega\mu_o k_{yB} \sin(k_{xB} x) \exp[-k_{yB}(y - a)],$$

$$H_{yB} = j \sum_r B_r \beta k_{yB} \sin(k_{xB} x) \exp[-k_{yB}(y - a)],$$

$$E_{yB} = j \sum_r B_r \omega\mu_o k_{xB} \cos(k_{xB} x) \exp[-k_{yB}(y - a)],$$

$$H_{xB} = -j \sum_r B_r \beta k_{xB} \cos(k_{xB} x) \exp[-k_{yB}(y - a)].$$

C. Characteristic Equations of the TE_0^s Modes

Although the conformal mapping technique gives results that are close to experimental results, it requires calculations that are complicated and lengthy. The transverse resonance technique is more readily amenable to computation, especially if the number of terms in the precisely required series is limited. The approach is outlined here. The equations are given for solution of the guide fields and characteristics using two alternative boundary conditions and with first- and second-order approximations.

The boundary conditions imposed by Nakahara and Kurauchi (1964) between the grooved region A and the evanescent side region B, i.e., at $y = a$, are given by the following:

$$H_{zA} = H_{zB} \qquad \text{for} \quad |x| \le c,$$

$$\partial H_{zA}/\partial y = \partial H_{zB}/\partial y \qquad \text{for} \quad |x| \le c,$$

$$\partial H_{zA}/\partial y = 0 \qquad \text{for} \quad c \le |x| \le b.$$

The same boundary conditions were imposed by Schiek and Schlosser (1964) and Snurnikova (1970, 1971).

The first-order approximation H_{zA} and H_{zB} in Eqs. (1) and (2) (taken for $p = r = 1$ only) for the solution of the TE_{11} mode is then given by the solution of the following simultaneous equations to obtain u and v:

$$v = \frac{\pi^2 b(c/b - b/c)^2}{16c \cos^2(\pi c/2b)} [u \tan(u)] \tag{3}$$

and

$$u^2 + v^2 = \pi^2[(a/2c)^2 - (a/2b)^2], \tag{4}$$

where $u = k_{yA} a$ ($< \pi/2$ for TE_{11} mode) and $v = k_{yB} a$.

The first equation comes from the characteristic equation obtained by applying the boundary conditions at $|y| = a$ and is approximated by the terms of $p = r = 1$ only. The second equation comes from the assumed solution of the wave equation. The values of k_{xA} and k_{xB} are obtained directly from $p\pi/2b$ and $r\pi/2c$, respectively.

Thus we can analyze the characteristics of this mode to the first-order approximation by Eqs. (3) and (4), but this approximation is obtained by neglecting the discontinuity at $y = a$ for the propagation of a wave in the y direction. However, the higher-order waves are a result of this discontinuity, and the boundary condition at this plane must be satisfied. The effect of this discontinuity should therefore be considered.

Considering the first two terms of Eqs. (1) and (2) in this way, i.e., p and r given by both 1 and 3, the second-order approximation of the characteristic equation (Choi, 1982) is given by

$$X_1 \tan(X_1) = \frac{K_{11} - K_{13}K_{31}}{K_{33} - X_3 \tan(X_3)},$$ (5)

where

$$K_{mn} = \sum_{t=1}^{3} U_{mt} W_{tn} Y_t,$$

$$X_p = k_{yA} a = [X_1^2 - \pi^2(a/2b)^2(p^2 - 1)]^{1/2},$$

$$Y_r = k_{yB} a = \{-X_1^2 + \pi^2(a/2b)^2[(br/c)^2 - 1]\}^{1/2},$$

$$U_{pr} = \frac{k_{xA} \cos(k_{xA} c) \sin(k_{xB} c)}{\frac{1}{2}c(k_{xB}^2 - k_{xA}^2)},$$

$$W_{rp} = (c/b)U_{pr},$$

and m, n, p, and r are the subscripts of K, X, Y, U, and W.

Now if the guide dimensions, i.e., the parameters c/b and $a/2c$, are defined, X_p ($=k_{yA} a$) can be calculated from Eq. (5), and then the cutoff wavelength can be obtained from the relations

$$k_c^2 = k_{xA}^2 + k_{yA}^2,$$ (6)

$$k_c^2 = k_{xB}^2 + (jk_{yB})^2,$$ (7)

which are obtained directly from the wave equation. These second-order expressions are not too complicated, and they give results within 1% of the measured values for both cutoff wavelengths and field-decay constants.

Variations for the electric field component in the y direction of this mode at $x = 0$ using both first- and second-order approximations are shown in Fig. 3, where the improvement in the discontinuity is clear. Values of basic distribution parameters and relative component magnitudes for the particular guide geometry and wavelength are also given.

Alternative boundary conditions, imposed by Griemsmann (1964) between the grooved region A and the evanescent side region B, i.e., at $|y| = a$, are given by

$$(H_{zA})_{x=b} = (H_{zB})_{x=c}$$

and

$$(E_{xA})_{x=b}(b) = (E_{xB})_{x=c}(c).$$

The first expression comes from the assumption of equal "current" density and the second from equal "voltage" at the interfaces. These boundary conditions are only suitable for guide with relatively shallow grooves.

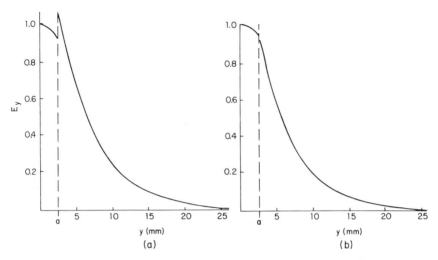

FIG. 3   Variation of E_y with y at $x = 0$ for $2c = 10$ mm, $d/2c = \frac{1}{3}$; $2a/2c = \frac{1}{2}$, $\lambda_0 = 3.0$ mm. (a) First-order solution: $\lambda_c/2c = 26.05$; $k_{yA1}a = 0.376$; $k_{yB1}a = 0.503$. (b) Second-order solution: $\lambda_c/2c = 26.52$; $k_{yA1}a = 0.359$; $k_{yB1}a = 0.516$; $jk_{yA3}a = -1.284$; $k_{yB3}a = 2.281$; $A_3 = -0.056A_1$; $B_1 = 0.551A_1$; $B_3 = -0.090A_1$.

The solution of the TE_{11} mode is then given by the solution of the following simultaneous equations:

$$v = (b/c)[u \tan(u)], \tag{8}$$

$$u^2 + v^2 = \pi^2[(a/2c)^2 - (a/2b)^2], \tag{9}$$

where u and v have the same meaning as given earlier in this section.

Equation (3) reduces to Eq. (8) when the value of $\cos(\pi c/2b)$ approaches zero, i.e., c/b approaches 1 (relatively shallow grooves). These expressions are the simplest to use, but experimental and theoretical cutoff wavelengths are not as close as those obtained by using the former expressions. Results can be obtained by using the second-order expressions that are as accurate as is likely to be required under any practical conditions.

Having found k_{yA} and k_{yB} from the simultaneous equations, we can determine other characteristics of the guide, i.e., λ_c from k_c and β from $k_c^2 = \omega^2 \mu_0 \varepsilon_0 - \beta^2$.

D.   AMPLITUDE COEFFICIENTS OF THE TE_o^s MODES

The relative amplitudes of the field components for $p = r = 3$, compared with those of the basic first-order mode given by $p = r = 1$, can be obtained from Choi (1982):

$$\frac{A_3}{A_1} = \frac{X_1 \sin(X_1) - \cos(X_1)(\sum_{r=1}^{3} U_{1r} W_{r1} Y_r)}{\cos(X_3)(\sum_{r=1}^{3} U_{3r} W_{r1} Y_r)} \tag{10}$$

and

$$B_r = \sum_{p=1}^{3} A_p U_{pr} \cos(X_p). \tag{11}$$

E. MODING CHARACTERISTICS

1. Transmission Modes

There are two types of wave propagation in groove guide. One corresponds to power flow in both the y and z directions (Fig. 4) and is characterized by a real wave number in the y direction in region B; it is defined as a non-transmission mode because energy is lost transversely from the guide. The other type, which has power flow in the z direction only, is characterized by an imaginary wave number in the y direction in region B, giving an exponential transverse decay. Only the second type propagates in the z direction without loss, and this is designated the transmission mode.

When TE_{pq} modes are excited in region A, the TE_{rs} modes, where $r \leq p$, are induced in region B. The relations among the wave numbers of these modes are given by

$$k_o^2 = k_{xA}^2 + k_{yA}^2 + k_{zA}^2,$$

$$k_o^2 = k_{xB}^2 + k_{yB}^2 + k_{zB}^2.$$

Therefore,

$$k_{yA}^2 - k_{yB}^2 = k_{xB}^2 - k_{xA}^2,$$

where k_o, k_{xA}, k_{xB}, k_{yA}, k_{yB} have the meaning given previously. Note that k_{yB} is real for a wave propagating in the y direction.

The wave numbers k_{yA} and k_{yB} can be related to transverse rank numbers q and s by $k_{yA} \leq q\pi/2a$ and $k_{yB} \leq s\pi/2h$. The wave number in the z direction is the same for regions A and B, i.e., $k_{zA} = k_{zB}$.

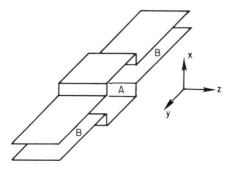

FIG. 4   Cavity regions for the transverse resonance approach.

The dimensions of the cavity are $2a$, $2b$, $2c$, and $2h$, as shown in Fig. 2. The overall mode type in the groove guide is denoted by the TE_{pq}–TE_{rs} mode pair. The conditions for resonance (low-loss transmission) in groove guide are that power should not flow out into the y direction, i.e., k_{yB} must be imaginary; and the main mode in the groove region should be above cutoff to satisfy the impedance matching at the junction.

2. Condition for Diverging Higher-Order Modes

For modes to diverge (i.e., non-transmission of the modes), k_{yB} must be real. For k_{yB} to be real, the following condition must be satisfied:

$$k_{xA}^2 - k_{xB}^2 + k_{yA}^2 \geq 0. \tag{12}$$

Provided that k_{yA} is real, $k_{xA}^2 - k_{xB}^2 \geq 0$, i.e.,

$$(2c) \geq (r/p)(2b), \qquad \text{where} \quad p \geq r, \tag{13}$$

will satisfy the condition for diverging higher-order modes. If $p = r$, it is not possible for condition (13) to be satisfied, and the mode is a transmission mode, e.g., the TE_{1q}–TE_{1s} and TM_{1q}–TM_{1s} mode pairs are transmission modes. Such mode pairs ($p = r$) are called the self-coupled modes. However, if $p > r$ it is possible that condition (13) is satisfied, and the mode will then propagate transversely out of the guide and will be removed from the transmission system.

3. Condition for Cutting Off Higher-Rank Modes (Modes with More than One Transverse Periodicity)

Condition (12) can be written as $k_{yA}^2 \geq (r\pi/2c)^2 - (p\pi/2b)^2$. Now,

$$q(\pi/2a) \geq k_{yA} \geq (q - 1)(\pi/2a),$$

where q is an odd number for the odd-rank mode and an even number for the even-rank mode. Therefore,

$$(1/2a)^2 \geq [r/(q - 1)]^2(1/2c)^2 - [p/(q - 1)]^2(1/2b)^2 \tag{14}$$

is the general condition for diverging higher-rank modes. For the case when $p = r \geq 1$, condition (14) reduces to

$$(1/2a)^2 \geq [p/(q - 1)]^2[(1/2c)^2 - (1/2b)^2].$$

The special case of $q \geq 2$ and $p = r = 1$ reduces condition (14) to

$$(1/a)^2 \geq (1/c)^2 - (1/b)^2, \tag{15}$$

which is the condition for diverging the first-order higher-rank modes and is the same as Nakahara and Kurauchi's (1965) expression.

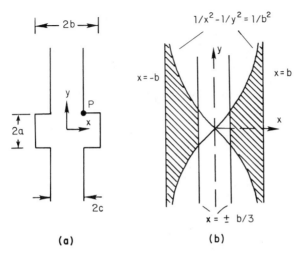

FIG. 5   (a) Guide cross section and (b) edge point $P(x = c; y = a)$ conditions for divergence of higher-order modes, shown for $p = 3$. (From Nakahara and Kurauchi, 1965).

4.   *Nakahara and Kurauchi's Graphical Representation of Divergence Conditions for Higher Modes*

When the coordinates (c, a) of the edge point P in Fig. 5 are in the hatched area, the guide propagates only the first-order modes independently of the frequency, and any other modes diverge (propagate out of the guide) in the y direction. The hatched area is bounded by the following conditions:

$$c < b, \tag{16}$$

$$c \geq b/p, \tag{17}$$

$$1/x^2 - 1/y^2 \leq 1/b^2. \tag{18}$$

Condition (16) corresponds to the physical existence of a groove. For the condition (17), the TE_{pq} and TM_{pq} modes, where $p \geq 3$ in region A and $p = 1$ in region B, diverge. This is derived from the wave equation that gives propagation in both y and z directions, i.e., the wave numbers in the y and z directions are real. Condition (18) corresponds to expression (15), and under this condition TE_{1q} and TM_{1q} modes ($q \geq 2$) diverge.

F.   CHOICE OF GROOVE-GUIDE DIMENSIONS

In the design of groove guide with relatively low-loss, low-dispersion, and with single-mode propagation, we have chosen a plane separation $2c$ that is from 3 to 3.4 times the operating wavelength. Such a plane separation can result in low loss and low dispersion, and yet, because the number of allowable modes ($p < 7$) is small, mode control can readily be achieved by a suitable choice of the groove depth d.

This guide can support three symmetrical self-coupled mode pairs, i.e., the dominant TE_{1q} self-coupled mode in addition to the higher-order self-coupled TE_{3q} and TE_{5q} modes, when the guide is excited symmetrically about the center of the guide. To diverge the higher-order modes, condition (13) must be satisfied, i.e., the groove depth $d \leq (p/r - 1)c$, where $p > r$.

Suppose that the plane separation $2c$ is chosen to be $3\lambda_0$.

(1) To diverge the self-coupled TE_{3q} mode, $d \leq 2c$. To ensure that the TE_{3q} mode in region A can be coupled effectively to the TE_{1s} mode in region B, d should be greater than about $\frac{1}{4}(2c)$.

(2) To diverge the self-coupled TE_{5q} mode, $d \leq \frac{1}{3}(2c)$. To ensure that the TE_{5q} mode in region A can be coupled effectively to the TE_{3s} and TE_{1s} modes in region B, d should be $> \frac{1}{8}(2c)$.

(3) If the groove depth $d > \frac{1}{5}(2c)$, the TE_{7q}–TE_{5s} mode pair becomes a transmission mode because condition (13) is not satisfied, and if $d \leq \frac{1}{5}(2c)$ the TE_{3q} self-coupled mode is less effectively rejected. To diverge the TE_{7q}–TE_{5s} mode pair, $d \leq \frac{2}{3}(2c)$. To ensure that the TE_{7q} mode in region A can be coupled more effectively to the TE_{3s} and TE_{1s} modes in region B, d should be $> \frac{3}{8}(2c)$.

Therefore, for effective rejection of higher-order modes, $2c \geq d \geq \frac{1}{4}(2c)$ for the rejection of the TE_{3q} self-coupled mode, and $\frac{1}{3}(2c) \geq d \geq \frac{1}{8}(2c)$ for the rejection of the TE_{5q} self-coupled mode. A logical choice of the groove depth d would be $\frac{1}{3}(2c) \geq d \geq \frac{1}{4}(2c)$. With this groove depth, the TE_{7q}–TE_{5s} mode pair is being less effectively rejected. To have this mode pair being more effectively rejected, the upper limit is chosen, i.e., $d = \frac{1}{3}(2c)$.

To diverge higher-rank modes, condition (14) must be satisfied. Now if d is chosen to be $\frac{1}{3}(2c)$, the groove width should be $2a < [(q - 1)/p]\frac{5}{4}(2c)$, where $p \geq 1$ for effective rejection of self-coupled higher-rank modes. Hence, for cutting off the TE_{12} and TE_{32} self-coupled modes, $2a < \frac{5}{12}(2c)$, e.g., $2a = 0.4(2c)$. Because the even-rank modes are not excited with symmetrical excitation, the groove width can be chosen so that the TE_{13} and TE_{33} self-coupled modes are cut off, i.e., $2a < \frac{5}{6}(2c)$.

From these considerations, it has been concluded that for low-loss, low-dispersion, and single-mode propagation at millimeter wavelengths, the groove dimensions will be optimized by the following conditions: $2c = 3$–$3.4\lambda_0$, $d = \frac{1}{3}(2c)$, $2a = 0.4(2c)$. It should be noted that these recommended dimensional relationships are not critical to a small fraction of a wavelength. The effectiveness of these relationships has been shown experimentally, as discussed in Section III.

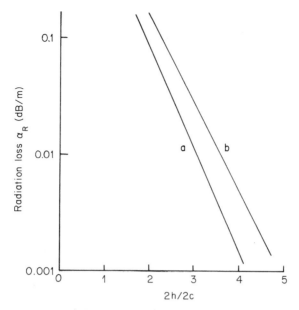

FIG. 6   Variation of radiation loss (approximate method) with normalized waveguide height for the lowest TS_0^s mode; curve a, $2a/2c = \frac{1}{2}$, $d/2c = \frac{1}{3}$; curve b, $2a/2c = \frac{1}{2}$, $d/2c = \frac{1}{4}$.

Section II.G gives a simplified basis for calculation of loss by radiation α_r. Figure 6 shows this loss, which is independent of frequency, versus the normalized guide height for the lowest TE_0^s mode. It can be seen that if $2h/2c \geq 5$, the radiation loss is negligible. The guide attenuation constant α_g is then dominated by the dissipation losses in the metal walls.

In view of the above discussion, suitable selections for the groove-guide parameters under normal conditions are as follows: plane separation, approximately $3-3.4\lambda_0$; groove depth, one-third the plane separation; groove width, four-tenths the plane separation; guide height, greater than five times the plane separation.

G.   ATTENUATION CONSTANT OF TE_0^s MODES

Let P_A and P_B represent the power transmitted through the regions A and B, respectively. Because of the finite conductivity of the metal walls of the guide, there is power dissipation in the guide and a corresponding attenuation constant. When these power losses are small, we can determine the attenuation constant using the field values for a guide constructed of perfect conductors. The dissipation in the metal walls is designated W_{A1}, W_{A2}, and W_B, and the radiated power is W_R, as shown in Fig. 7. The transmitted power P_z

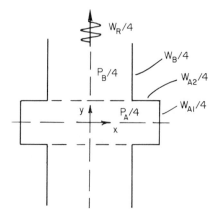

Fɪɢ. 7   Notation for propagating power, conductor power loss, and radiation power loss.

can be obtained in terms of the field components by integrating the z component of the Poynting vector over the cross-sectional area.

The average dissipated power in the guide wall W_m is given by

$$W_m = W_{A1} + W_{A2} + W_B,$$

where

$$W_{A1}/4 = \tfrac{1}{2}R_s \int_0^a (|H_{yA}(x = b)|^2 + |H_{zA}(x = b)|^2)\, dy,$$

$$W_{A2}/4 = \tfrac{1}{2}R_s \int_b^c (|H_{xA}(y = a)|^2 + |H_{zA}(y = a)|^2)\, dx,$$

$$W_B/4 = \tfrac{1}{2}R_s \int_a^h (|H_{yB}(x = c)|^2 + |H_{zB}(x = c)|^2)\, dy,$$

R_s is the surface resistance of the metal wall $(\omega\mu_0/2\sigma)^{1/2}$, and σ is the conductivity of the metal wall.

A very approximate loss of power at the guide edges can be determined by integrating the axial component of the Poynting vector over the unguided area, i.e.,

$$W_R/4 = \int_h^\infty \int_0^c \tfrac{1}{2}\,\mathrm{Re}(E_{xB}H_{yB}^* - E_{yB}H_{xB}^*)\, dx\, dy.$$

It is assumed here simply that the power that would have been transmitted in the guide section, $y > h$, is lost by radiation. Consequently, the attenuation constant α_g of the guide in nepers per meter is given by

$$\alpha_g = W/2P_z = (W_{A1} + W_{A2} + W_B + W_R)/[2(P_A + P_B)]$$
$$= \alpha_{A1} + \alpha_{A2} + \alpha_B + \alpha_R. \tag{19}$$

H. COMPUTED CHARACTERISTICS

1. *Cutoff Wavelength*

The computed results of the normalized cutoff wavelength $\lambda_c/2c$ versus the normalized groove half-width $a/2c$ using Eq. (5) are shown in Fig. 8. For a groove depth of one-third the plane separation and a groove width of two-fifths the plane separation, the cut-off wavelength is about 25% greater than for the parallel-plane guide without grooves. Greater widths can add up to 50% total on the cutoff wavelength. A comparison between the results obtained using the first-order approximation, Eqs. (3) and (4), and the second-order approximation, Eq. (5), is shown in Table I. As seen from this table, the difference between the two approximations is small ($<6.5\%$) even for a very deep groove depth ($d = 2c$), and the cutoff wavelength can

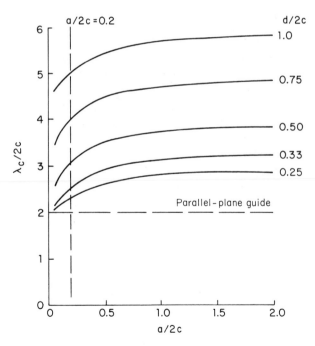

FIG. 8 Variation of normalized cutoff wavelength $\lambda_c/2c$ with normalized groove half-width $a/2c$. $d/2c$ is the groove depth–plane separation.

TABLE I

COMPARISON OF APPROXIMATE CUTOFF WAVELENGTHS[a]

	First-order approximation			Second-order approximation			Difference between the two λ_c values (%)
c/b	$\lambda_c/2c$	$k_{yA1}a$	$k_{yB1}a$	$\lambda_c/2c$	$k_{yA1}a$	$k_{yB1}a$	
0.333	4.77	0.200	0.713	5.10	0.163	0.722	6.5
0.400	3.90	0.252	0.674	4.14	0.214	0.687	5.7
0.500	3.09	0.323	0.599	3.22	0.289	0.616	4.1
0.600	2.60	0.376	0.503	2.65	0.359	0.516	1.8
0.667	2.39	0.398	0.430	2.40	0.391	0.436	0.6
0.800	2.12	0.391	0.263	2.12	0.392	0.262	0.1
0.900	2.03	0.317	0.130	2.03	0.317	0.129	0.0

[a] $a/c = \frac{1}{2}$.

therefore be obtained with sufficient precision by using only the first-order approximation.

2.   Wave Number and Field Decay Constant in the y Direction

Figures 9–12 show the product of the groove half-width a and the wave numbers in the y direction for both regions. These figures give the "fraction of a sinusoid" across the groove half-width and the "exponential decay" in a transverse distance equal to the groove half-width, respectively. As expected, the product of a and the wave number, and the product of a and the field decay constants in the y direction of the TE_0^s mode tend to zero when the groove half-width a tends to zero, i.e., the guide no longer confines the power at the center. As seen from Fig. 10, the field of this mode is a surface wave field, because of the real value k_{yB1}. Also, because of the pure imaginary value k_{yA3}, it can be seen that the fields expressed by the terms of $p = r = 3$ store reactive energy in the neighborhood of $y = a$. The full field components along the z direction can be expressed by the sum of the $p = r = 1$ and $p = r = 3$ fields:

Region A, $[0 \leq y \leq a]$,

$$H_{zA} = A_1 k_c^2 \sin(k_{xA1}x) \cos(k_{yA1}y)$$
$$+ A_3 k_c^2 \sin(k_{xA3}x) \cosh(k_{yA3}y); \qquad (20)$$

Region B, $[a \leq y \leq h]$,

$$H_{zB} = B_1 k_c^2 \sin(k_{xB1}x) \exp[-k_{yB1}(y - a)]$$
$$+ B_3 k_c^2 \sin(k_{xB3}x) \exp[-k_{yB3}(y - a)]; \qquad (21)$$

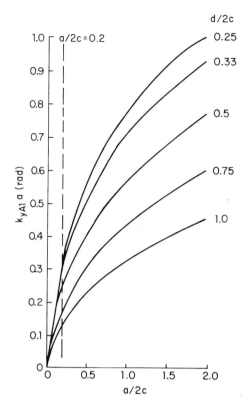

FIG. 9 Variation of the product of dimension a and the wave number in the y direction, $k_{yA1}a$, for groove region A, with normalized groove half-width $a/2c$.

where A_1, A_3 are the amplitude coefficients of field components in region A; B_1, B_3 are the amplitude coefficients of field components in region B; k_c is the cutoff wave number; k_{xA1}, $k_{xA3} = (\pi/2b)$ and $(3\pi/2b)$, respectively, are the wave numbers in the x direction in region A; k_{xB1}, $k_{xB3} = (\pi/2c)$ and $(3\pi/2c)$, respectively, are the wave numbers in the x direction in region B; k_{yA1}, $jk_{yA3} \leq (\pi/2a)$ are the wave numbers in the y direction in region A; jk_{yB1}, $jk_{yB3} \leq (\pi/2h)$ are the wave numbers in the y direction in region B.

3. Relative Amplitude Coefficients

The magnitudes of the fields in region A and region B are given in terms of amplitude coefficients A_p and B_r, respectively, with spatial variations as shown in Eqs. (20) and (21) and Figs. 9–12 for the various mode components.

The relative amplitude coefficients B_1/A_1, B_3/A_1, and A_3/A_1 calculated by Eqs. (10) and (11) are shown in Figs. 13 and 14. As seen from Fig. 14,

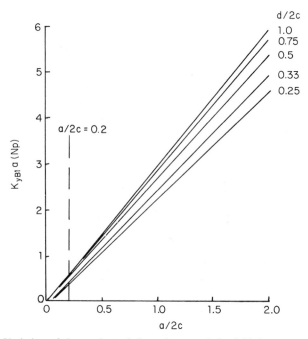

FIG. 10 Variation of the product of dimension a and the field decay constant in the y direction, $k_{yB1}a$, for region B, with normalized groove half-width $a/2c$.

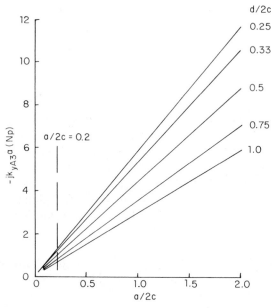

FIG. 11 Variation of the product of dimension a and the field decay constant for $p = r = 3$ in the y direction, $jk_{yA3}a$, for groove region A, with normalized groove half-width $a/2c$.

116

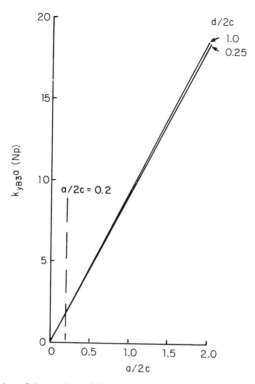

FIG. 12   Variation of the product of dimension a and the field decay constant for $p = r = 3$ in the y direction, $k_{yB3}a$, for region B, with normalized groove half-width $a/2c$.

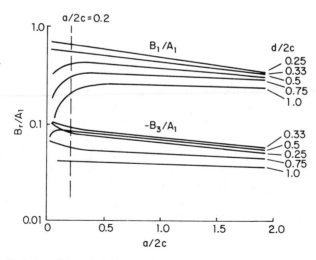

FIG. 13   Variation of the relative amplitude coefficient B_1/A_1 and B_3/A_1 with normalized groove half-width $a/2c$.

117

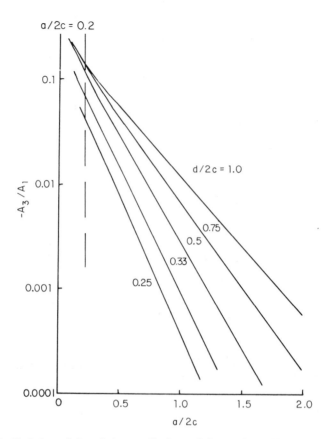

FIG. 14   Variation of the relative amplitude coefficient A_3/A_1 with normalized groove half-width $a/2c$.

$|A_3/A_1|$ decreases rapidly as the normalized groove half-width $a/2c$ increases, showing that the effects of the discontinuity become small at the center of the guide when the normalized groove half-width $a/2c$ is large, e.g., if $a/2c > 2$, $A_3 < 0.001A_1$ even for a deep groove depth ($d \to 2c$). It can also be noted that, even when the normalized groove half-width $a/2c$ tends to zero (i.e., when region A vanishes), the amplitude coefficients A_1 and A_3 in region A still remain. This means that the field with amplitude coefficients B_1 and B_3, which exists in region B in the limit of $a \to 0$, is continuous with the superposed field of A_1 and A_3 for the same limit.

4.   Theoretical Attenuation Constant

Figure 15 shows the variation of attenuation constant with plane separation for 1, 1.5, and 3 mm wavelengths. It can be seen that for low-loss opera-

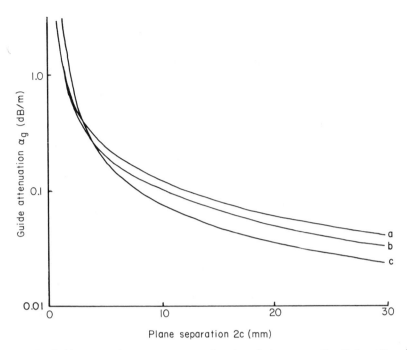

FIG. 15   Guide attenuation constant α_g as a function of plane separation $2c$ for $d/2c = \frac{1}{3}$; $2a/2c = \frac{1}{2}$; and $h/c = 6$: curve a, $\lambda_0 = 1$ mm; curve b, $\lambda_0 = 1.5$ mm; curve c, $\lambda_0 = 3$ mm.

tion, e.g., less than 0.1 dB/m at 3 mm wavelength, the plane separation $2c$ must be large compared with the wavelength, e.g., three times greater than the operating wavelength. Larger plane separations are required for shorter wavelengths to achieve such a low-loss characteristic, although in practice the separation will be kept to a few wavelengths and a higher attenuation will be accepted.

I.   POWER-HANDLING OF THE GUIDE

With the development of high-power sources such as gyrotrons, giving kilowatt mean power levels even at 300 GHz, the problem of transmission becomes important. There are two important limitations: waveguide heating caused by conductor loss and breakdown of the gaseous insulation. For the former low attenuation is required, and this is a criterion for which groove guide is well suited. The relatively large cross-section dimensions are also advantageous for meeting the second criterion. Assuming a plane separation of about 3λ leads to a peak power capacity of about 1 MW at 100 GHz and 100 kW at 300 GHz.

III. Experimental Measurements of the Propagation
Characteristics of Single Rectangular-Groove Guide at 100 GHz

Measurement of the guide characteristics has been made using a resonance technique (Harris *et al.*, 1978a, b). A length of the guide under test was terminated at one end by a fixed conducting plate with an input aperture and at the other end by a moving short-circuit plate with an output aperture on which was mounted a crystal detector. The position of the moving plate was registered by a linear displacement transducer giving a dc output voltage proportional to displacement. The source was an IMPATT oscillator of nominal frequency 100 GHz, and the output from the crystal detector was connected, via a phase-locked amplifier, to an $x-y$ plotter. The resultant spectrum obtained on moving the short-circuit plate gives the guide wavelength from resonance peak separation, the moding characteristics from the overall pattern, and the attenuation from the Q-factor measurements.

The measured guide was constructed from aluminum plate of 8-cm width, with a plane separation of 10 mm and groove dimensions $2a \times d$, where $2a$ is the groove width and d the groove depth in millimeters. The longitudinal grooves were milled by conventional workshop techniques. The outer edges of the guide were left open to allow unwanted unconfined modes to leave the resonator system.

The guide wavelength λ_g can be determined approximately by doubling the distance between any two adjacent resonances of the same mode, and from this value the number of standing waves (which must be an integer) in the waveguide resonator can be estimated. If m is the number of standing waves in the waveguide resonator with resonant length l_m, then the average value of the guide wavelength λ_g is given by $\lambda_g = 2l_m/m$.

The experimental and theoretical guide wavelengths λ_g of the guides under test are given in Table II. The value of m is large, and the accuracy is

TABLE II

GUIDE WAVELENGTHS

λ_0 (mm)	Guide dimension d^a (mm)	Experimental results			Theoretical calculations	
		l_m (mm)	m	λ_g (mm)	λ_c (mm)	λ_g (mm)
3.080	2.50	399.084	257	3.106	24.037	3.106
2.990	3.33	479.860	319	3.009	26.518	3.009
2.778	3.33	428.945	307	2.794	26.518	2.793

^{a}Other guide dimensions: $2c = 10$ mm; $2a = 5$ mm; $2h = 60$ mm.

largely given by the accuracy in measurement of l_m, provided that the right integer m is distinguished. The correlation between theoretical and experimental values of λ_g is very good indeed. When the measured guide wavelengths λ_g are compared with the operating wavelength λ_0, it is seen that as expected the guides under test will have relatively low dispersion, because both wavelengths are within 1 % of each other.

The measured moding spectra for different groove dimensions at 100 GHz were examined and found to correspond to the expected behavior, thus confirming the conditions for diverging higher-order–rank modes. Hence with proper choice of groove dimensions, the guide is shown to transmit only the first-order mode, and higher-order modes leak transversely from the central region. The choice of optimum guide dimensions was given in

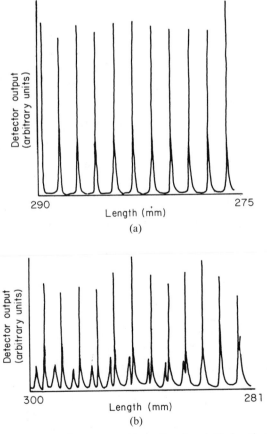

FIG. 16 Moding spectrum for a waveguide cavity of variable length at 100 GHz. (a), $2c = 10$ mm, $2a = 4$ mm, $d = 3\frac{1}{3}$ mm; (b), $2c = 10$ mm, $2a = 10$ mm, $d = 2.5$ mm.

Section II.F. An example of the moding spectrum for the optimized guide dimensions at 100 GHz is shown in Fig. 16a. For a guide with groove dimensions outside the allowable range, additional modes become apparent in the spectrum. These additional modes were identified and correlated with the guide dimensions. An example is shown in Fig. 16b.

The attenuation of the guide can be determined from the resonance Q factor for different positions of the moving short-circuit plate, to eliminate the effect of end-plate and coupling-aperture losses. Comparisons of experimental and theoretical guide attenuation constants are given in Table III. The experimental loss is shown to be from 1.2 to 7.5 times the theoretical value. No attempt has been made to minimize this loss by guide surface treatment.

Our conclusion from a wide range of experimental measurements at 100 GHz is that the guide characteristics are very close to those predicted for

TABLE III

ATTENUATION CONSTANTS OF SINGLE RECTANGULAR-GROOVE GUIDE

2a (mm)	λ_g (mm)	l_o (mm)	Δl_o (mm)	Q_1	α_g (dB/m)	λ_g (mm)	α_g (dB/m)
		Experimental results				Theoretical calculations	
Outer edges open[a]							
5	2.9052	489.239	0.08347	6029	0.44	2.9121	0.063
	2.9179	344.291	0.07663	4578			
Outer edges closed by lossy polystyrene blocks[a]							
5	2.9047	498.155	0.07825	6429	0.48	2.9121	0.063
	2.9043	342.715	0.07031	4920			
Outer edges open[b]							
4	3.0173	479.750	0.0679	7172	0.10	3.0106	0.081
	3.0058	290.055	0.0659	4435			
5	3.0069	470.605	0.0702	6759	0.37	3.0092	0.074
	3.0053	280.980	0.0624	4537			
6	3.0039	473.130	0.0720	6617	0.45	3.0082	0.068
	3.0026	306.265	0.0636	4838			
7	3.0192	397.040	0.0701	5757	0.29	3.0074	0.062
	3.0087	282.810	0.0664	4299			
8	2.9955	479.272	0.0599	8006	0.43	3.0067	0.058
	3.0024	289.730	0.0509	5720			

[a] The operating wavelength λ_0 is 2.891 mm. Guide dimensions: $2c$, 10 mm; d, 2.5 mm; $2h$, 60 mm.

[b] The operating wavelength λ_0 is 2.990 mm. Guide dimensions: $2c$, 10 mm; d, $3\frac{1}{3}$ mm; $2h$, 60 mm.

dispersion and moding behavior. The attenuation measurements are also encouraging in that the ratio of experimental to theoretical loss is relatively low, even for a guide constructed from aluminum by conventional workshop techniques.

IV. Characteristics of Double Rectangular-Groove Guide

A. FIELD CONFIGURATIONS

Double-groove guide, as shown in Fig. 17, can support two modes, with field distributions as shown in Fig. 18. The TE_{11} mode is symmetrical about the guide center, whereas the TE_{12} mode is asymmetrical with zero field at the guide center. The two modes will have different phase velocities, and hence power transfer will occur from one groove to the other with an effective transfer wavelength λ_{gb} related to the difference in phase constant $\Delta\beta$. There is thus the potential for a groove-guide power coupler. Additionally, the loss of the TE_{11} mode may be lower than that of the corresponding single-groove guide. For these reasons the characteristics of double-groove guide have been investigated theoretically and experimentally using 100 GHz as the design frequency. A basic theory for double-groove guide was developed (Harris and Lee, 1981; Lee, 1978) using a first-order approximation,

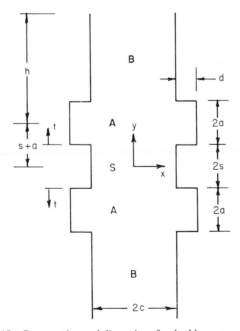

FIG. 17 Cross section and dimensions for double-groove guide.

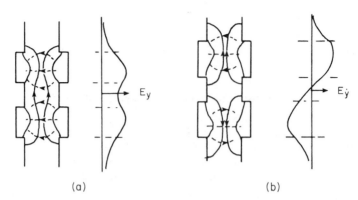

FIG. 18   Field pattern and transverse field E_y amplitude distribution for double-groove guide: (a), TE_{11} (odd) mode; (b), TE_{12} (even) mode; ——, E; - - -, H.

but the theory was then extended and a more precise calculation was made possible using a second-order approximation (Choi, 1982), because this is required to obtain a value of $\Delta\beta$ and λ_{gb} with sufficient precision.

B.   First- and Second-Order Solutions of the Characteristic Equations of the TE (Symmetrical and Asymmetrical) Modes, TE_{11} and TE_{12}

The equations are obtained in a way similar to that for single rectangular-groove guide by applying the Nakahara and Kurauchi inter-region boundary conditions, and are given as follows.

For the first-order approximation,

$$X_1 \tan(X_1) = U_{11} W_{11} Y_1, \tag{22}$$

$$X_{1(t)} \tan(X_{1(t)}) = U_{11} W_{11} Y_{1(t)} \begin{bmatrix} \tanh(Y_{1(t)}(s/t)) \\ \coth(Y_{1(t)}(s/t)) \end{bmatrix}, \tag{23}$$

where tanh is appropriate for the TE_{11} and coth for the TE_{12} modes, and U, W, and Y have the same meaning as for the single-groove analysis in Eq. (5).

For the second-order approximation,

$$X_1 \tan(X_1) = K_{11} - \frac{K_{13} K_{31}}{K_{33} - X_3 \tan(X_3)}, \tag{24}$$

where

$$K_{mn} = \sum_{r=1}^{3} U_{mr} W_{rn} Y_r,$$

$$X_p = \{X_1^2 - (p^2 - 1)[\pi(2a - t)/2b]^2\}^{1/2},$$

$$Y_r = \{-X_1^2 + [(rb/c)^2 - 1][\pi(2a - t)/2b]^2\}^{1/2},$$

and

$$X_{1(t)} \tan(X_{1(t)}) = K_{11(t)} - \frac{K_{13(t)}K_{31(t)}}{K_{33(t)} - X_{3(t)} \tan(X_{3(t)})}, \tag{25}$$

where

$$K_{mn(t)} = \sum_{r=1}^{3} U_{mr} W_{rn} Y_{r(t)} \begin{bmatrix} \tanh(Y_{r(t)} s/t) \\ \coth(Y_{r(t)} s/t) \end{bmatrix},$$

$$X_{p(t)} = [X_{1(t)}^2 - (p^2 - 1)(\pi t/2b)^2]^{1/2},$$

$$Y_{r(t)} = \{-X_{1(t)}^2 + [(rb/c)^2 - 1](\pi t/2b)^2\}^{1/2}.$$

In these equations, $X_p = k_{yA}(2a - t)$, $X_{p(t)} = k_{yA}(t)$, $Y_r = k_{yB}(2a - t)$, and $Y_{r(t)} = k_{yB}(t)$. The H_z fields of the TE modes in the three regions can be written as follows:

(1) in region A,

$$H_{zA} = \sum_p A_p k_c^2 \sin(k_{xA}x) \cos[k_{yA}(y - s - t)]; \tag{26}$$

(2) in region B,

$$H_{zB} = \sum_r B_r k_c^2 \sin(k_{xB}x) \exp[-k_{yB}(y - s - 2a)]; \tag{27}$$

(3) in region S, for the TE_{11} mode,

$$H_{zS} = \sum_r S_r k_c^2 \sin(k_{xS}x) \cosh(k_{yS}y), \tag{28}$$

and for the TE_{12} mode,

$$H_{zS} = \sum_r S_r k_c^2 \sin(k_{xS}x) \sinh(k_{yS}y). \tag{29}$$

In the foregoing, A_p, B_r, and S_r are the amplitude constants; t is the value of y for maximum field strength; $k_{xA} = p\pi/2b$, the wave number in the x direction in region A; $k_{xB} = r\pi/2c$, the wave number in the x direction in region B; $k_{xS} = r\pi/2c$, the wave number in the x direction in region S; k_{yA} is the wave number in the y direction in region A; jk_{yB} is the wave number in the y direction in region B; and jk_{yS} is the wave number in the y direction in region S.

Also, for the three regions A, B, and S,

$$k_c^2 = k_{xA}^2 + k_{yA}^2, \tag{30}$$

$$k_c^2 = k_{xB}^2 - k_{yB}^2, \tag{31}$$

$$k_c^2 = k_{xS}^2 - k_{yS}^2, \tag{32}$$

$$k_c^2 = k_o^2 - \beta^2, \tag{33}$$

where $k_c = 2\pi/\lambda_c$, the cutoff wave number; $k_o = 2\pi/\lambda_o$ is the free-space wave number; and $\beta = 2\pi/\lambda_g$ is the propagation constant in the z direction. It follows that because $k_{xB} = k_{xS}$, then $k_{yB} = k_{yS}$. Therefore,

$$k_{yS} = k_{yB} = (k_{xB}^2 - k_{xA}^2 - k_{yA}^2)^{1/2}. \tag{34}$$

Now if the guide dimensions are defined, X_p or $X_{p(t)}$ can be calculated from Eqs. (22) and (23), or (24) and (25). The cutoff wavelength can then be obtained. Other characteristics, such as the amplitude coefficients and attenuation constants, can be determined using the same approach as for the single-groove guide.

C. Computed Characteristics and Experimental Behavior

The propagation characteristics of double-groove guide having each groove with the normalized dimensions $d/2c = \frac{1}{3}$ and $a/c = \frac{1}{2}$ have been computed for various normalized groove separations s/c, using both the first- and second-order approximations.

The normalized cutoff wavelength of the TE_{11} and TE_{12} modes is shown in Fig. 19 as a function of groove separation. When this separation is greater than twice the plane separation, each groove behaves effectively as an

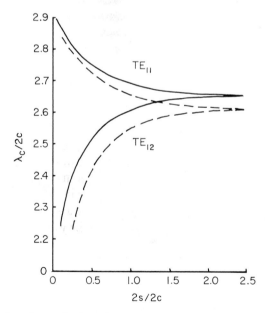

Fig. 19   Variation of normalized cutoff wavelength of TE_{11} and TE_{12} modes with normalized groove separation $2s/2c$: ———, second-order solution; ———, first-order solution.

independent separate groove. The propagation phase constant for each mode, the difference in phase constant between the two modes, and the subsequent beat wavelength for interference between them can then be readily determined; these are shown in Figs. 20 and 21. It can be seen that because of the small difference in phase constants, the beat wavelength is long, which would make a coupler based on this property inconvenient in length. The other parameters depicted in Section II for single-groove guide can also be computed for this double-groove guide structure. The attenuation constant would be significant if it were appreciably lower than the single-groove equivalent. A computed result for the two modes is shown in Fig. 22, from which it can be noted that, although the TE_{11} mode can have a lower attenuation than the single-groove case, the reduction is not a significant one.

The experimental behavior was determined by the resonant technique as outlined in Section III. Double-resonant spectra were obtained, as illustrated in Fig. 23, the mode being identified by a perturbation method. A similar display could be shown by detection at either groove. The corresponding experimental points are shown in Figs. 20 and 21, good correspondence being obtained with the second-order approximation computation in each case.

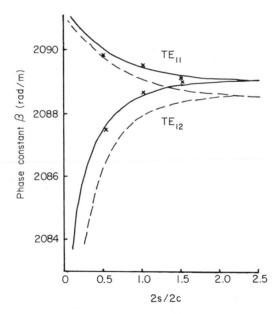

FIG. 20 Variation of phase constants of the TE_{11} and TE_{12} modes with normalized groove separation $2s/2c$: ——, second-order solution; ---, first-order solution; ×, experimental values, $\lambda_0 = 2.99$ mm; $2c = 10$ mm; $2a = 5$ mm; $d = 3\frac{1}{3}$ mm.

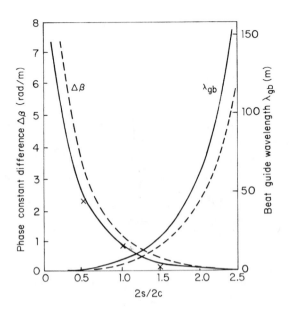

FIG. 21 Variation of difference in phase constant between the TE_{11} and TE_{12} modes and the beat guide wavelength between them, with normalized groove separation $2s/2c$: $\lambda_0 = 2.99$ mm; $2c = 10$ mm; $2a = 5$ mm; $d = 3\frac{1}{3}$ mm; ——, second-order solution; ----, first-order solution; ×, experimental values.

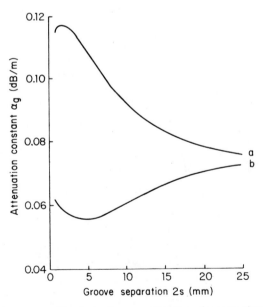

FIG. 22 Double-groove guide attenuation α_g as a function of grooves separation $2s$: curve a, TE_{12}; curve b, TE_{11}. The plane separation $2c = 10$ mm.

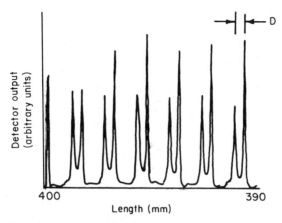

FIG. 23 Moding spectrum for a double-groove guide cavity of variable length at 100 GHz: $2s = 5$ mm, $d = 3\frac{1}{3}$ mm, $2c = 10$ mm, $2a = 5$ mm. D is the separation of the TE_{11} and TE_{12} mode resonances having the same known numbers of standing waves in the waveguide resonator.

Measurements on the transfer of energy between the two grooves in a transmission experiment also gave good correlation with the theoretical beat-wavelength calculations. The measured attenuation of the two modes for a groove separation of half the plane separation was 0.2 and 0.6 dBm^{-1}, respectively, factors of 4 and 6 times greater than the theoretical prediction.

D. APPLICABILITY OF DOUBLE-GROOVE GUIDE

Although the double-groove guide has interesting properties, its application will be limited. The attenuation of the TE_{11} mode is shown to be lower than that for single-groove guide, but the gain is marginal. The presence of the two modes has been demonstrated, and the second-order approximation theory agrees well with experimental values. The beat wavelength between them is inconveniently long, however, and this is likely to prohibit the use of the two modes for an effective power coupler.

V. Components in Groove Guide

For any waveguide possibility to find application to a practical system, a range of appropriate components must be developed. Among the early papers on groove guide, that by Nakahara and Kurauchi (1964) included reference to an application for communication with moving trains, using the open nature of the guide to allow coupling along its length. No results were included for this system, however. During recent years attention has been given at Cardiff to the development of some basic active and passive devices to enable groove-guide systems to be used at frequencies of 100 GHz and above, to take advantage of the low-attenuation characteristics. There has also been a proposal for a groove-guide radiator (Oliner and Lampariello, 1982). A description of components we have fabricated and their performance at 100 GHz is given below, but they have not yet been optimized.

A. GUIDE STRAIGHT JUNCTIONS AND TRANSITIONS

The low tolerances on groove-guide dimensions make it easy to construct even at short millimeter wavelengths, but this advantage would be nullified if the junction between two guide sections required high precision. However, experiments have shown that this is not the case (Harris and Choi, 1979). Straight butt ends have been used, made by conventional machine-shop techniques. The loss at an individual junction is approximately 0.2 dB at 100 GHz even with a gap of up to 1 mm, and a misalignment by 0.5 mm can be tolerated without introducing an additional loss of more than 0.1 dB. These desirable features follow from the relatively large cross-sectional dimensions and the low wall currents in the direction of propagation. The only requirements for guide junctions, therefore, are good flat end surfaces, a means of alignment to, say, ± 0.5 mm, and a method for mechanical coupling. Such an arrangement is shown in Fig. 24.

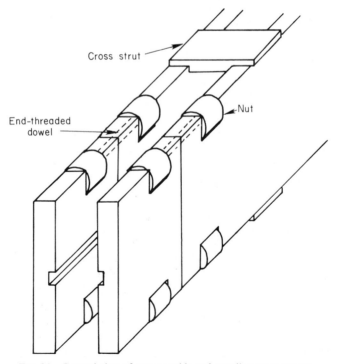

FIG. 24   General view of groove guide and coupling arrangement.

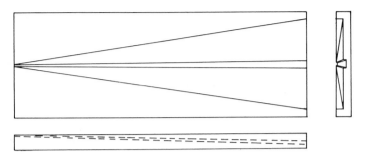

FIG. 25 Transition from dominant-mode rectangular guide to groove guide.

At present, most components available for 100–300 GHz use are in dominant-mode rectangular guide. A transition to such guide is therefore required and can be readily constructed. Ruddy (1965) quotes a transition for 27–40 GHz with a loss less than 0.6 dB, and Nakahara and Kurauchi (1965) give a design but without experimental performance. We have considered several different ways of gradually transforming from one cross section to the other. A simple technique that gives good performance and can be readily milled from solid material is shown in Fig. 25. A pair of transitions milled from aluminum for operation at 100 GHz has a total loss less than 1 dB, and the VSWR is estimated to be about 1.2.

B. *E*- AND *H*-PLANE BENDS

Cylindrical bends in both *E* and *H* planes have been attempted but with very limited success. The loss was found to be unacceptably high. The *E*-plane bend loss was more than 2 dB even with a cylindrical outer reflector, and the *H*-plane bend was even poorer with a loss of 5 dB. The *E*-plane situation can be remedied by use of a 90° corner with a 45° reflection plate, as in Fig. 26.

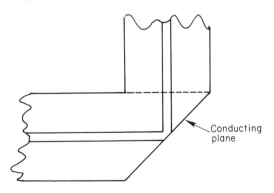

Conducting plane

FIG. 26 An *E*-plane corner bend.

This is easily made, and the loss at 100 GHz was 0.6 dB. The H-plane bend presents somewhat greater difficulty. The high loss was not expected and may be due to nonsymmetrical transverse field distribution with transverse radiation of harmonics. The loss can be reduced significantly by tapering down to a narrower plane separation, e.g., to one wavelength, but a satisfactory solution has yet to be reached.

C. GROOVE-GUIDE COUPLERS

Two approaches have been investigated for coupling out part of the main guide power: the use of a section of double-groove guide and the use of a quasi-optical coupler with a 45° partial reflector to couple power into an orthogonal groove-guide section.

It was shown in Section IV that two modes can exist on double-groove guide, one being symmetrical and the other asymmetrical across the two grooves, with a consequent transfer of energy from one groove to the other. A series of experiments on transfer of power for various lengths of the double-groove section has confirmed the theoretically expected transfer behavior. The power coupled into the second groove was extracted by a 90° reflector bend. Two difficulties limiting this approach are the required length for a reasonable transfer of power, e.g., 80 cm for 10-dB coupling, and the loss due to intrusion of the 45° reflecting plane.

The quasi-optical directional coupler is simple in concept and design. Two groove guides intersect at 90°, and a partial reflector is set at 45°, as in Fig. 27. This partial reflector can be a thin dielectric film for low coupling,

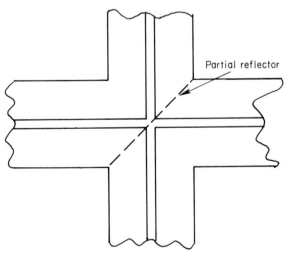

FIG. 27   A quasi-optic directional coupler.

or a partially reflecting grid for coupling out a large part of the incident power. Melinex film of thickness 175, 125, 75, and 40 μm gave experimental coupled powers of 13, 17, 20, and 26 dB respectively. The use of line or mesh grids for partial reflectors is well documented. They can be conveniently made by photoetching Duroid (e.g., RT Duroid 5880). Grids with strip widths and line spacings of 0.05 and 0.6 mm, 0.1 and 1.1 mm, and 0.1 and 1.4 mm gave coupled powers of 3.5, 7.1, and 9.6 dB, respectively. This is close to calculated design values. The overall insertion loss of the coupler is appreciable, but is less than 1 dB. However, it is expected that improvement in design, e.g., by tapering the groove region towards the partial reflector, will reduce this loss.

D. ATTENUATORS, MATCHED TERMINATIONS, AND PHASE SHIFTERS

The open nature of the guide makes it relatively simple to introduce lossy materials for attenuation or terminations, or to introduce higher permittivity materials for a change of phase along the guide. Simple but practical attenuators have been made using a disk of Tufnol coated with microwave absorbent, pivoted near the circumference, and mounted near the outer edge of the guide so that it could be moved into the more intense field region by a rotation. Typically a 60° rotation will readily give up to 10 dB attenuation. Thus, only a simple construction and physical measurement is required. Likewise, the large cross-sectional dimensions lead to simple matched termination possibilities. A V-shaped card coated with microwave absorbent placed parallel to the guide wall appeared to be satisfactory, although careful measurement of VSWR was not carried out.

Two approaches that could be used to give a change of phase along the guide are a change of guide wall separation and insertion of a dielectric with relative permittivity greater than unity. The first is more easily carried out for groove guide than for rectangular guide, because of the absence of a need for a transverse conducting surface. A simple "squeeze section" could be made and would suffice. As for rectangular guide, a dielectric insertion could be used, but with the added limitation that a disturbance of field distribution will produce transverse space harmonics that will radiate and cause loss. The former technique is therefore probably preferable.

E. DESIGN OF GROOVE-GUIDE DETECTORS

To take full advantage of groove-guide characteristics, it is essential to be able to design and make video detectors and mixers directly in the guide without the need to transform to dominant-mode rectangular guide. Beam-lead Schottky diodes are available, such as the GEC diode type DC1346, with a good detection capability at 100 GHz. A combination of low capacitance and low series resistance (e.g., 0.02 pF and 5 Ω, respectively) with good mechanical strength is achieved by using low-loss glass to support the connecting strips. The overall size is very small (< 1 mm), but they can be bonded directly into a miniature microwave integrated-circuit configuration.

A suitable mount has been designed and tested using an etched circuit on Duroid 5880 (Harris and Mak, 1981). It consists essentially of a triangular dipole with a centrally mounted diode placed transversely across the groove guide, as in Fig. 28. The dipole length and flare angle were determined by a scaled-up model at 10 GHz, an overall length of 0.4 λ with a flare angle of 40° giving good results. The beam-lead diode was bonded to the circuit by thermal compression, and the dc conditions were checked through the connecting lines. Because the dipole intercepts only a fraction of the incident wave, a concentrating reflector is essential for good sensitivity. Flat, corner, cylindrical, and parabolic reflectors were studied. The latter gave the highest sensitivity, but this was only some 20% better than the cylindrical reflector, which was adopted to make manufacturing easier. The cylindrical reflector is mounted behind the diode on a simple micrometer mount, adjusted for maximum sensitivity. The diode bias current was also adjusted for best results.

The voltage sensitivity at 100 GHz increased with bias current to a maximum of about 400 V/W at 60 μA, and the tangential sensitivity was estimated

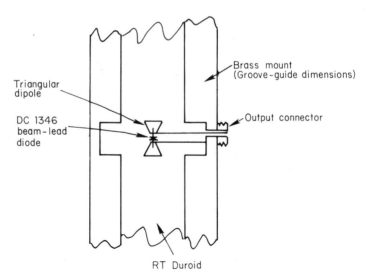

FIG. 28   A detector in groove-guide mount, using a beam-lead diode.

to be -42 dBm. The power levels had to be estimated because of the absence of accurate power-measuring devices, but we assume that the levels are known to within ± 3 dB. This type of groove-guide detector is simple to construct, given the availability of a suitable diode, and its performance as a video detector is superior to many of the available rectangular-guide detectors. The mount could be readily scaled down for operation at higher frequencies. No measurements have been made on the performance of the component as a mixer, but there is no reason to believe that it would not be equally effective.

F. SOLID-STATE OSCILLATORS

It would be valuable if oscillators could be made for direct groove-guide use, but it is not so critical because oscillators do not usually proliferate in a system. There are several possible configurations for introducing an oscillator active device such as a Gunn diode or IMPATT diode directly into groove guide. For example, a transverse conducting plate can be tapered into groove guide at the midpoint because the electric field is parallel to the guide walls at that point. Two diodes mounted back to back could be connected to the plate. We have taken the simpler approach of mounting a diode (in our case an IMPATT diode) in an over-moded cavity whose width is equal to the distance between the guide at the groove position (Harris and Mak, 1982). The height of the over-moded cavity had to be reduced to half that of the groove width because of the need to restrict the length of the biasing circuit in contact with the diode to maintain mechanical rigidity. The oversize cavity dimensions, approximately 6.5 and 0.75 times the operating wavelength, enabled a simple taper transformer to be used to link the cavity to the groove guide, with no dominant mode-guide sections in the system. The frequency of operation was primarily controlled by the diameter of the resonant cap. A schematic diagram of the oscillator is shown in Fig. 29. The IMPATT diode used for the work was of the Si double-drift ($p^+-p-n-n^+$) type supplied by Plessey (Caswell). These devices are fabricated by epitaxial techniques; they use substrates with high-conductivity arsenic doping to minimize both the substrate series resistance and the contact resistance of the substrate metallization. Both the cavity and the transformer section were machined from aluminum using standard workshop techniques. The resonant cap for the diode had a taper angle of $14°$ and a diameter close to a half free-space wavelength, i.e., 1.5 mm in diameter. This cap is an integral part of the biasing network, which consists of a four-section high–low impedance low-pass filter insulated from the main oscillator block by a thin sheet of Mylar.

The oscillator readily produced power at 95.6 GHz for a diode current of 60 mA. The variable back short-circuit was adjusted for maximum power, which was estimated to be 15 mW at 90 mA, an encouraging figure in view

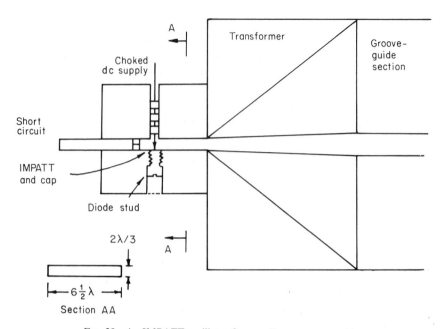

FIG. 29   An IMPATT oscillator for coupling to groove guide.

of the lack of extensive optimization, the use of the aluminum circuit, and the absence of high-surface-finish techniques used in fabrication. A frequency spectrum analyzer enabled the oscillator output linewidth to be measured. The $f_0/\Delta f$ value at -30 dB was measured to be 10^6, a value slightly greater than that for an oscillator with a dominant-mode rectangular cavity. Some frequency tuning resulted from a change of bias current, with a tuning range of about 100 MHz.

It has thus been shown that a convenient oscillator and transition can be made for groove-guide systems without the need to introduce dominant-mode rectangular guide, but the problem of providing an isolator remains.

G.   MEASUREMENT OF VSWR

At lower microwave frequencies, VSWR is measured by moving a probe with detector at great precision along a slot in the broad face of rectangular guide. This becomes exceedingly difficult at 100 GHz or above because of the minute dimensions involved. A simple technique at 100 GHz has been used, however, taking advantage of the open nature of the guide and relatively cheap linear transducers. The basic system is shown in Fig. 30. A small transverse dipole can be moved along the guide above the groove region by a micrometer to scatter a fraction of the energy in the guide. A horn and a

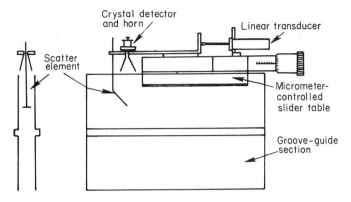

FIG. 30   A scattering-type standing-wave indicator.

detector moving with the dipole give a measure of the scattered radiation and, therefore, of the field intensity at the dipole. The output from the detector is passed, via a phase-locked amplifier, to the y deflection of an $x-y$ plotter. The output from the linear transducer, which is proportional to the distance along the guide, is connected directly to the x deflection of the plotter, and the voltage standing wave is plotted directly. An example of the results obtained is shown in Fig. 31. This is the case of a short-circuit termination with a small aperture. There are more efficient ways of coupling power from the dipole to the detector; these are now being investigated, but the overall technique seems promising for an effective standing-wave detector for 100 GHz or more at a relatively low cost.

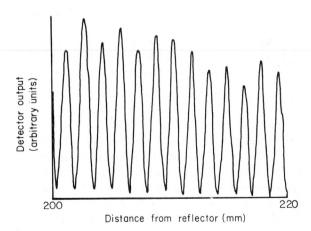

FIG. 31   Example of a standing wave in groove guide: reflection from a near-short-circuit.

H. Groove-Guide Antennas and Radiators

In normal groove guide, the power is distributed over an area a few wavelengths in each transverse direction. An open-ended guide is therefore a relatively good match and will radiate directly. The gain can be conveniently increased by diverging the guide walls in one direction and reducing the groove depth or width to spread out the field in the other transverse direction.

A proposal has recently been made to use transverse leaky modes to produce a linear-array radiator (Oliner and Lampariello, 1982). A variation in the symmetry from a single half-sinusoid between the guide walls will produce space harmonics that will radiate freely transversely from the outer edge of the guide. This asymmetry can be produced by introducing additional conducting surfaces between the guide walls, e.g., a strip parallel to the groove direction and perpendicular to one of the guide walls, and initial work shows that the approach is an encouraging one. A beam that can be scanned from about 15° to nearly 60° from the normal can result from a scan of frequency.

VI. Summary of Groove-Guide Possibilities

Groove guide has the valuable property of being a practical single-mode propagation waveguide even though cross-sectional dimensions can be considerably greater than the wavelength. This characteristic is a feature of the guide itself and does not rely upon the introduction of a mode filter, but the outer edges of the guide must be open or lossy to absorb any higher-order modes that might be generated. No dimensions of the guide are critical to a fraction of a wavelength, and it is therefore simple and easy to construct by conventional techniques. The loss of the guide is at least an order of magnitude less than that of dominant-mode rectangular guide for the same frequency. The simple form, low loss, and relatively large cross section allows the guide to be scaled down conveniently for frequencies in excess of 100 GHz, and it is believed that operation at 300 GHz is very feasible. The low loss and large dimensions also allow high power to be transferred by the guide, a factor that could be valuable with the development of high-power sources such as gyrotrons.

The theoretical behavior of the guide is now well documented. The optimum dimensions are a compromise between lower losses on the one hand and higher-order mode propagation on the other. For the work at Cardiff, the following dimensions are used: plane separation, 3–3.4 λ; groove depth, one-third the plane separation; groove width, four-tenths the plane separation; transverse dimension, six times the plane separation. The experimental attenuation at 100 GHz is around 0.2 dBm^{-1} for aluminum guide produced by milling from sheet material.

Some progress has been made toward an overall groove-guide system. It is easy to join sections of the guide together with a simple butt joint. A gap of 0.5 mm and a transverse misalignment by a similar amount can be tolerated without a serious increase in attenuation. Transitions can also be made with dominant-mode rectangular guide. It has been found that reflection-corner E-plane bends can be readily made with low loss, but difficulty has been encountered so far with H-plane bends. The most convenient couplers made are quasi-optic devices, using partial reflecting films and grids to divert power to an orthogonal groove, with coupling from 3 to 20 dB. Because the field in the groove region is similar to that for rectangular guide, similar approaches can be used for some components. The open edges of the guide also allow material to be easily inserted, e.g., lossy vanes, and thus variable attenuators and loads are simply made. Phase change with low loss is probably best achieved, however, by a variation of plane separation. Groove guide also lends itself to fairly simple radiators, either directed out from the guide end with suitable expansion of the field-occupation area or by transverse radiation through controlled-field asymmetry.

There have been successful attempts to generate and detect radiation at 100 GHz in structures associated with groove guide. The oscillator was an IMPATT in an over-moded cavity, with a transverse dimension to fit the groove region of the guide so that only a simple transition was required. The performance was similar to that for normal rectangular guide. The detector consisted of a beam-lead diode mounted in a triangular dipole in Duroid, situated across the guide with a cylindrical rear reflector. A good voltage sensitivity and tangential sensitivity were achieved.

For measurements, a simple but effective standing-wave detector using a movable dipole scattering element has been used.

The possible use of double-groove guide to give lower loss as a coupler has been investigated, but there seems to be little significant advantage to this approach.

It is not expected that groove guide will find use at frequencies below 100 GHz, where strip-line and dominant-mode guide will suffice except for special conditions, e.g., high-power transmission or low-loss, long-distance guide requirements. At frequencies above 100 GHz, however, alternative techniques look increasingly unattractive as the frequency increases, although the transverse dimensions of groove guide become more convenient. The guide, and systems derived from it, could be particularly valuable for the 100–300-GHz frequency range.

ACKNOWLEDGMENTS

The authors wish to acknowledge the part played by other research workers in the development of this program, and particularly Andrew Doswell, Kin Wah Lee, and Simon Mak.

Yat Man Choi is grateful to the University of Wales Institute of Science and Technology for the award of a research studentship for the period of his work.

The program has received substantial help from the Science and Engineering Research Council through the award of several research grants. The assistance of the GEC Hirst Research Centre with the provision of beam-lead diodes, and of the Plessey Allan Clark Research Laboratory with IMPATT diodes, is also gladly acknowledged.

REFERENCES

Bava, G. P., and Perona, G. (1966). *Electron. Lett.* **2**, 13–15.

Choi, Y. M. (1982). "Groove-Guide as a Short-Millimetric Waveguide System." Ph. D. Thesis, University of Wales, Cardiff, United Kingdom.

Griemsmann, J. W. E. (1964). *Proc. Symp. Quasi-Opt.* **14**, 565–578.

Harris, D. J. (1980). *Microwave System News* **10**, 62.

Harris, D. J., and Choi, Y. M. (1979). *Electron. Lett.* **15**, 687–688.

Harris, D. J., and Lee, K. W. (1981). *IEE Proc.* **128H**, 6–10.

Harris, D. J. and Mak, S. (1981). *Electron. Lett.* **17**, 516–517.

Harris, D. J., and Mak, S. (1982). *Electron. Lett.* **18**, 399–400.

Harris, D. J., Lee, K. W., and Batt, R. J. (1978a). *Infrared Phys.* **18**, 741–747.

Harris, D. J., Lee, K. W., and Reeves, J. M. (1978b). *IEEE Trans. Microwave Theory Tech.* **MTT-26**, 998–1001.

Lee, K. W. (1978). "An Investigation of Rectangular-Groove Guide as a Waveguide for the 100–1000 GHz Frequency Range." Ph. D. Thesis, University of Wales, Cardiff, United Kingdom.

Nakahara, T., and Kurauchi, N. (1964). *J. Inst. Electr. Commun. Eng. Japan* **47**, 1029–1036.

Nakahara, T., and Kurauchi, N. (1965). *Sumitomo Electr. Tech. Rev.* **5**, 65–71.

Oliner, A. A., and Lampariello, P. (1982). *Electron. Lett.* **18**, 1105–1106.

Ruddy, J. M. (1965). *IEEE. Trans. Microwave Theory Tech.* **MTT-13**, 880–881.

Schiek, B., and Schlosser, W. (1964). *AEU* **18**, 481–486.

Snurnikova, G. K. (1970). *Radio Eng. Electron. Phys.* **15**, 509–511.

Snurnikova, G. K. (1971). *Radio Eng. Electron. Phys.* **16**, 352–354.

Tischer, F. J. (1963a). *IEEE Trans. Microwave Theory Tech.* **MTT-11**, 291–296.

Tischer, F. J. (1963b). *Proc. IEEE* **51**, 1050–1051.

Yee, H. Y., and Audeh, N. F. (1965). *Proc. Nat. Electron. Conf.* **21**, 18–23.

CHAPTER 4

The Application of Oversized Cavities for Millimeter-Wave Spectroscopy

F. Kremer, A. Poglitsch, D. Böhme, and L. Genzel

Max-Planck-Institut für Festkörperforschung
Stuttgart, Federal Republic of Germany

I. Introduction

An oversized cavity (with typical dimensions being much larger than the wavelength) is an untuned cavity with high reflectivity walls within which an isotropic and homogeneous field is generated on a time-averaged basis. The uniqueness of this configuration has allowed the device to find applications in many areas, including dielectric measurements of gaseous, liquid, and solid materials, power measurements, studies of electromagnetic compatibility, analyses of industrial heating and microwave ovens, and determinations of the effect of exposing biological systems to microwave radiation. The spectral extent of these applications covers the microwave through the submillimeter region to wavelengths as short as 0.25 mm. In the guise of an integrating sphere, it is widely used for many spectroscopic applications in the infrared and visible spectra.

The usefulness of the device arises from several advantages. For absorption measurements of gaseous samples, the oversized cavity serves as a long-path absorption cell, thus providing a many-fold increase in sensivity compared to a single-path setup. The photon-storing capability of the device can also be used to advantage for measuring the resistivity of metals. In this case the walls of the oversized cavity are partly composed of the material being studied. For dielectric measurements of inhomogeneous samples, the main advantage is that such measurements are only sensitive to the absorption loss in the specimen and not to scattering and reflection losses. Thus, specimen preparation in a plane-parallel form is not necessary, as it is for the established optical methods. It therefore allows the quantitative study of a substantial body of materials, including powders, thin films, and composites. This advantage applies to all spectral regions. At microwave and millimeter-wave frequencies, a further advantage is that, in principle, the device is broadband. So, in contrast to resonant methods such as the cavity perturbation technique or the use of open resonators, the untuned cavity can provide a broadband spectrometer when combined with a swept frequency source. For low-temperature absorption measurements at microwave and millimeter-wave frequencies, the device can be used with a cryostat such that the total sample chamber inside the cavity is composed of fused silica that is nearly transparent. In this way serious problems can be circumvented by heat conduction in waveguide systems or by standing waves between the windows of the cryostat in quasi-optical systems.

The main disadvantage of the oversized cavity is the loss of phase information. Thus, only indirect measurements of the real part n of the complex index of refraction $\tilde{n} = n + ik$ are possible, as will be discussed later. Furthermore, the oversized cavity technique requires larger amounts of sample material than quasi-optical or waveguide methods. This is especially limiting for studies of biological samples that are available only in small quantities.

The theory of the oversized cavity as applied to absorption measurements was originally developed by Lamb (1946) for gases and vapors that entirely fill the cavity, and therefore no interface effects have to be considered. This theory was employed by Becker and Autler (1946) to determine the profile of the 0.744 cm^{-1} rotational absorption line of water vapor using a number of single-frequency magnetron sources. Richards and Tinkham (1960) made the first truly broadband measurements with an oversized cavity in their studies of the difference between the surface resistance in the normal and super-conducting states of several metals between 4 and 45 cm^{-1}. Gebbie and Bohlander (1972) subsequently made broadband determinations of the power transmission spectrum of gaseous carbon monoxide from 12 to 45 cm^{-1}, using a spherical cavity having a specularly reflecting internal surface. Pinkerton and Sievers (1982) used a dual nonresonant cavity setup to measure the far-infrared absorptivity of metals. Oversized cavities were also used as an integrating cavity in the visible region to measure quantitatively the absorption of low absorbing materials (Elterman, 1970; Schultz, 1960).

The theory developed by Lamb (1946) for measurement of gaseous samples cannot be directly applied to solid or contained-liquid samples that fill only a small fraction of the cavity. Llewellyn-Jones et al. (1980) developed the first theory for the absorption of low-loss samples in an oversized cavity. Their results demonstrate that for bulk samples with the dimensions on the order of ten wavelengths or more, the absorption coefficient is only weakly dependent on the sample geometry. They established an empirical factor to correct for the effect of reflection at the sample boundaries. The dependence of this correction factor on the sample's refractive index was sufficiently weak under their experimental conditions that precise knowledge of the real part n of the refractive index was not required to determine the absorption coefficient α with acceptable accuracy. Absorption measurements using this theory have been reported at a single frequency, 156 GHz, for some low-loss polymers (Llewellyn-Jones et al., 1980) and also, using broadband radiation from 3 to 15 cm^{-1}, on polytetrafluoroethylene (PTFE) (Willis et al., 1981). An alternative approach to the study of solids and liquids by the use of an oversized cavity has been developed by Kremer and Izatt (Izatt and Kremer, 1981; Kremer and Izatt, 1981). They consider only specimens of a well-defined lamellar form. By making complete allowance for all interface effects, they showed that for weakly absorbing samples ($k \leq 0.05$) the influence of the real part n of the complex index of refraction on the determination of the absorption coefficient is small. For strongly absorbing samples ($k \geq 0.05$), they demonstrated both theoretically and experimentally that simultaneous measurement of the real and imaginary parts of the refractive index can be accomplished by measuring the absorption of lamellar samples of different thicknesses. Using this approach, measurements were reported at 70 GHz for PTFE, polyethylene, and Plexiglas (Kremer and Izatt, 1981), for water and glycerine (Izatt and Kremer, 1981), and for biological macro-molecules such as RNA, collagen, and lysozyme (Kremer, 1983; Kremer and Genzel, 1981). The temperature dependence of the millimeter-wave absorp-tion of biological cells was measured (Kremer et al., 1980; Genzel et al., 1981) and, furthermore, measurements of the temperature dependence of the absorption coefficient of hemoglobin and poly-L-alanine from 4 K to room temperature and at frequencies from 40 to 150 GHz were carried out (Genzel et al., 1983). Furthermore, the dielectric properties of hydrated lysozyme were measured in this frequency and temperature range (Poglitsch et al., 1984).

Birch et al. (1983) performed a systematic study of the oversized cavity and its application to absorption measurements using broadband near-milli-meter-wavelength radiation from a thermal source. It was found that the necessary isotropic internal radiation flux could be established in a cavity of regular shape without a mode stirrer. For coherent radiation this was not the case, thus proving the need for a mode stirrer.

In this chapter the application of oversized cavities for millimeter-wave spectroscopy using coherent radiation sources (backward-wave oscillators) is described. The theory is presented for the measurement of the complex index of refraction of strongly absorbing samples ($k \geq 0.05$), for the measurement of the absorption coefficient of weakly absorbing samples ($k \leq 0.05$), and for the measurement of the temperature-dependent absorption coefficient of weakly absorbing samples ($k \leq 0.05$). The experimental setup and its limitations are described in detail, and examples of the measurements are given.

II. Theoretical Approach

A. BASIC PROCEDURE

It is convenient to associate a quality factor Q with each of the principal cavity loss mechanisms. The Q factor is defined generally by the following relationship.

$$Q^{-1} = \frac{\text{(rate of loss of photons from cavity)}}{\omega(\text{number of photons stored in cavity})}, \tag{1}$$

where ω is the angular frequency of the radiation in the cavity. The Q factor associated with absorption in the sample, Q_s, can be evaluated by comparing the absorption loss with the loss of a known load. The Q factor associated with this load is Q_c. Such a known loss can be realized by opening a hole of a certain area in the cavity (Lamb's hole method) or by placing a sample of known geometry and known dielectric properties in the cavity. The Q factor associated with the remaining resonant losses, such as absorption in the walls and losses through the aperture that is used to feed radiation into the cavity, is denoted by Q_R.

It is convenient to formulate the description in terms of rate equations rather than to use the totally equivalent procedure of summing inverse Q factors, as is often done. A detector placed inside the cavity produces, within its linear range, a signal proportional to the local photon density, which is taken to be homogeneous and isotropic throughout the cavity. Therefore, it suffices to consider the evolution of the total number of photons in the cavity. Let N_0 represent the number of photons in the cavity when no sample is present, and let N_s be the number of photons after the sample is introduced. (Correspondingly, N_c will denote the number of photons when a calibration load of known dielectric properties is in the cavity). If radiation is fed into the empty cavity at a constant rate M, an equilibrium population will be established that satisfies the following equation:

$$dN_0/dt = 0 = M - \ell_R N_0; \qquad (2)$$

and with the sample in the cavity,

$$dN_s/dt = 0 = M - \ell_R N_s - \ell_s N_s. \qquad (3)$$

The loss coefficient ℓ_R represents all losses inherent in the oversized cavity and satisfies the condition

$$\ell_R = \omega/Q_R. \qquad (4)$$

The loss coefficient ℓ_s is entirely due to the absorption in the sample; ℓ_c is entirely due to losses in the calibration load.

To avoid problems associated with absolute calibration of the detector, it is expedient to employ only ratios of detector signals. The ratio of the detector signal with the sample present to that measured with the cavity empty is denoted by $R_s = N_s/N_0$ (correspondingly, $R_c = N_c/N_0$). Application of Eqs. (2) and (3) for the sample and the known load in succession, then, leads to the following result:

$$\ell_s/\ell_c = R_c(1 - R_s)/R_s(1 - R_c). \qquad (5)$$

To calculate the absorption loss in the sample, the radiation field inside the cavity is treated as a photon gas. Consider an elementary area dA of the sample that receives radiation incident upon it between angles of incidence θ and $\theta + d\theta$ and of azimuth ϕ and $\phi + d\phi$ angles, where ϕ is referred to some arbitrary azimuth angle. In a time dt, all photons traveling toward the area dA within the solid angle $\sin \theta \, d\theta \, d\phi$ about the incident direction defined by θ and ϕ, and within a distance $c\,dt$ of the elementary area dA, will strike that area (c is the speed of light). Thus, because of the varying aspect angle of the sample one is concerned with all photons within a volume $c|\cos \theta|dA\,dt$ that are traveling towards the elementary area dA.

For N photons in the cavity of volume V, a photon density of N/V results owing to the assumed homogeneity of the radiation field. The number of photons that will strike this sample area A_s from both sides during the time interval dt is given by

$$(N/V)c\,dt\,A_s\left(\int_0^{2\pi}\int_0^{\pi}\sin\theta|\cos\theta|\,d\theta\,d\phi\right)\bigg/\left(\int_0^{2\pi}\int_0^{\pi}\sin\theta\,d\theta\,d\phi\right)$$

$$= (N/V)c\,dt\,A_s\int_0^{\pi/2}\sin\theta\cos\theta\,d\theta\,d\phi. \tag{6}$$

It is convenient to express the dielectric properties of the sample in terms of the complex index of refraction $\tilde{n} = n + ik$. The real part n is just the refractive index of Snell's law, and the absorption coefficient in Lambert's law is related to k by $\alpha = 2\omega k/c$. The sample's reflection factor, which specifies the fraction of the radiation incident at angle θ that is reflected from the sample back into the cavity field, will be denoted by $R = R(\theta, n, k, d, \lambda)$, where λ is the vacuum wavelength of the radiation and d the sample thickness. The transmission factor, which is defined in an analogous fashion, will be denoted by $T = T(\theta, n, k, d, \lambda)$. These quantities are indicated in Fig. 1. Both coefficients are also polarization dependent, and polarizations normal and parallel to the plane of incidence will be denoted respectively by subscripts $\perp$ and $\|$. Of the photons striking the sample, the fraction A that is absorbed is given in terms of R and T by the energy conservation relationship, $R + T + A = 1$, neglecting scattering and refraction. Because the radiation field in the cavity is isotropic, the total absorption for a thin, flat lamellar sample is obtained by averaging over the two perpendicular polarization components. Thus the number of the photons absorbed in the sample per time unit is given by

$$N\ell_s = (NcA_s/V)\int_0^{\pi/2}A_T(\theta, n, k, d, \lambda)\sin\theta\cos\theta\,d\theta, \tag{7}$$

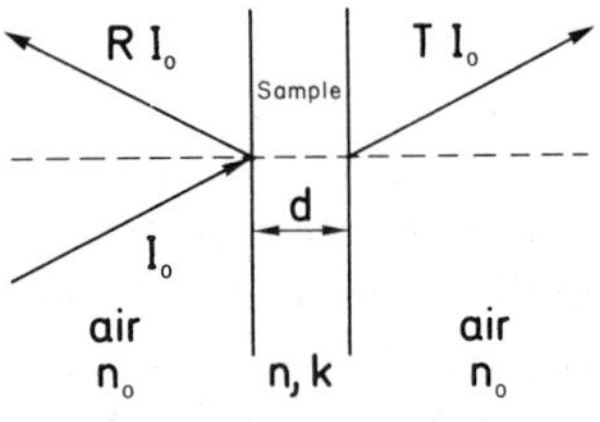

FIG. 1   Scheme of a lamellar sample: d, sample thickness; n, real part of the complex index of refraction; k, imaginary part of the complex index of refraction; n_0, real part of the refractive index of air; I_0, incident radiation; RI_0, reflected radiation; TI_0, transmitted radiation.

where

$$A_T(\theta, n, k, d, \lambda) = \tfrac{1}{2}[A_\perp(\theta, n, k, d, \lambda) + A_\parallel(\theta, n, k, d, \lambda)]. \tag{8}$$

The integral

$$I_s(n, k, d, \lambda) = \int_0^{\pi/2} A_T(\theta, n, k, d, \lambda) \sin\theta \cos\theta \, d\theta \tag{9}$$

is defined as the integral absorption, i.e., the normalized absorption of a thin, flat, double-sided sample in the isotropic radiation field of the cavity averaged over all angles of incidence.

For a hole of area A_H one finds, neglecting diffraction effects, the photon loss of a single-sided sample with $A_T = 1$. This gives, with Eq. (7),

$$N\ell_H = NA_H c/4V \tag{10}$$

in accordance with Lamb (1946). Incorporating this expression and Eq. (9) in Eq. (5) yields

$$I_s(n, k, d, \lambda) = (A_s/4A_H)R_c(1 - R_s)/R_s(1 - R_c). \tag{11}$$

For a calibration sample of known dielectric properties (n_c, k_c) and of area A_c with thickness d_c, Eqs. (5), (8), and (9) yield

$$I_s(n, k, d, \lambda) = (A_c/A_s)R_s(1 - R_s)/R_s(1 - R_c)[I_c(n_c, k_c, d_c, \lambda)]. \tag{12}$$

B. DETERMINATION OF THE COMPLEX INDEX OF REFRACTION

For a lamellar sample with $n = 1.5$, $k = 0.5$, and a thickness of $d = 0.1$ cm (Fig. 1), reflection, transmission, and absorption versus angle of incidence are shown in Fig. 2. (For the detailed formulas see Appendix I.) Integrating over all angles of incidence delivers the integral absorption of the sample. The dependence of the integral absorption on the sample thickness is shown in Fig. 3. Comparing two samples of the same dielectric properties and equal sample areas but different thicknesses yields, according to Eq. (12),

$$I(n, k, d_1, \lambda)/I(n, k, d_2, \lambda) = R_2(1 - R_1)/R_1(1 - R_2), \tag{13}$$

where R_1 (R_2) denotes the ratio of the detector signal when the sample with thickness d_1 (d_2) is in the cavity over the detector signal when no sample is in the cavity. The right-hand side [Eq. (13)] is given by experiment and the integral equation has to be solved to determine the complex index of refraction. The result (Fig. 4) gives different trajectories in the n, k space for the various sample comparisons. Besides mathematical intersections of just

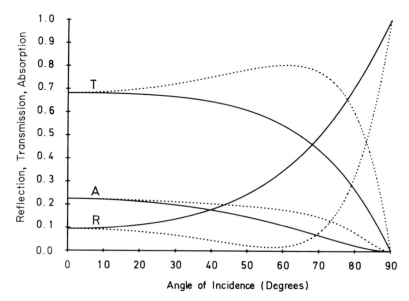

FIG. 2 Reflection (R), transmission (T), and absorption (A) of a lamellar sample with $n = 1.5$, $k = 0.5$, $v = 70$ GHz, and $d = 0.1$ cm: ——, polarization normal to the plane of incidence; $\cdots$, polarization parallel to the plane of incidence.

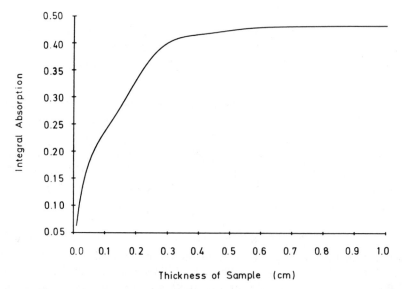

FIG. 3 Integral absorption, defined in Eq. (9), versus sample thickness for a lamellar sample with $n = 1.5$, $k = 0.5$, $v = 70$ GHz, and $d = 0.1$ cm. The undulations are caused by interference within the sample.

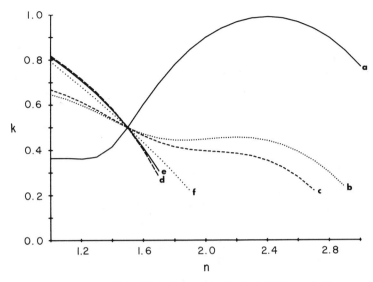

FIG. 4 Space loci (n, k) of the integral equation $I(n, k, d_i, \lambda)/I(n, k, d_j, \lambda) = $ constant at 70 GHz for various intersample comparisons between samples with thicknesses of $d_1 = 0.1$ cm, $d_2 = 0.05$ cm, $d_3 = 0.01$ cm, and $d_4 = 0.005$ cm. Curve a is the trajectory corresponding to an intersample comparison between samples of thicknesses d_1 and d_2; curve b, d_1 and d_3; curve c, d_1 and d_4; curve d, d_2 and d_3; curve e, d_2 and d_4; and curve f, d_3 and d_4.

two trajectories (curves d and e), there exists only one intersection of all trajectories. This intersection defines the physically meaningful (n, k) value of the material under study. Usually a comparison between samples of three different thicknesses is sufficient.

C. DETERMINATION OF THE ABSORPTION COEFFICIENT

For a lamellar sample with $n = 1.5$, $k = 0.05$ and a thickness $d = 1$ cm (Fig. 1), reflection, transmission, and absorption versus angle of incidence are shown in Fig. 5. Integrating over all angles of incidence delivers the integral absorption (Fig. 6). In Eqs. (11) and (12) the right-hand side is given by the experiment. So one has to solve an integral equation to determine the (n, k) values that fulfill the following equation:

$$I(n, k, d, \lambda) = \text{constant.} \tag{14}$$

This is done numerically and the results for several sample thicknesses d_i are shown in Fig. 7. As one can see directly, the influence of n on the determination of k using Eq. (14) is small in this case. So a relatively crude estimate of n (following FIR measurements, for instance) suffices to determine k and hence the absorption coefficient.

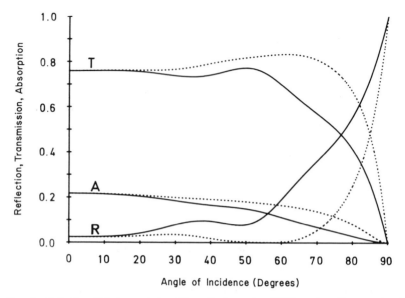

FIG. 5   Reflection (R), transmission (T), and absorption (A) for a lamellar sample with $n = 1.5$, $k = 0.05$, and $d = 1$ cm at 70 GHz: ——, polarization normal to the plane of incidence; $\cdots$, polarization parallel to the plane of incidence.

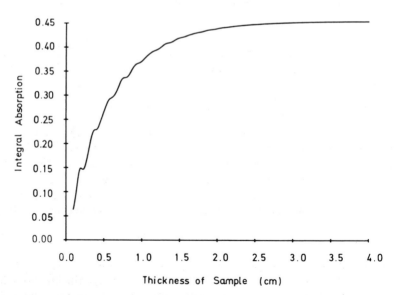

FIG. 6   Integral absorption, defined in Eq. (9), versus sample thickness for a lamellar sample with $n = 1.5$ and $k = 0.05$ at $v = 70$ GHz. The undulations are caused by interference within the sample.

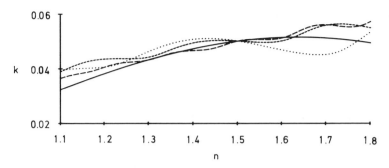

FIG. 7 Space loci (n, k) of the equation $I(n, k, d, \lambda)$ = constant for $n = 1.5$, $k = 0.05$, $v = 70$ GHz, and various sample thicknesses: ——, $d_1 = 0.050$ cm; $\cdots$, $d_2 = 0.100$ cm; ———, $d_3 = 0.500$ cm; ——, $d_4 = 1.00$ cm.

D. DETERMINATION OF THE TEMPERATURE-DEPENDENCE OF THE ABSORPTION COEFFICIENT

To provide for measurements at low temperatures, a special cryostat was constructed with the total sample chamber inside the cavity made of fused silica, which is nearly transparent for millimeter waves. The fact that the cryostat itself has a small but nonnegligible temperature-dependent absorption requires a modification of the rate equations (2) and (3) and hence of Eq. (12). If radiation is fed into the cavity at a constant rate M, the following relation holds for the empty cavity at room temperature T_0:

$$M = \ell_R(T_0)N_0(T_0); \tag{15}$$

for the cavity with a calibration sample we have

$$M = \ell_R(T_0)N_c(T_0) + \ell_c(T_0)N_c(T_0). \tag{16}$$

For the empty cavity at a low temperature T,

$$M = \ell_R(T)N_0(T); \tag{17}$$

for a sample at the temperature T in the cavity,

$$M = \ell_R(T)N_s(T) + \ell_s(T)N_s(T). \tag{18}$$

Incorporating Eq. (15) into Eqs. (16) and (18) yields

$$\ell_R(T) = [N_0(T_0)/N_0(T)]\{N_c(T_0)/[N_0(T_0) - N_c(T_0)]\}\ell_c(T_0). \tag{19}$$

Incorporating Eq. (15) into Eq. (16) and the result into Eq. (19) gives

$$\ell_s(T) = \frac{N_c(T_0)}{N_s(T)} \frac{1 - N_s(T)/N_0(T)}{1 - N_c(T_0)/N_0(T_0)} \ell_c(T_0). \tag{20}$$

Assuming the detector signals to be proportional to the number of photons in the cavity, one finds with Eq. (9) that

$$I_s[n(T), k(T), d_s, \lambda] = \frac{D_c(T_0)}{D_s(T)} \frac{1 - D_s(T)/D_0(T)}{1 - D_c(T_0)/D_0(T_0)}$$

$$\times I_c[n_c(T_0), k_c(T_0), d_c, \lambda], \qquad (21)$$

where $D_0(T_0)$ is the detector signal of the empty cavity at room temperature T_0, $D_c(T_0)$ the detector signal with calibration sample in the cavity at room temperature T_0, $D_0(T)$ the detector signal of the empty cavity at a temperature T, and $D_s(T)$ the detector signal with the unknown sample in the cavity at a temperature T. Note that absolute calibration of the detector is not necessary because only ratios of detector signals are involved.

The right-hand side of Eq. (21) is given by experiment. If we assume that the real part n of the refractive index exhibits no strong temperature dependence, the temperature dependence of the imaginary part k, and hence of the absorption coefficient $\alpha(T) = 4\pi k(T)/\lambda$, can be determined by solving the integral equation (21).

III. Experimental System

A. DESCRIPTION OF THE CAVITY

The overall experimental system includes a broadband millimeter-wave source that can be stabilized in frequency (Fig. 8) and the oversized cavity.

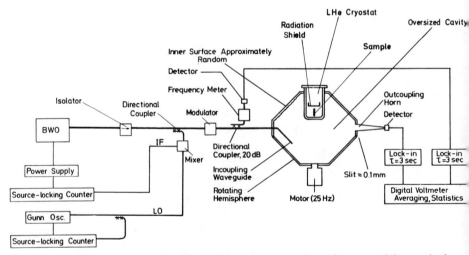

FIG. 8   Schematic description of the millimeter-wave experimental system and the oversized cavity (with cryostat).

The latter is made out of aluminum and is nearly spherical (Fig. 9). It is constructed out of two hemispheres, one of which is fixed whereas the other rotates at about 25 Hz (large mode stirrer). Because of a very good bearing on the mode stirrer, it was possible to reduce the slit between the two hemispheres to less than 0.10 mm. The inner surface of the cavity is nearly randomly pitted with a spherical pit (radius 6 mm). This inner structure and the rotating mode stirrer are essential to minimize the degree of degeneracy of the modes

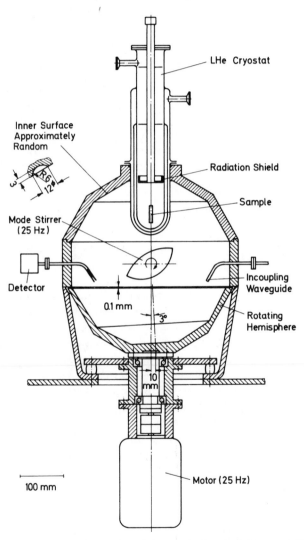

FIG. 9 Schematic description of the oversized cavity (with cryostat).

in the resonator. To further increase the mode mixing, two additional smaller mode stirrers are mounted in the equatorial plane of the oversized cavity perpendicular to each other (one mode stirrer is shown in Fig. 9).

The millimeter-wave power enters the cavity through a waveguide that is directed at the rotating mode stirrer. One percent of the input power is coupled out with a directional coupler and measured with a crystal detector. A small fraction of the field intensity inside the resonator is coupled out through a waveguide with a tapered longitudinal slit and measured by a second crystal detector. Broadband fin-line detectors were employed (AEG-Telefunken, Ulm). Because of the temperature-dependent responsivity of the detectors, their housing was temperature-controlled by a flow of water ($32 \pm 0.2°C$). Both detector signals were amplified by lock-in amplifiers (PAR 124 A) and averaged over a time (3 or 10 sec) that was long compared with the period of the mode stirrers, to integrate over the modes of the resonator. With a stabilized frequency, the short-term stability (10 min) of the whole system is $\pm 0.15\%$, measured as half of the maximum difference in ten consecutive measurements (each being a mean value of 100 single measurements taken in 60 sec). The long-term stability (10 h) of the whole system is $\pm 0.5\%$.

B.   ISOTROPY OF THE RADIATION INSIDE THE CAVITY

An essential requirement of the oversized cavity is a high degree of isotropy for the radiation inside. This was determined by two methods. First, the change in the signals transmitted through the cavity was measured for different polarizations of the outcoupling horn [see also Corona et al. (1980a, b)]. The result for various types of mode mixing is shown in Table I. For a stabilized frequency of 50 ± 0.001 GHz, the relative change (last column in Table I) caused by changing the polarization of the outcoupling horn with respect to the incoupling horn is minimized when all three mode stirrers are rotating. The isotropy of the cavity can be enhanced further by applying a frequency modulation (± 100 MHz). The use of two mode stirrers allows a further, but now small, improvement.

A second method to determine the degree of isotropy of the radiation inside the cavity is to measure the frequency dependence of the transmission of the device. Figure 10 shows the influence of mode mixing on the transmission. When the mode stirrers are fixed, the transmission exhibits sharp peaks that correspond to the cavity modes. Application of the mode stirrers smears out these structures.

C.   HOMOGENEITY OF THE RADIATION FIELD INSIDE THE CAVITY

To check the homogeneity of the internal radiation field, a small cube ($6 \times 6 \times 6$ mm^3) of a strongly absorbent material (microwave absorber

TABLE I

EFFECTIVENESS OF MODE MIXING ON ISOTROPY OF RADIATION IN AN OVERSIZED CAVITY[a]

Type of mode mixing	Detector signal for parallel polarization of in- and outcoupling horn (Arbitrary units)	Detector signal for perpendicular polarization of in- and outcoupling horn (Arbitrary units)	Relative change (%)
One mode stirrer (lower hemisphere)	0.7717	0.7494	2.88
Two mode stirrers (lower hemisphere and one side mode stirrer)	0.7774	0.7661	1.46
Three mode stirrers (lower hemisphere and two side mode stirrers)	0.7761	0.7694	0.85
One mode stirrer (lower hemisphere) and FM ±100 MHz	0.7839	0.7817	0.284
Two mode-stirrers (lower hemisphere and one side mode-stirrer) and FM ±100 MHz	0.7838	0.7820	0.229

[a] For the measurements, the frequency was stabilized at 50 ± 0.001 GHz using a phase-locked loop. For measurements with frequency modulation (FM), a stabilization was not possible. But the center of frequency f_0 was also adjusted to 50 GHz. The width of the FM was measured by harmonic mixing to ± 100 MHz. The modulation frequency was 333 Hz. Each value in column two and three is the mean value of 500 single measurements taken in 300 sec.

AN72 supplied by Emerson and Cuming, Inc.) was systematically positioned on a grid of points in the cavity, and the resulting signal level transmitted through the cavity was recorded. The absorber will absorb a constant fraction of the radiation flux incident on it. Thus variations in the detected signal will correspond to field inhomogeneities. The absorber was supported with a Teflon PTFE strand on a PTFE rod and could be positioned following cylindrical coordinates in a volume 19 cm high and 14 cm in diameter, including nearly all of the upper half of the cavity. A larger variation in the radius was not possible because of the side mode stirrers.

The number of photons absorbed in the absorber per unit time can readily be deduced from Eq. (5):

$$\ell_s = (\text{constant})(D_0 - D_s)/D_s, \tag{22}$$

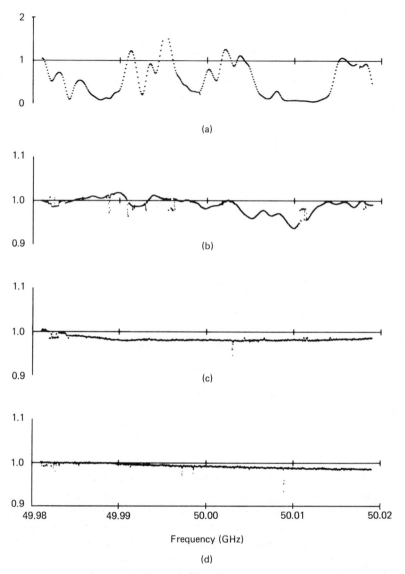

FIG. 10 Transmission of the oversized cavity between 49.981 and 50.019 GHz normalized by the value of the transmission at 49.981 GHz. The millimeter waves were stabilized with a phase-locked loop and swept in steps of 0.100 MHz. Normalized transmission is shown: (a) when the mode stirrers are switched off (note that the scale covers values from 0 to 2); (b) when the large mode stirrer (lower hemisphere) rotates; (c) when the large mode stirrer and one side-mode stirrer are rotating; (d) shows the transmission when all three mode stirrers are switched on. Note that for (b)–(d) the scale goes from 0.9 to 1.1. The sharp peaks are caused by momentary failures of the phase-lock system.

where D_0 is the detector signal of the empty cavity and D_s that of the cavity with the sample. The absorber was positioned in 1-cm steps at points along the central axis (Table II) and in 30° steps in circles (radii: 3, 5, 7 cm) at a distance of 16 cm from the upper wall (Table III). For each position, the signal of the cavity $D(x)$ was measured as the mean value of 100 measurements taken in 100 sec. The value of $P(x) = [D_0 - D_x(x)]/D_x(x)$ for the different positions of the absorber is shown in Table II. Its relative variability is smaller than $\pm 10\%$. Because in the experiments the samples measured in the cavity were 50 mm in diameter, the size of sample itself averaged over the small residual inhomogeneities of the internal radiation field. Nevertheless, the sample was always placed in the same position for the experiments [see also D'Ambrosio and Ferrara (1983) and D'Ambrosio et al. (1980)].

D. Q VALUE OF THE CAVITY

To determine the Q value of the cavity, two methods were employed: the hole method described by Lamb (1946) and a method using a calibration load of known dielectric properties (Kremer and Izatt, 1981). Modifying Eq. (3) for the photon loss through a hole in the cavity wall (instead of the loss of a sample placed inside the cavity) yields, with Eqs. (4) and (10),

$$\ell_R = cA_H/4V[R_H/(1 - R_H)], \qquad (23)$$

where A_H denotes the area of the hole, c the speed of light, V the cavity volume, and R_H the ratio of the detector signals with hole open and closed. For a calibration load of known dielectric properties, Eqs. (2-4) and (9) yield

$$\ell_R = (A_c c/V)[R_c/(1 - R_c)]I_c(n_c, k_c, d_c, \lambda), \qquad Q_R = \omega/\ell_R, \qquad (24)$$

where A_c denotes the area of one side of the lamellar calibration sample and I_c its absorption integral. Plexiglas (PMMA) was used as the material for the calibration sample. Because it is highly homogeneous, it can easily be measured optically. This was done in the frequency range from 4 to 20 cm^{-1} by J. R. Birch (National Physical Laboratory, Teddington) using Fourier transform spectroscopy. A further measurement at 60 GHz was carried out by M. Lemke (Philips Laboratories Hamburg) with an open resonator. The results of the Q value measurements are shown in Table IV.

The measurement of the Q value using the hole method delivers a smaller value than that measured using a calibration sample. This probably has two causes: the photon density at the walls is smaller compared to that at the center of the cavity (where the calibration sample was placed); and the walls of the resonator have a finite thickness (15 mm), thus decreasing the effective hole area.

TABLE II

ENERGY ABSORPTION OF A STRONG ABSORBER AT VARIOUS CENTRAL POSITIONS IN THE CAVITY[a]

	Distance of sample from upper wall (cm)																	
	4	5	6	7	8	9	10	11	12	13	14	15	16	17	18	19	20	21
$P(x)$[b]	6.5	6.8	6.8	6.6	6.6	6.5	6.8	6.8	6.6	6.9	6.8	6.6	6.8	7.1	7.2	6.5	7.4	6.9
$100[\overline{P(x)} - P(x)]/\overline{P(x)}$	4.5	0.1	0.1	2.3	2.3	4.5	0.1	0.1	2.3	−2.1	0.1	2.3	0.1	−4.3	−6.5	4.5	−8.7	−2.1

[a] AN 72, size: $6 \times 6 \times 6$ mm³. The sample was moved along the central axis of the cavity in 1-cm steps.

[b] The value $P(x) \equiv [D_0 - D_s(x)]/[D_s(x)]$ is proportional to the number of photons absorbed per unit time in the absorber. Its mean value is denoted by $\overline{P(x)}$. The variability in percent of the $P(x)$ values is presented in the second row. For the experiment, the microwaves were phase-locked at 50 0.001 GHz and all three mode stirrers were used.

158

TABLE III

ENERGY ABSORPTION OF A STRONG ABSORBER AT VARIOUS ANGULAR POSITIONS IN THE CAVITY[a]

		Angular position of sample (deg)											
		0	30	60	90	120	150	180	210	240	270	300	330
I	$P(x)$	6.5	5.9	6.2	6.4	6.4	6.5	6.2	6.2	5.9	6.2	6.5	6.4
	$100[\overline{P(x)} - P(x)]/\overline{P(x)}$	−2.0	7.3	2.7	0.3	0.3	−2.0	2.7	2.7	7.3	2.7	−2.0	0.3
II	$P(x)$	6.1	6.4	6.2	6.5	6.5	6.9	6.4	6.2	6.2	6.5	6.6	6.5
	$100[\overline{P(x)} - P(x)]/\overline{P(x)}$	5.0	0.3	2.7	−2.0	−2.0	−9.0	0.3	2.7	2.7	−2.0	−4.3	−2.0
III	$P(x)$	6.6	6.1	6.2	6.4	6.2	6.8	6.5	6.5	6.4	6.6	6.6	6.1
	$100[\overline{P(x)} - P(x)]/P(x)$	−4.3	5.0	2.7	0.3	2.7	−6.7	−2.0	−2.0	0.3	−4.3	−4.3	5.0

[a] AN72, size: $6 \times 6 \times 6$ mm^3. The sample was placed 17 cm from the upper wall and moved in 30° steps along circles of radius 3 cm (I), 5 cm (II), and 7 cm (III). (See Table II.)

TABLE IV

METHODS FOR MEASUREMENT OF THE Q VALUE

Type of measurement	Q value
Hole method	
Hole diameter 50 mm	$33,000 \pm 1,000$
Hole diameter 100 mm	$32,000 \pm 1,000$
Use of a calibration sample	
(Plexiglas, diameter 50 mm,	
thickness 6 mm)	$45,000 \pm 1,000$

[a] The frequency was stabilized at $50,000 \pm 0.001$ GHz and all three mode stirrers were used (see Fig. 9).

Assuming that the only loss mechanism is absorption in the cavity walls, Lamb (1946) derived the following expression:

$$Q = \tfrac{3}{2}(V/A_w \delta), \tag{25}$$

where V is the cavity volume, A_w the area of the cavity walls, and δ the skin depth of the walls. The definition of δ is

$$\delta = \sqrt{2/\omega \mu_0 \sigma}. \tag{26}$$

In this expression σ is the wall conductivity, ω the angular frequency of the radiation, and μ_0 the permeability of free space. For a Q value of 45,000 a value for the skin depth of 2.0×10^{-4} cm is found. Assuming for standard grade aluminum a conductivity of $(2 \times 10^3)/\Omega \cdot$ cm one calculates a value of 8.8×10^{-5} cm for the skin depth. This difference between the measured and the calculated skin depth is due to the assumption that the only loss mechanism in the cavity is absorption in the cavity walls.

An estimation of the Q value of the cavity is also possible from Fig. 10a. The cavity modes at 50 GHz have a half-width of about 1.5 MHz. This results in a Q value of about 35,000.

E. ACCURACY OF THE MEASUREMENT

Because there exist no analytic expressions for n and k as a function of the measured detector signals, one is restricted to a numerical evaluation of the error propagation in this case. For example, to achieve $\pm 0.5\%$ accuracy in both n and k for liquid H_2O by comparing the integral absorption of samples of different thicknesses, sample cells of uniform thickness within a maximum deviation of ± 1 μm are required, and the detector signals must be reproducible to within about $\pm 0.1\%$ (Izatt and Kremer, 1981).

On the other hand, accuracy in determining the integral absorption is readily obtained from Eq. (21):

$$\frac{\Delta I_s(T)}{I_s(T)} = \frac{\Delta D_c(T_0)}{D_c(T_0)} + \frac{\Delta D_c(T_0)}{D_0(T_0)[1 - D_c(T_0)/D_0(T_0)]}$$

$$-\frac{\Delta D_s(T)}{D_s(T)} - \frac{\Delta D_s(T)}{D_0(T)[1 - D_s(T)/D_0(T)]}$$

$$+\frac{\Delta D_0(T)}{D_0(T)} \frac{D_s(T)}{D_0(T)[1 - D_s(T)/D_0(T)]}$$

$$-\frac{\Delta D_0(T_0)}{D_0(T_0)} \frac{D_c(T_0)}{D_0(T_0)[1 - D_c(T_0)/D_0(T_0)]}. \tag{27}$$

Assuming $D_0(T) \simeq D_0(T_0) = D_0$ and replacing $D_c(T_0)$ by D_c and $D_s(T)$ by D_s, we obtain

$$\frac{\Delta I_s(T)}{I_s(T)} = \frac{\Delta D_c}{D_c} \frac{D_0}{D_0 - D_c} - \frac{\Delta D_s}{D_s} \frac{D_0}{D_0 - D_s} + \frac{\Delta D_0}{D_0} \left(\frac{D_s}{D_0 - D_s} - \frac{D_c}{D_0 - D_c} \right). \tag{28}$$

It is obvious that accuracy is strongly decreased if D_c or D_s approaches D_0, i.e., in the case of weak absorption and small sample amounts.

For the case of $D_c = D_s = 0.9$ and $D_0 = 1$, a value of $\Delta D_s = 0.01$ results in a variability $\Delta I_s(T)/I_s(T)$ of 0.11. This corresponds to a variability in the absorption coefficient of less than $\pm 15\%$.

F. LOW-TEMPERATURE CRYOSTAT FOR THE CAVITY

To provide for measurements at liquid helium temperature and above, a special cryostat was constructed with the total sample chamber inside the cavity made of fused silica, which is nearly transparent at millimeter-wave frequencies (Fig. 9). The fused silica walls on the upper part of the cryostat inside the cavity were gold-plated to decrease the microwave leakage. But in spite of that, the Q value of the cavity was decreased by a factor of two. To check the extent by which the isotropy and homogeneity of the radiation field inside the cavity is disturbed by the regularly shaped cryostat, the absorption of a sample was measured at room temperature in the cavity with and without the cryostat. For low absorbing samples ($k \leq 0.05$) no measurable difference was found. For strongly absorbing samples ($k > 0.05$), a difference in the integral absorption of about 20% was measured due to some disturbance of the field inside the cryostat.

IV. Comparison of the Oversized Cavity Technique with Other Methods

A. COMPARISON WITH THE CAVITY-PERTURBATION TECHNIQUE

The well-known cavity-perturbation technique is widely used at frequencies up to 30 GHz. Applications at 75 GHz are described (Amrhein and Schulze, 1972). With further increasing frequency, the strongly decreasing dimensions of waveguides and cavities compromise the accuracy of the method. Compared with the oversized-cavity technique, the main disadvantage arises from the fact that the small cavities for the cavity-perturbation technique are very narrow banded.

B. COMPARISON WITH OPEN-RESONATOR MEASUREMENTS

An open resonator is a very accurate device for measurements of weakly absorbing materials ($k \leq 0.05$) at microwave frequencies (Cullen and Yu, 1971; Cullen and Davies 1978; Clarke and Rosenberg, 1982). Beside the fact that it is a resonant technique as well and therefore narrow banded, it requires very good homogeneity of the material under study (Birch and Clark, 1982). Furthermore, it is not applicable to strongly absorbing materials, and temperature-dependent measurements cannot easily be performed.

C. COMPARISON WITH WAVEGUIDE TECHNIQUES

Waveguide techniques such as bridges or reflectometers are used up to frequencies of about 170 GHz (Stumper, 1973). Temperature-dependent measurements, especially at low temperatures cannot easily be performed because of heat conduction in waveguides.

D. COMPARISON WITH FOURIER TRANSFORM SPECTROSCOPY

Fourier transform spectroscopy used at frequencies below twenty wave numbers is difficult because of the low power levels emitted by the thermal sources (Bell, 1967; Birch and Parker, 1979). When combined with low-temperature bolometers, the method can be applied to frequencies as low as 2–3/cm. As an optical method it requires that the sample has a high degree of homogeneity. So, in contrast to the oversized-cavity technique, scattering is a more serious problem. For high-density polyethylene a direct comparison between the oversized cavity method and Fourier transform spectroscopy was carried out (Birch *et al.*, 1983). The comparison demonstrated a good agreement between the two techniques. For highly reflecting and weakly absorbing samples, e.g., films of doped polyacetylene, (see below) application of Fourier transform spectroscopy is difficult because of reflection losses. The oversized-cavity technique can be used advantageously in this case, because in principle it measures only absorptive losses. Furthermore the spectral resolution of microwave methods is much higher than that of Fourier transform spectroscopy.

V. Results

A. MEASUREMENT OF THE TEMPERATURE-DEPENDENT MILLIMETER-WAVE ABSORPTION OF HEMOGLOBIN AND HYDRATED LYSOZYME

Information is sparse concerning the optical properties of biological macromolecules such as proteins in the submillimeter region (Vergoten et al., 1980). The situation is even worse at millimeter wavelengths, where almost nothing is known about the dynamical processes that determine the dielectric properties of the biomolecules. This spectral regime could be of particular interest for the understanding of molecular dynamics such as vibrations or relaxations (McCammon et al., 1979; Frauenfelder et al., 1979). As an example, the dielectric properties of lyophylized hemoglobin and lysozyme were studied in the frequency range from 50 to 150 GHz and at temperatures between 4.2 and 300 K (Genzel et al., 1983; Poglitsch et al., 1984).

For the experiment, about 20 g of sample material were used. In the case of lyophylized hemoglobin, the material was dried over P_2O_5 at 80°C for one week and then pressed into disk-shaped pellets 50 mm in diameter and 12 mm thick (density 0.99 g/cm^3).

The observed integral absorption for dried hemoglobin at 70 GHz is shown in Fig. 11a. Assuming a value of 1.6 for n (obtained from far-infrared measurements) the absorption coefficient α could be determined (Fig. 11b). Note that a variation in n between 1.4 and 1.8 changes the calculated value of α by only $\pm 5\%$. It is found that the absorption coefficient α increases exponentially with temperature and as $v^{1.2}-v^2$ with frequency (Fig. 12). This frequency and temperature dependence is quantitatively described (solid lines in Fig. 12) as being caused by three distinct relaxation processes on a picosecond time scale occurring in asymmetric double-well potentials. These processes are most probably assigned to the NH···OC hydrogen bonds of the peptide backbone (Genzel et al., 1983).

To study the influence of adsorbed water on the dynamics of the biomolecules, lysozyme was chosen as a particularly stable protein that can be completely dehydrated except for approximately three tightly bound water molecules per lysozyme molecule (Imoto et al., 1972). For dried lysozyme (water content $\leq 0.5\%$ by weight) a nearly linear frequency dependence and an exponential temperature dependence of the absorption coefficient are observed between 50 and 300 K (Fig. 13). Hydrating the sample results in a nearly frequency-independent contribution to the absorption; this occurs

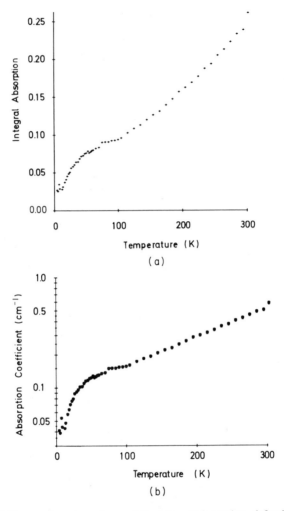

FIG. 11   (a) Temperature dependence of the integral absorption, defined in Eq. (9), for lyophylized hemoglobin at 70 GHz. (b) Temperature dependence of the absorption coefficient of lyophylized hemoglobin at 70 GHz as determined from the data shown in (a).

only at temperatures above 150 K (Fig. 14). This frequency and temperature dependence of the absorption coefficient is described for the dried material by a minimum set of three relaxation processes occurring in asymmetric double-well potential with picosecond relaxation times. Owing to the high frequencies involved, these relaxations can most probably be assigned to relaxation of the NH$\cdots$OC hydrogen bond of the peptide backbone. The contribution of the adsorbed water is assigned to relaxation of the bound

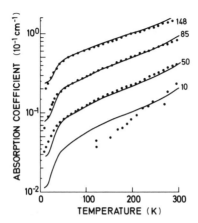

FIG. 12 Temperature and frequency dependence of the microwave absorption coefficient of hemoglobin. The numbers on the curves are the frequencies in GHz: ●, the experimental data; ——, absorption coefficient due to the theoretical model. The experiments between 50 and 148 GHz were carried out with the oversized-cavity technique. At 10 GHz the cavity-perturbation technique was employed. (From Genzel et al., 1983. © 1983 John Wiley & Sons, Inc.)

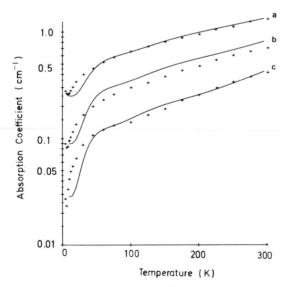

FIG. 13 Temperature and frequency dependence of the millimeter-wave absorption of lysozyme: +, experimental data; ——, absorption due to the theoretical model. Curve a, 150 GHz; curve b, 90 GHz; curve c, 50 GHz. (From Poglitsch et al., 1984.)

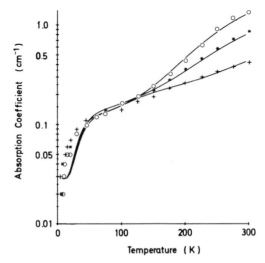

FIG. 14   Millimeter-wave absorption at 50 GHz of lysozyme at various degrees of hydration. Experimental data: +, 0.5% g(H$_2$O)/g (protein); *, 8% g/g; O, 18% g/g; ——, absorption due to the theoretical model. (From Poglitsch *et al.*, 1984.)

water, which exhibits much slower relaxation rates. (Poglitsch *et al.*, 1984; Bone and Pethig, 1982.)

B.   TEMPERATURE-DEPENDENT DIELECTRIC STUDIES OF DOPED POLYACETYLENE

Polyacetylene is a substance of widespread theoretical and practical interest (Baeriswyl *et al.*, 1982). Its conductivity can be increased over ten orders of magnitude by doping it with I or AsF$_5$. The mechanism of this conductivity is still a highly controversial topic (Roth *et al.*, 1982). Nothing is known about the dielectric properties at millimeter-wave frequencies and information at submillimeter wavelength is sparse. (Montaner *et al.*, 1981; Hoffman *et al.*, 1981.)

For this experiment polyacetylene films (thickness about 100 μm, diameter 50 mm) were prepared and doped with I or AsF$_5$. The observed temperature dependence of the integral absorption is shown in Fig. 15 for measurements at various frequencies and with different dopants and concentrations. Because the sample is strongly absorbing and the real part n of the complex refractive index exhibits a strong temperature dependence, it is not possible to determine from the measurement in a straightforward way the temperature dependence of the absorption coefficient α. This would have been possible if an additional measurement of the temperature-dependent transmission had been carried out.

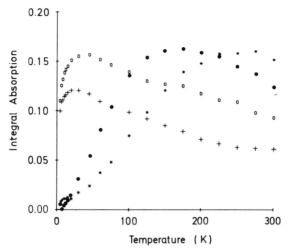

FIG. 15   Temperature and frequency dependence of the integral absorption I, defined in Eq. (9), for doped polyacetylene. (cis)-polyacetylene doped with 0.7% [mol % (poly-acetylene)] A_sF_5: +, at 50 GHz; 0 , at 111 GHz. (trans)-polyacetylene doped with :*, 1.1% I at 50 GHz; ●, 1.9% I at 50 GHz.

For doped polyacetylene, a maximum in the integral absorption versus temperature is observed that depends on the frequency, the dopant, and the dopant concentration (Fig. 15). For undoped polyacetylene (thickness 100 μm, diameter 50 mm), no measurable absorption was found. Thus the strong millimeter-wave absorption was induced by the dopants. At low temperatures not many charge carriers are activated; the sample is nearly transparent and hence shows only a small absorption. At room temperature the dopants are thermally activated, resulting in high reflectivity of the material and medium absorptivity. In between, the maximum in the integral absorption is observed. So the transition from the insulating to the semi-conducting state can be directly studied with temperature-dependent millimeter-wave spectroscopy. From the frequency and temperature dependence of the dielectric function and for given models for conductivity (Nagels, 1979), conclusions can be drawn about the number of effective charge carriers, their mobility, and the activation energies involved (Genzel et al., 1984).

C.   MEASUREMENT OF THE COMPLEX INDEX OF REFRACTION OF A STRONGLY ABSORBING POLYMER

A strongly absorbing polymer (an eccosorb absorber purchased from Emerson and Cuming, Inc.) was given to us by Professor P. L. Richards (University of California at Berkeley). It is used as a calibration blackbody in

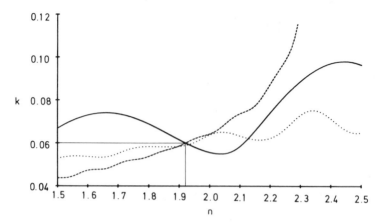

FIG. 16 Space loci (n, k) of the integral equation $I(n, k, d_i, \lambda) = I_{exp}$. I_{exp} is the measured integral absorption at 111 GHz. Three sample thicknesses were used: ——, trajectory for $d_1 = 0.147$ cm; $\cdots$, trajectory for $d_2 = 0.396$ cm; ---, trajectory for $d_3 = 0.901$ cm.

millimeter and submillimeter telescopes to measure the 3-K cosmic background radiation. For the experiment, lamellar samples (diameter 50 mm) of different thicknesses ($d_1 = 0.147$ cm, $d_2 = 0.396$ cm, $d_3 = 0.901$ cm) were measured. The numerical solution of the equation $I(n, k, d_i, \lambda) = $ constant is graphically shown for different sample comparisons in Fig. 16. Values of $n = 1.92 \pm 0.02$ and $k = 0.059 \pm 0.0025$ were obtained at room temperature.

The temperature dependence of the absorption coefficient α was determined assuming that the real part n of the refractive index exhibits no temperature dependence (Fig. 17). For strongly absorbing samples such as water or glycerin, the method of the intersample comparison was used to determine the complex index of refraction (Izatt and Kremer, 1981).

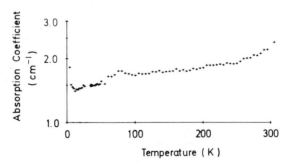

FIG. 17 Temperature dependence of the absorption coefficient for the eccosorb absorber at 111 GHz. For the determination of the absorption coefficient α it was assumed that the real part n of the complex refractive index exhibits no temperature dependence; hence a value of $n = 1.9$ was assumed.

VI. Conclusion

In this study the oversized-cavity technique is described for broadband millimeter-wave spectroscopy using coherent radiation sources. An exact calculation of the absorption of lamellar samples as a function of the complex index of refraction is shown to enable measurements of both the real and imaginary parts (n, k) for strongly absorbing samples and the absorption coefficient α for weakly absorbing samples. Temperature-dependent measurements between 5 and 300 K are easily obtained using a fused-silica cryostat inserted in the cavity. Experimental applications on samples of current interest are given for all types of measurements. The oversized cavity proves to be particularly advantageous for absorption measurements of weakly absorbing scattering samples and of highly reflecting samples, because the detected change in the quality factor Q of the resonator is caused by absorptive losses only, regardless of scattering and reflection.

Appendix I: Lamellar Sample with Windows

A lamellar sample with windows is shown schematically in Fig. 18. The reflection factor R represents the fraction of the radiation incident upon either face of the sample that is reflected back. The transmission factor T represents the fraction of the incident radiation that passes through the sample. Each of these factors depends on the polarization of the radiation. We denote the component corresponding to $\mathbf{E}$ normal to the plane of incidence with $\perp$ and the parallel component with $\parallel$. The absorption factor will then be given by energy conservation:

$$A_\perp = 1 - R_\perp - T_\perp, \tag{29a}$$

$$A_\parallel = 1 - R_\parallel - T_\parallel. \tag{29b}$$

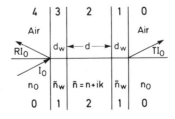

FIG. 18 Schematic diagram of sample with windows. All the interfaces are normal to the x axis: d, sample thickness; d_w, thickness of windows.

Because the radiation in the untuned cavity is unpolarized, the total absorption factor is obtained by averaging over the two perpendicular polarization components:

$$A_T = \tfrac{1}{2}(A_\perp + A_{\parallel}). \tag{30}$$

Two methods will be presented to calculate R and T. The first one is based on a matrix method described by Abelès (1967) and has been published by Izatt and Kremer (1981). Typographic errors in the latter publication have been corrected here. The windows are assumed to be lossless, and hence their refractive index is real: $\tilde{n}_w = n_w$. The reflection and transmission factors can be written as follows:

$$R_\perp = (A^2 + B^2)/(C^2 + D^2), \tag{31a}$$

$$T_\perp = 4p_0^2/(C^2 + D^2), \tag{31b}$$

where

$A = -a \sin \alpha \sinh \gamma + b \cos \alpha \sinh \gamma - c \sin \alpha \cosh \gamma,$

$B = a \cos \alpha \cosh \gamma + b \sin \alpha \cosh \gamma + c \cos \alpha \sinh \gamma,$

$C = d \cos \alpha \cosh \gamma - e \sin \alpha \sinh \gamma - f \sin \alpha \cosh \gamma - g \cos \alpha \sinh \gamma,$

$D = d \sin \alpha \sinh \gamma + e \cos \alpha \cosh \gamma + f \cos \alpha \sinh \gamma - g \sin \alpha \cosh \gamma,$

$a = [(p_0^2 - p_w^2)/p_w] \sin 2\beta_w,$

$b = p_s\{[(p_0^2 - \rho_s^2)/\rho_s^2] \cos^2 \beta_w + [(p_w^4 - p_0^2 \rho_s^2)/\rho_s^2 p_w^2] \sin^2 \beta_w\},$

$c = q_s\{[(p_0^2 - \rho_s^2)/\rho_s^2] \cos^2 \beta_w + [(p_w^4 + p_0^2 \rho_s^2)/\rho_s^2 p_w^2] \sin^2 \beta_w\},$

$d = 2p_0 \cos 2\beta_w,$

$e = [(p_0^2 + p_w^2)/p_w] \sin 2\beta_w,$

$f = p_0 p_s[(p_w^2 + \rho_s^2)/\rho_s^2 p_w] \sin 2\beta_w - q_s\{-[(p_0^2 - \rho_s^2)/\rho_s^2] \cos^2 \beta_w + [(p_w^4 - p_0^2 \rho_s^2)/\rho_s^2 p_w^2] \sin^2 \beta_w\},$

$g = p_0 q_s[(p_w^2 - \rho_s^2)/\rho_s^2 p_w] \sin 2\beta_w - p_s\{[(p_0^2 + \rho_s^2)/\rho_s^2] \cos^2 \beta_w - [(p_w^4 + p_0^2 \rho_s^2)/\rho_s^2 p_w^2] \sin^2 \beta_w\}.$

If the directions of propagation relative to the x axis in Fig. 18 are denoted by θ_0, θ_w, and θ, respectively, for the air, window, and sample regions, then

$p_0 = n_0 \cos \theta_0,$

$p_w = n_w \cos \theta_w = (n_w^2 - n_0^2 \sin^2 \theta_0)^{1/2}.$

For $p_s + iq_s = \tilde{n}\cos\theta$, we have the following:

$$p_s = (1/\sqrt{2})[\rho_s^2 + (n^2 - k^2 - n_0^2\sin^2\theta_0)]^{1/2},$$

$$q_s = (1/\sqrt{2})[\rho_s^2 - (n^2 - k^2 - n_0^2\sin^2\theta_0)]^{1/2},$$

$$\rho_s^2 = p_s^2 + q_s^2 = [(n^2 - k^2 - n_0^2\sin^2\theta_0)^2 + 4n^2k^2]^{1/2},$$

$$\beta_w = (2\pi/\lambda)n_w d_w\cos\theta = (2\pi/\lambda)d_w p_w.$$

For $\alpha + i\gamma = (2\pi/\lambda)\tilde{n}d\cos\theta$, we have the following:

$$\alpha = (2\pi/\lambda)p_s d, \qquad \gamma = (2\pi/\lambda)q_s d.$$

The reflection and transmission factors for $\mathbf{E}$ parallel to the plane of incidence are obtained from the formulas given for normal polarization as follows where p_0, p_w, β_w, α, and γ are left unchanged; replace

$$q_s \quad \text{by} \quad q_s' = q_s[1 - (n_0^2\sin^2\theta_0/\rho_s^2)],$$

$$p_s \quad \text{by} \quad p_s' = p_s[1 + (n_0^2\sin^2\theta_0/\rho_s^2)],$$

and

$$p_i \quad \text{by} \quad p_i' = p_i[1 + (n_0^2\sin^2\theta_0/p_i^2)], \qquad i = 0, w.$$

Calculate $a \cdots g$, $A \cdots D$, and $R_\parallel$, $T_\parallel$ using the same formulas as in the case of normal polarization.

The second method, described by Birch (pers. comm.), offers the possibility of calculating the transmission and reflection factors for an arbitrary number of dielectric (absorbing or nonabsorbing) layers in a repetitive way. We start with one layer. Figure 19 shows medium 2 immersed in media 1 and 3, which extend infinitely to the left and right, respectively. The incident radiation is considered as a plane, coherent wave. Instead of directly solving the boundary problem, one can treat it as a multiple reflection from boundaries between two semi-infinite media. Adding all partial waves then yields the total reflected and transmitted wave. According to Fig. 19, the complex amplitude reflectivity will be

$$\tilde{r}_{31} \equiv \tilde{r}_{321} = \tilde{r}_{32} + \tilde{t}_{32}\tilde{a}_2^2\tilde{r}_{21}\tilde{t}_{23}(1 + \tilde{a}_2^2\tilde{r}_{21}\tilde{r}_{23} + \tilde{a}_2^4\tilde{r}_{21}^2\tilde{r}_{23}^2 + \cdots)$$

$$= \tilde{r}_{32} + [\tilde{t}_{32}\tilde{a}_2^2\tilde{r}_{21}\tilde{t}_{23}/(1 - \tilde{a}_2^2\tilde{r}_{21}\tilde{r}_{23})]. \tag{32}$$

The complex amplitude transmittance will be

$$\tilde{t}_{31} \equiv \tilde{t}_{321} = \tilde{t}_{32}\tilde{t}_{21}\tilde{a}_2(1 + \tilde{a}_2^2\tilde{r}_{21}\tilde{r}_{23} + \tilde{a}_2^4\tilde{r}_{21}^2\tilde{r}_{23}^2 + \cdots)$$

$$= [\tilde{t}_{32}\tilde{t}_{21}\tilde{a}_2/(1 - \tilde{a}_2^2\tilde{r}_{21}\tilde{r}_{23})], \tag{33}$$

where $\tilde{r}_{ik}$ and $\tilde{t}_{ik}$ are the complex reflection and transmission factors given by the Fresnel formulas for a single boundary between medium i and medium k.

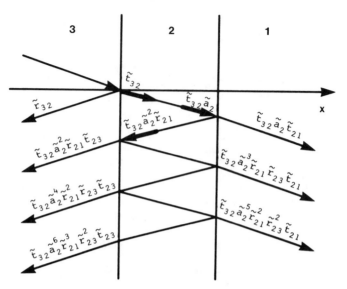

FIG. 19 Multiple reflection of a plane wave in medium 2 that is measured in media 1 and 3. All interfaces are normal to the *x* axis.

Note that the summation over all partial waves has to be carried out at the same point, i.e., at the intersection of the *x* axis and the boundary plane. Thus for a plane wave (Fig. 20), the complex propagation factor $\tilde{a}_j$ in medium j will be $\tilde{a}_j = \exp[(2\pi/\lambda)i\tilde{n}_j d_j \tilde{\cos}\,\theta_j]$, where $\tilde{n}_j = n_j + ik_j$ is the complex refractive index and λ the wavelength. The complex $\tilde{\cos}\,\theta_j$ is obtained by Snell's law:

$$\tilde{\sin}\,\theta_j = (\tilde{n}_k/\tilde{n}_j)\,\tilde{\sin}\,\theta_k,$$

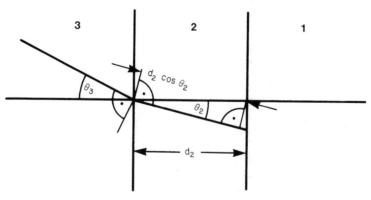

FIG. 20 Distance between planes of equal phase that intersect the boundaries on the *x* axis.

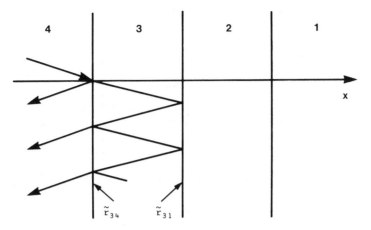

FIG. 21 Calculation of the reflectivity with one layer being added.

therefore

$$\tilde{\cos} \theta_j = [1 - (\tilde{n}_k/\tilde{n}_j)^2 \tilde{\sin}^2 \theta_k]^{1/2}$$

for any combination j, k.

This method can now be extended to more than two interfaces, taking the effective reflectivity $\tilde{r}_{31}$ for the overall reflection on boundary 3–2 (Fig. 21). Equation (32) then transforms to

$$\tilde{r}_{41} \equiv \tilde{r}_{4321} = \tilde{r}_{43} + [\tilde{t}_{43}\tilde{a}_3^2\tilde{r}_{31}\tilde{t}_{34}/(1 - \tilde{a}_3^2\tilde{r}_{31}\tilde{r}_{34})]. \tag{34}$$

For the transmission, the extension has to be in the other direction (Fig. 22). The radiation entering medium 1 is given by the effective transmittance $\tilde{t}_{31}$. The overall reflection on boundary 1–2 is given by the effective reflectivity $\tilde{r}_{13}$, which has to be calculated as shown in Eq. (32). Equation (33) then transforms to

$$t_{30} \equiv t_{3210} = [\tilde{t}_{31}\tilde{t}_{10}\tilde{a}_1/(1 - \tilde{a}_1^2\tilde{r}_{10}\tilde{r}_{13})]. \tag{35}$$

With an arbitrary number N of layers, we can generalize Eqs. (34) and (35) to the recursion formulas

$$\tilde{r}_{n,0} = \tilde{r}_{n,n-1} + [\tilde{t}_{n,n-1}\tilde{a}_{n-1}^2\tilde{r}_{n-1,0}\tilde{t}_{n-1,n}/(1 - \tilde{a}_{n-1}^2\tilde{r}_{n-1,0}\tilde{r}_{n-1,n})],$$
$$n = 2 \cdots N + 1 \tag{36}$$

and

$$\tilde{t}_{N+1,m} = [\tilde{t}_{N+1,m+1}\tilde{t}_{m+1,m}\tilde{a}_{m+1}/(1 - \tilde{a}_{m+1}^2\tilde{r}_{m+1,m}\tilde{r}_{m+1,N+1})],$$
$$m = N - 1 \cdots 0. \tag{37}$$

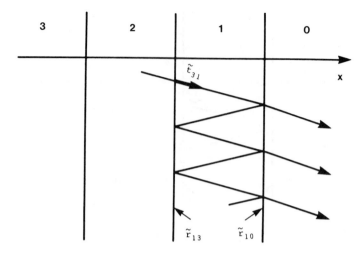

FIG. 22   Calculation of the transmittance with one layer being added.

The power reflection and transmission will be

$$R = |\tilde{r}_{N+1,0}|^2 \tag{38a}$$

and

$$T = |\tilde{t}_{N+1,0}|^2. \tag{38b}$$

The complex reflection and transmission amplitude factors $\tilde{r}_{ik}$ and $\tilde{t}_{ik}$ for adjacent layers ($k = i \pm 1$) are given by the Fresnel formulas in the form (Stratton, 1941)

$$\tilde{r}_{ik_\perp} = (\tilde{n}_i \, c\tilde{o}s \, \theta_i - \tilde{n}_k \, c\tilde{o}s \, \theta_k)/(\tilde{n}_i \, c\tilde{o}s \, \theta_i + \tilde{n}_k \, c\tilde{o}s \, \theta_k), \tag{39a}$$

$$\tilde{t}_{ik_\perp} = (2\tilde{n}_i \, c\tilde{o}s \, \theta_i)/(\tilde{n}_i \, c\tilde{o}s \, \theta_i + \tilde{n}_k \, c\tilde{o}s \, \theta_k) \tag{39b}$$

for polarization normal to the plane of incidence, and in the form

$$\tilde{r}_{ik_\parallel} = (\tilde{n}_k \, c\tilde{o}s \, \theta_i - \tilde{n}_i \, c\tilde{o}s \, \theta_k)/(\tilde{n}_k \, c\tilde{o}s \, \theta_i + \tilde{n}_i \, c\tilde{o}s \, \theta_k), \tag{40a}$$

$$\tilde{t}_{ik_\parallel} = (2\tilde{n}_k \, c\tilde{o}s \, \theta_i)/(\tilde{n}_k \, c\tilde{o}s \, \theta_i + \tilde{n}_i \, c\tilde{o}s \, \theta_k) \tag{40b}$$

for polarization parallel to the plane of incidence. The magnetic permeability of all layers is assumed to be $\mu = 1$. For both polarizations we have from Eqs. (39a) and (40a) the following:

$$\tilde{r}_{ik} = -\tilde{r}_{ki}. \tag{41}$$

One should note that only the formulas (39) hold directly for the electric-field vector, whereas formulas (40) are valid for the corresponding magnetic-field vector. For the overall reflection and transmission, however, this need not be taken into account, because the multilayer arrangement is assumed to be immersed in the same medium on both sides, i.e., $\tilde{n}_{N+1} = \tilde{n}_0$. We now apply this method to our problem (Fig. 1). By symmetry, we have

$$\tilde{r}_{21} = \tilde{r}_{23}, \qquad \tilde{r}_{10} = \tilde{r}_{34}, \qquad \tilde{r}_{24} = \tilde{r}_{20}, \qquad \tilde{r}_{14} = \tilde{r}_{30},$$

$$\tilde{t}_{21} = \tilde{t}_{23}, \qquad \tilde{t}_{10} = \tilde{t}_{34},$$

$$\tilde{a}_1 = \tilde{a}_3.$$

Thus, using Eqs. (36), (37), and (41), we get the following recursion, which has to be calculated for both polarizations:

$$\tilde{r}_{20} = \tilde{r}_{21} + \frac{\tilde{t}_{21}\tilde{a}_1^2\tilde{r}_{10}\tilde{t}_{12}}{1 + \tilde{a}_1^2\tilde{r}_{10}\tilde{r}_{21}}, \qquad \tilde{t}_{42} = \frac{\tilde{t}_{01}\tilde{t}_{12}\tilde{a}_1}{1 + \tilde{a}_1^2\tilde{r}_{21}\tilde{r}_{10}},$$

$$\tilde{r}_{30} = -\tilde{r}_{21} + \frac{\tilde{t}_{12}\tilde{a}_2^2\tilde{r}_{20}\tilde{t}_{21}}{1 - \tilde{a}_2^2\tilde{r}_{20}\tilde{r}_{21}}, \qquad \tilde{t}_{41} = \frac{\tilde{t}_{42}\tilde{t}_{21}\tilde{a}_2}{1 - \tilde{a}_2^2\tilde{r}_{21}\tilde{r}_{20}},$$

$$\tilde{r}_{40} = -\tilde{r}_{10} + \frac{\tilde{t}_{01}\tilde{a}_1^2\tilde{r}_{30}\tilde{t}_{10}}{1 - \tilde{a}_1^2\tilde{r}_{30}\tilde{r}_{10}}, \qquad \tilde{t}_{40} = \frac{\tilde{t}_{41}\tilde{t}_{10}\tilde{a}_1}{1 - \tilde{a}_1^2\tilde{r}_{10}\tilde{r}_{30}} = \tilde{t}_{04},$$

$$R = |\tilde{r}_{40}|^2, \qquad T = |\tilde{t}_{40}|^2.$$

The complex propagation factors given here are polarization independent:

$$\tilde{a}_1 = \exp(2\pi i\tilde{n}_1\bar{v}d_1\cos\theta_1), \qquad \cos\theta_1 = [1 - (n_0/\tilde{n}_1)^2\sin^2\theta_0]^{1/2},$$

$$\tilde{a}_2 = \exp[2\pi i\tilde{n}_2\bar{v}d_2\cos\theta_2), \qquad \cos\theta_2 = [1 - (n_0/\tilde{n}_2)^2\sin^2\theta_0]^{1/2},$$

where θ_0 is the angle of incidence in the surrounding medium. The reflection and transmission factors for each polarization will be as follows.

For polarization normal to the plane of incidence,

$$\tilde{r}_{10\perp} = \frac{\tilde{n}_1\cos\theta_1 - n_0\cos\theta_0}{\tilde{n}_1\cos\theta_1 + n_0\cos\theta_0}, \qquad \tilde{t}_{10\perp} = \frac{2\tilde{n}_1\cos\theta_1}{\tilde{n}_1\cos\theta_1 + n_0\cos\theta_0},$$

$$\tilde{t}_{01\perp} = (2n_0\cos\theta_0)/(\tilde{n}_1\cos\theta_1 + n_0\cos\theta_0).$$

For polarization parallel to the plane of incidence,

$$\tilde{r}_{10\parallel} = \frac{n_0\cos\theta_1 - \tilde{n}_1\cos\theta_0}{n_0\cos\theta_1 + \tilde{n}_1\cos\theta_0}, \qquad \tilde{t}_{10\parallel} = \frac{2n_0\cos\theta_1}{n_0\cos\theta_1 + \tilde{n}_1\cos\theta_0},$$

$$\tilde{t}_{01\parallel} = (2\tilde{n}_1\cos\theta_0)/(n_0\cos\theta_1 + \tilde{n}_1\cos\theta_0).$$

For polarization normal to the plane of incidence,

$$\tilde{r}_{21_\perp} = \frac{\tilde{n}_2 \cos \theta_2 - \tilde{n}_1 \cos \theta_1}{\tilde{n}_2 \cos \theta_2 + \tilde{n}_1 \cos \theta_1}, \qquad \tilde{t}_{21_\perp} = \frac{2\tilde{n}_2 \cos \theta_2}{\tilde{n}_2 \cos \theta_2 + \tilde{n}_1 \cos \theta_1},$$

$$\tilde{t}_{12_\perp} = (2\tilde{n}_1 \cos \theta_1)/(\tilde{n}_2 \cos \theta_2 + \tilde{n}_1 \cos \theta_1).$$

For polarization parallel to the plane of incidence,

$$\tilde{r}_{21_\parallel} = \frac{\tilde{n}_1 \cos \theta_2 - \tilde{n}_2 \cos \theta_1}{\tilde{n}_1 \cos \theta_2 + \tilde{n}_2 \cos \theta_1}, \qquad \tilde{t}_{21_\parallel} = \frac{2\tilde{n}_1 \cos \theta_2}{\tilde{n}_1 \cos \theta_2 + \tilde{n}_2 \cos \theta_1},$$

$$\tilde{t}_{12_\parallel} = (2\tilde{n}_2 \cos \theta_1)/(\tilde{n}_1 \cos \theta_2 + \tilde{n}_2 \cos \theta_1).$$

The advantage of this method is that one can also take into account the absorption of the windows. On the other hand, complex arithmetic (addition, multiplication, division, square, root, and modulus) is required.

REFERENCES

Abelés, F. (1967). *In* "Advanced Optical Techniques" (A. C. S. van Heel, ed.), pp. 144–288. North-Holland Publ., Amsterdam.

Amrhein, E. M., and Schulze, H. W. (1972). *Kolloid-Zeitschrift und Zeitschrift für Polymere* **250** (10), 921–926.

Baeriswyl, D., Harbeke, G., Kiess, H., and Meyer, W. (1982). *In* "Electronic Properties of Polymers" (J. Mort and G. Pfister, eds.), pp. 267–326. Wiley, New York.

Becker, G. E., and Autler, S. H. (1946). *Phys. Rev.* **70**, 300–307.

Bell, E. E. (1967). *In* "Hanbuch der Physik," Vol. 25/2a, pp. 1–58. Springer-Verlag, Berlin and New York.

Birch, J. R., and Clarke, R. N. (1982). *Radio Electron. Eng.* **52** (11/12), 565–584.

Birch, J. R., and Parker, T. (1979). *In* "Infrared and Millimeter Waves," Vol. 2 (K. J. Button, ed.), pp. 137–271. Academic Press, New York.

Birch, J. R., Bechtold, G., Kremer, F., and Poglitsch, A. (1983). "The Use of an Untuned Cavity for Broad Band Near Millimetre Wavelength Absorption Measurements on Low Loss Solids." National Physical Laboratory, Teddington, Middlesex, United Kingdom.

Bone, S., and Pethig, R. (1982). *J. Mol. Biol.* **157**, 572–575.

Clarke, R. N., and Rosenberg, C. B. (1982). *J. Phys. E.* **15**, 9–24.

Corona, P., Latmiral, G., and Paolini, E. (1980a). *IEEE Trans. Electromagn. Compat. Trans.* **EMC-22**, 2–5.

Corona, P., A. De Bonitatibus, Ferrara, G., and Paolini, E. (1980b). *Proc. 5th* Int. Wroclaw Symp. *Electromagn. Compat.*, Wroclaw, Poland, pp. 829–837.

Cullen, A. L., and Davies, J. A. (1978). *IEE. J. Microwaves Opt. Acoust.* **2**, 77–84.

Cullen, A. L., and Yu, P. K. (1971). *Proc. R. Soc. Lond., Ser. A.* **325**, 483–509.

D'Ambrosio, G., and Ferrara, G. (1983). *Proc. 5th Symp. Tech. Exhibition Electromagnetic Compatibility*, Zurich, Switzerland, March 1983, 403–406.

D'Ambrosio, G., F. Di Meglio, and Ferrara, G. (1980). *Alta Freq.* **49**, 89–94.

Elterman, P. (1970). *Appl. Opt.* **9**(9), 2140–2142.

Frauenfelder, H., Petsko, G. A., and Tsernoglov, D. (1979). *Nature* **280**, 558–568.

Gebbie, H. A., and Bohlander, R. A. (1972). *Appl. Opt.* **11**, 723–728.

Genzel, L., Kremer, F., Santo, L., and Shen, S. C. (1981). *In* " Biological Effects of Nonionizing Radiation" (K. H. Illinger, ed.), 84–93. American Chemical Society Symposium Series, No. 157.

Genzel, L., Kremer, F., Poglitsch, A., and Bechtold, G. (1983). *Biopolymers* **22**, 1715–1729.

Genzel, L., Kremer, F., Poglitsch, A., Bechtold, G., Menke, K., and Roth, S. (1984). *Phys. Rev. B*, in press.

Hoffman, D. M., Tanner, D. B., Epstein, A. J., and Gibson, H. W. (1982). *Mol. Cryst. Liq. Cryst.* **83**, 143–150.

Imoto, T., Johnson, L. N., North, A. C. T., Phillips, D. C., and Ruppley, J. A. (1972). *In* "The Enzymes" (P. D. Boyer, ed.), pp. 665–868. Academic Press, New York.

Izatt, J. R., and Kremer, F. (1981). *Appl. Opt.* **20**, 2555–2559.

Kremer, F., Genzel, L., and Drissler, F. (1980). "Ondes Electromagnetiques et Biologie." L'Union Radio Scientifique International Symp., Juiy-en-Josas, Paris, France, 31–32.

Kremer, F., and Izatt, J. R. (1981). *Int. J. Infrared Millimeter Waves* **2**, 675–694.

Kremer, F. (1983). *In* "Biological Effects and Dosimetry of Nonionizing Radiation" (M. Grandolfo, S. M. Michaelson, and A. Rindi, eds.), pp. 233–250. Plenum, New York.

Kremer, F., and Genzel, L. (1981). *Dig. Int. Conf. Infrared Millimeter Waves*, Miami, Florida, Paper T-1-4, IEEE Catalog No. 81 CH 1645-1 MTT.

Lamb, W. E., Jr. (1946). *Phys. Rev.* **70**, 308–317.

Llewellyn-Jones, D. T., Knight, R. J., Moffat, P. H., and Gebbie, H. A. (1980). *Proc. IEE* **127(A)**, 535–540.

McCammon, J. A., Gelin, B. R., and Karplus, M. (1979). *Nature* **262**, 325.

Montaner, A., Galtier, M., Benoit, C., and Aldissi, M. (1981). *Phys. Status Solidi A* **66**, 267–269.

Nagels, P. (1979). *In* "Topics in Applied Physics," vol. 36 (M. H. Brodsky, ed.), pp. 138–158. Springer-Verlag, Berlin and New York.

Pinkerton, F. E., and Sievers, A. J. (1982). *Infrared Phys.* **22**, 377–392.

Poglitsch., A., Kremer, F., and Genzel, L. (1984). *J. Mol. Biol.* **173**, 137–142.

Richards, P. L., and Tinkham, M. (1960). *Phys. Rev.* **119**, 575–590.

Schultz, G. (1960). *Ann. Phys. (Leipzig)* 7. Folge, Bd. 6, H. 7–8, 345–354.

Roth, S., Ehinger, K., Menke, K., Peo, M., and Schweizer, R. J. (1982). *Proc. Les Arcs Conf.*, Les Arcs, France.

Stratton, J. A. (1941). "Electromagnetic Theory." McGraw-Hill, New York.

Stumper, U. (1973). *Rev. Sci. Instrum.* **44**(2), 165–169.

Vergoten, G., Fleury, G., and Moschetto, I. (1980). *In* "Advances in Infrared and Raman Spectroscopy," vol. 4 (R. J. H. Clark, ed.), pp. 195–269. Heyden, London.

Willis, H. A., *et al.* (1981). *Polymer* **22**, 20–22.

CHAPTER 5

Powerful Gyrotrons

V. A. Flyagin, A. L. Gol'denberg, and G. S. Nusinovich

Institute of Applied Physics
Academy of Sciences of the USSR
Gorky, USSR

I. Introduction

By the mid-1950s powerful microwave electronics covered the centimeter-wavelength range as well as longer wavelengths. However, descending along the wavelength scale to the millimeter range, traditional microwave devices lose their maximum power quickly. The main cause of this is heat loss from small-scale elements of electrodynamic structures, such as slow-wave structures of devices based on Cherenkov radiation (traveling-wave tubes, backward-wave oscillators, magnetrons) and resonator cavities of klystrons based on transient radiation.

As a result, by the end of the 1950s there arose a particular interest in *Bremsstrahlung* generators, in which electrons can resonate with the fast waves of space-extended electrodynamic systems devoid of small-scale elements. Rotating in the external homogeneous magnetic field, the electrons

179

undergo phase bunching because of the relativistic dependence of cyclotron frequency (electron mass) on electron energy. This is one of the mechanisms of electron bunching that results in coherent *Bremsstrahlung*. The mechanism, discovered independently by Twiss (1958), Schneider (1959), and Gaponov (1959a), forms the basis of cyclotron resonance masers (CRMs).

The maser effect was observed in the very first CRM experiments (Gaponov, 1959b; Hirshfield and Wachtel, 1964; Pantell, 1959). Subsequently the power of various CRMs was essentially increased (Antakov *et al.*, 1966; Gaponov, Petelin, and Yulpatov, 1967; Gaponov *et al.*, 1965) until at 8 mm it amounted to 0.8 kW (Antakov *et al.*, 1966).

An important step in the escalation of CRM power throughout the millimeter and even part of the submillimeter ranges was the invention of the gyrotron (Gaponov *et al.*, 1976). The main components of the gyrotron are a magnetron-type electron gun and an open high-selectivity cavity represented by a section of slightly irregular waveguide (see Fig. 1).

The electron gun forms a hollow cylindrical beam in which the energy of electron gyration is a larger (or at least significant) part of the total energy. In the resonator cavity, which is immersed in a maximum magnetic field, the electron beam, operating under the cyclotron resonance condition, excites one of the cavity eigenmodes at the frequency near cutoff. When interacting with the rf field, electrons uniformly distributed over the gyration phases at the cavity input behave as follows. First (see Fig. 2, in which the ring of electrons is shown to gyrate around one of the guiding centers) the orbital momentum of an electron is modulated by the rf electric field. Half of the electrons (those that enter the accelerating phase of the synchronous component of the rf field) increase their energy, and the other half (those that enter the decelerating phase) decrease their energy. Owing to the relativistic mass–energy dependence, the variation of the particle energy results in a change in the electron cyclotron frequency. As a consequence, inertial phase bunching of particles occurs. The electrons bunch into the decelerating phase if the rf field frequency is slightly higher than the cyclotron frequency. This may result in coherent electron radiation. The rf field consumes only the rotational energy of the electrons. The fact that the axial wave number of the excited mode is small weakens the negative influence of the axial-velocity spread on the Doppler broadening of the cyclotron resonance band, i.e., on the conditions of gyrotron excitation.

Because of diffraction effects, the rf field of the cavity penetrates into an output waveguide with a smoothly enlarging cross section (Fig. 1). The narrowing of the cavity on the output side defines the diffraction Q factor as well as the Q selection of modes that differ in the axial structure of the rf field. The diameter and the length of the cavities of powerful gyrotrons are both significantly larger than the wavelength. This, together with the mitiga-

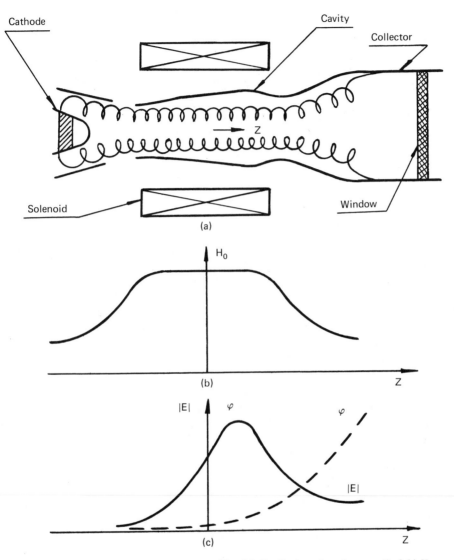

FIG. 1   (a) The scheme of the gyrotron; (b) axial distribution of static magnetic field H_0; (c) alternating rf field $E = |E|\,\mathrm{Re}\{\exp[i(\omega t - \varphi)]\}$.

tion of the heat problems, makes it possible to use electron optical systems having a very large perveance.

Present-day pulsed gyrotrons (with pulse durations less than 1 msec) operating at frequencies up to 100 GHz (see Fig. 3) have a power of more than 1 MW, and at 100 kW the frequency range increases to 500 GHz. The

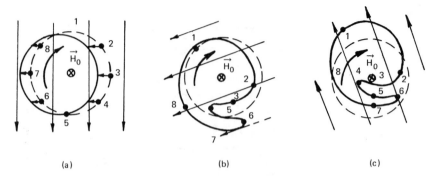

(a)   (b)   (c)

FIG. 2   Evolution of the electron ring interacting with the rf field: (a) initial modulation, displacement of the ring as a whole in the decelerating region; (b) inertial orbital bunching caused by the relativistic dependence of electron cyclotron frequency on its energy; (c) deceleration of the bunch for $\omega > \omega_H$.

output power of cw and long-pulsed gyrotrons reaches ~ 200–400 kW at frequencies up to 100 GHz and up to tens of kilowatts at higher frequencies. [The data given in Fig. 3 are taken from works by Andronov et al. (1978), Carmel et al. (1982), Faillon and Mourier (1982), Jory et al. (1982), Jory (1984), Luchinin et al. (1982), Temkin et al. (1982), and Zaytsev et al. (1974).]

Powerful gyrotrons are of interest for various fields of physics. Long-pulsed and cw millimeter gyrotrons have been used successfully in experiments on electron-cyclotron plasma heating in tokamaks [see, for example, Alikaev (1981), Alikaev et al. (1972), and Uckan (1981)], and further development of gyrotrons is stimulated by prospects of experiments on future generations of these and other plasma assemblies.

The gyrotron is a member of a big family of gyro tubes using magnetron-type electron guns and having similar electrodynamic system performance. This group also includes amplifiers of the waveguide type: gyro TWT, sectional amplifiers, two- and multicavity gyroklystrons, and combined amplifiers–gyrotwistrons, wherein the output waveguide section is excited by the beam of electrons that are bunched because of modulation in the input cavity. An interesting member of this family is a device in which the direction of the electromagnetic-wave radiation is perpendicular to the direction of the external magnetic field (CRM with a transverse current). Each of these devices has its own advantages: high efficiency and gain in gyroklystrons (Andronov et al., 1978; Jory et al., 1978), a large bandwidth in gyro TWTs (Lau et al., 1981), a combination of these qualities in gyrotwistrons (Bratman et al., 1973; Flyagin et al., 1977; Moiseev, 1977), and a high power level in gyrotrons with a transverse current (Sprangle et al., 1981; Zarnitsyna and Nusinovich, 1978), which is combined with a large band-

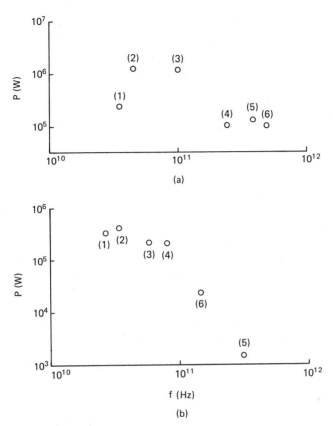

FIG. 3 State-of-the-art pulsed and cw gyrotrons: (a) Short-pulsed gyrotrons ($\tau < 1$ msec): (1) $\tau = 6$ μsec (Thompson-CSF, France, 1982); (2) $\tau = 50$ μsec [Institute of Applied Physics (IAP), USSR 1973]; (3) $\tau = 100$ μsec (IAP, USSR, 1977); (4) $\tau = 1$ μsec (Massachusetts Institute of Technology, 1982); (5), (6) $\tau = 100$ μsec (IAP, USSR, 1982). (b) Long-pulsed and cw gyrotrons: (1) cw (Varian, 1978); (2) $\tau = 40$ msec (Naval Research Laboratory, 1981); (3) $\tau = 100$ msec (Varian, 1982), cw (Varian, 1984); (4) $\tau = 150$ msec (IAP, USSR, 1981); (5), (6) cw (IAP, USSR, 1974).

width in amplifiers of this type (Bykov *et al.*, 1974; Zhurakhovsky, 1969). Here we restrict our consideration mainly to problems essential for powerful oscillators: gyrotrons.

In this chapter we will examine some of the principal ideas and methods of the gyrotron theory that are most current and fruitful. We will present a number of relatively old but still significant results from rather rare publications, so that researchers need not waste time deriving them afresh. In Section II we will give a general theory of the interaction between electrons rotating in the external magnetic field and the rf field of the resonators and waveguides. In Section III, together with simplified drift adiabatic description of the operating gyrotron gun we will present a number of new numerical methods of computation, and results based on them, for guns with enhanced current. The problems of mode selection essential for gyrotrons with extended interaction space are discussed in Section IV. These problems have been partially reviewed in previously published papers (Gaponov *et al.*, 1980, 1981). In the concluding section we will discuss the state of the art and possible further increases in the microwave power and operating frequency of gyrotrons.

II. Theory of Electron–rf Field Interaction

Gyrotron theory is similar to the theory of other microwave tubes. To obtain a self-consistent set of equations, one can take either equations of electron motion in the external homogeneous magnetic field and the rf field or a kinetic equation and equations of the rf field excitation by a curved electron beam. The latter are equations for excitation of oscillations in gyrotrons with a high-Q cavity and equations of wave excitation in a waveguide for gyro-TWT devices.

We shall consider a linear and a nonlinear theory of the two gyrotron types: the amplifier (gyro-TWT) and the oscillator (gyromonotron).

A. EXCITATION OF THE WAVEGUIDE BY A HELICAL ELECTRON
   BEAM; ELECTRON BUNCHING IN THE WAVE FIELD

We take into account the influence of the rf field on the electron motion in the external magnetic field using the perturbation method. Within the framework of the linear theory, the electron shift from the stationary trajectory $\mathbf{r} = \mathbf{r}_0(\tau) + \mathbf{r}_\omega(\tau)e^{i\omega t}$ (τ is the electron transit time) owing to the high-frequency Lorentz force $\mathbf{F}_\omega$ can be presented in the form $\mathbf{r}_\omega \propto \mathbf{F}_\omega$. We present the electromagnetic-wave field as

$$\mathbf{E} = \text{Re}\{A\mathbf{e}(r, \theta) \exp[i(\omega t - hz)]\},$$
$$\mathbf{H} = \text{Re}\{A\mathbf{h}(r, \theta) \exp[i(\omega t - hz)]\}, \tag{1}$$

where A is the wave amplitude defined under the synchronism

$$|\omega - hv_\parallel - n\omega_H| \ll \omega_H \tag{2}$$

by an equation of excitation

$$dA/dz = -(I/N)(1/2\pi) \int_0^{2\pi} \mathbf{r}_\omega \mathbf{F}_\omega^* \, d\theta_0, \tag{3}$$

where I is the constant component of the beam current, N is the norm of the wave in the waveguide, $\mathbf{F}_\omega = \mathbf{e} + (1/c)[\mathbf{v}_0 \mathbf{h}]$, θ is the phase of electron gyration with respect to the wave field in the input cross section of the system, and the functions $\mathbf{e(h)}$ describe the electric (magnetic) field distribution of the wave in the transverse cross section of the waveguide.

Synchronous components of the electron shift vector $\mathbf{r}_\omega$ from a stationary trajectory can be defined by (Gaponov, 1961)

$$r_n = \frac{e}{m} A \frac{F_{rn}\Delta\omega_n - iF_{\theta n}\omega_H}{\Delta\omega_{n-1}\Delta\omega_n\Delta\omega_{n+1}},$$

$$r_0\theta_n = \frac{e}{m} A \left\{ \frac{iF_{rn}\omega_H + F_{\theta n}\Delta\omega_n}{\Delta\omega_{n-1}\Delta\omega_n\Delta\omega_{n+1}} - \frac{F_{zn}\beta_\parallel\beta_\perp + F_{\theta n}\beta_\perp^2}{(\Delta\omega_n)^2} \right\}, \qquad (4)$$

$$z_n = \frac{e}{m} A \frac{F_{zn}(1 - \beta_\parallel^2) - F_{\theta n}\beta_\parallel\beta_\perp}{(\Delta\omega_n)^2},$$

where $\beta_{\perp\,\parallel} = v_{\perp,\parallel}/c$ (v is the electron velocity, c the velocity of light), and $\Delta\omega_n = \omega - hv_{\parallel 0} - n\omega_{H0}$, i.e., taking into account the condition of synchronism (2),

$$|\Delta\omega_n| \ll \omega_H, \qquad \Delta\omega_{n-1} \simeq \omega_{H0} \simeq -\Delta\omega_{n+1}.$$

Synchronous components of the rf Lorentz force are

$$F_{rn} = \frac{1}{2\pi} \int_{-\pi}^{\pi} (e_r + \beta_\perp h_z - \beta_\parallel h_\theta)e^{-in\xi}\,d\xi,$$

$$F_{\theta n} = \frac{1}{2\pi} \int_{-\pi}^{\pi} (e_\theta + \beta_\parallel h_r)e^{-in\xi}\,d\xi, \qquad (5)$$

$$F_{zn} = \frac{1}{2\pi} \int_{-\pi}^{\pi} (e_z - \beta_\perp h_r)e^{-in\xi}\,d\xi,$$

where $\xi = \omega_H\tau$. In Eq. (4) the terms with the resonant denominators $\Delta\omega_n$ correspond to linear ($r_\omega \sim \tau$) electron bunching, and those with the resonant denominators $(\Delta\omega_n)^2$ to squared bunching ($r_\omega \sim \tau^2$). In this section we restrict our consideration to the cases when the linear electron bunching is negligible in comparison with squared bunching.

Under these assumptions, provided that the electron density is small (which makes it possible to consider the propagation constant h_s of normal waves close to the wave numbers h_{0s} of the cold system), one can substitute Eq. (4) into the equation of excitation (3) to obtain the dispersion equation (Gaponov, 1961)

$$\delta^2(\delta - \varepsilon) = -(|I|\omega)/(v_{\parallel 0}N)$$
$$\times \{|F_{zn}|^2(1 - \beta_\parallel^2) - |F_{\theta n}|^2\beta_\perp^2 - \beta_\parallel\beta_\perp(F_{\theta n}F_{zn}^* + F_{\theta n}^*F_{zn})\}, \qquad (6)$$

where

$$\delta = \left(h - \frac{\omega - n\omega_{H0}}{v_{\|0}}\right)\Big/h_{0s}, \qquad \varepsilon = 1 - \frac{\omega - n\omega_{H0}}{h_{0s}v_{\|0}}.$$

At a given frequency ω Eq. (6) is cubic with respect to h, and at small enough mismatch ε it always has complex conjugate roots for both fast and slow waves. This means that a growing-amplitude wave appears, owing to convective instability ($Imh > 0$). The first term in the right-hand side of Eq. (6) corresponds to axial electron bunching because of the z component of the Lorentz force. The second term defines phase (azimuthal) bunching caused by relativistic dependence of the electron-cyclotron frequency on energy. The last term describes the effect of additional axial bunching caused by the axial-velocity difference associated with the change in the relativistic mass of particles under the action of the wave field.

Let us express the Lorentz force components $F_{\theta n}$, F_{zn} via membrane functions of the waveguide ψ. For H waves the field components of interest are $e_{zH} = 0$, $e_{\theta H} = ik(\nabla_\perp\psi_H)_r$, and $h_{rH} = -ih(\nabla_\perp\psi_H)_r$, and for E waves they are $e_{zE} = \varkappa^2\psi_E$, $e_{\theta E} = -ih(\nabla_\perp\psi_E)_\theta$, $h_{rE} = -ih(\nabla_\perp\psi_E)_\theta$ (here $k = \omega/c$, $\varkappa = \sqrt{k^2 - h^2}$). Substituting these expressions into Eqs. (5) and (6) one can readily reduce the braces in the right-hand side of Eq. (6) to

$$\begin{aligned}
\{\cdots\}_H &= -\beta_\perp^2\varkappa^2|\Phi_{nH}|^2, \\
\{\cdots\}_E &= -(\beta_\| k - h_0)^2\varkappa^2|\Phi_{nE}|^2,
\end{aligned} \qquad (7)$$

where

$$\Phi_{nH} = \frac{1}{2\pi}\int_{-\pi}^{\pi}(\nabla_\perp\psi_H)_r e^{-in\xi}\,d\xi, \qquad \Phi_{nE} = \frac{1}{2\pi}\int_{-\pi}^{\pi}\psi_E e^{-in\xi}\,d\xi.$$

It follows from Eqs. (6) and (7), in particular, that the effect of squared bunching weakens with decreasing transverse wave number $\varkappa$. In the limit $h \to k$ ($\varkappa \to 0$), exact compensation of axial and azimuthal bunchings occurs that is caused by the effect of autoresonance (Davydovsky, 1962; Gaponov, 1960; Kolomensky and Lebedev, 1962). Hence, under conditions close to autoresonance, the effects of linear bunching of particles should be taken into account. [A more detailed analysis of bunching mechanisms is given by Gaponov (1961), Gaponov et al. (1967), and Gaponov and Yulpatov (1966).] If Eq. (6) has the form characteristic of TWT and BWO of the "O" type, inclusion of the linear bunching effects results in appearance in the dispersion equation of terms describing interaction of the "M" type (Gaponov, 1961; Gaponov et al., 1967). Note that the dispersion equation (6), as well as a more complete equation that takes into account linear bunching, can be obtained from a dispersion equation that describes wave propagation in an infinite magnetoactive plasma (Zheleznyakov, 1960) and in a plasma filling a cylindrical waveguide (Granatstein et al., 1975), if we neglect a nonsynchronous wave [see, e.g., Bratman et al. (1979b)].

In the case of weakly relativistic beams, the last term in the right-hand side of Eq. (6) can be neglected, and expressions for the azimuthal and axial components of the Lorentz force, omitting the terms of order $\beta_\parallel$, can be written in the form

$$|F_{\theta n}|^2 = k^2 |\Phi_{nHE}|^2, \qquad |F_{zn}|^2 \simeq \beta_\perp^2 h^2 |\Phi_{nHE}|^2$$

[compare with Eqs. (5), (7)]. From this it follows that in Eq. (6), for the case of fast waves ($k > h$), the term responsible for phase bunching ($\beta_\perp^2 |F_{\theta n}|^2$) dominates over the first term, $(1 - \beta_\parallel^2)|F_{zn}|^2$, which characterizes axial bunching. On the contrary, for the case of slow waves ($h > k$), axial bunching plays the main role (Chu and Hirshfield, 1978; Gaponov, 1961; Gaponov et al., 1967).

B. LINEAR GYROTRON THEORY

In the gyrotron a helical electron beam excites oscillations in a section of a slightly irregular waveguide at a frequency close to cutoff of the corresponding wave in the regular waveguide (waveguide of comparison) (Vlasov et al., 1969). These oscillations can be presented as superposition of fast waves with a large phase velocity $v_{ph} \gg c$ and, correspondingly, a small group velocity $v_{gr} \ll c$ in the axial direction. Condition $v_{ph} \gg c$, as mentioned above, means that phase bunching of particles has the main role in the gyrotron. This makes it possible to simplify the linear gyrotron theory, because one can omit the axial components of the Lorentz force in Eq. (4). At the same time, compared with wave excitation in a waveguide the theory of gyrotron-cavity excitation (Gaponov et al., 1967; Petelin and Yulpatov, 1975) is more complicated, because the eigenmode of the gyrotron cavity is a superposition of the waveguide waves with a characteristic spectrum width of the axial wave numbers $0 \leq h \lesssim \pi/L$, where L is the cavity length ($L \gg \lambda$).

From a known equation of excitation of the cavity's Sth mode (see, e.g., Vainstein and Solntsev (1972)],

$$\frac{dA_s}{dt} + i(\omega - \omega_s)A_s = -\frac{2\pi}{N_s} \int_v \mathbf{j}_\omega \mathbf{E}_s^* \, dV. \tag{8}$$

The conditions of self-excitation follow as

$$Q_s^{-1} = 4\pi\chi_{ss}'', \qquad (\omega - \omega_s')/\omega_s' = -2\pi\chi_s'. \tag{9}$$

The first equation defines the starting current of the oscillator; the second one, the electron shift of self-excitation frequency ω with respect to the eigen-frequency of the Sth mode $\omega_s = \omega'_s + i(\omega'_s/2Q_s)$ (Q_s is the Q factor of the Sth mode). Here the rf electric field is written in the form

$$\mathbf{E} = \mathrm{Re}\left\{e^{i\omega t} \sum_s A_s \mathbf{E}_s(\mathbf{r})\right\}.$$

The function $\mathbf{E}_s(\mathbf{r})$ describing the space structure of the Sth mode, because of the small axial wave number $h \ll k$, can be presented as

$$\mathbf{E}_s(\mathbf{r}) \simeq \mathbf{E}_s(\mathbf{r}_\perp)f_s(z),$$

where the function $f_s(z)$ is defined by the nonuniform string equation (Vlasov et al., 1969)

$$d^2 f_s/dz^2 + [k^2 - \varkappa_s^2(z)]f_s = 0, \tag{10}$$

with the corresponding boundary conditions.

We take into account that in the linear theory the complex amplitude of the alternating-beam current component $\mathbf{j}_\omega$ is proportional to small displacements $\mathbf{r}_\omega$ [Eq. (4)] of electrons from the stationary trajectory. Now, using corresponding integral representations for the rf field on the electron orbit, one can define the diagonal elements of the susceptibility tensor of the electron beam χ_{ss} [Eq. (9)] as follows (Petelin and Yulpatov, 1975):

$$\chi_{ss} = \frac{e|I|}{mN_s} \int \omega(\gamma)\left[\frac{M_{ss}}{\omega_H} + O_{ss}\frac{d}{d\omega}\right]\eta_{ss}\,d\gamma. \tag{11}$$

Here the function $\omega(\gamma)$ normalized to unity describes an arbitrary stationary distribution of electrons over coordinates and velocities. The quantity

$$\eta_{ss} = \frac{1}{\pi}\int_{-\infty}^{\infty}(|\tilde{f}_s|^2/\Delta\omega')\,d\Omega$$

describes the spectrum of the rf field intensity

$$\Omega = hv_\parallel, \qquad \Delta\omega' = \omega - n\omega_H + \Omega,$$

$$\tilde{f}_s(\Omega) = 1/2\pi v_\parallel \int_{-\infty}^{\infty} f_s(z)e^{-ihz}\,dz.$$

The coefficients O_{ss} and M_{ss} are defined in the case of H modes by [compare with Petelin and Yulpatov (1975)]

$$O_{ss}^{H} = 2\pi^2\beta_\perp^2(J'_n)^2|L_n|^2,$$

$$M_{ss}^{H} = \pi^2[(1 - \beta_\perp^2)(J_{n-1}^2 - J_{n+1}^2)|L_n|^2 + \beta_\perp^2(J'_n)^2(|L_{n-1}|^2 - |L_{n+1}|^2)],$$

$$\tag{12a}$$

and in the case of E modes by

$$O_{ss}^E = 2\pi^2 \beta_\parallel^2 J_n^2 |L_n|^2,$$

$$M_{ss}^E = \pi^2 \{\beta_\parallel^2 [J_n^2 (|L_{n-1}|^2 - |L_{n+1}|^2) + |L_n|^2 (J_{n-1}^2 - J_{n+1}^2)] - 4 J_n^2 |L_n|^2\},$$

$$(12b)$$

which characterize the efficiency of interaction of "O" or "M" types, respectively. In Eq. (12), J_n is the Bessel function with the argument $\varkappa_s a$ (a is the Larmor radius of electrons), the coefficient

$$L_n = \left(\frac{n}{|n|}\right)^n \left[\frac{1}{\varkappa_s}\left(\frac{\partial}{\partial X} + i \frac{n}{|n|} \frac{\partial}{\partial Y}\right)\right]^{|n|} \psi_s(X, Y)$$

describes the transverse structure of the rf Lorentz force acting on the electron with the guiding center coordinates X, Y; coordinates of the electron are $x = X + a \cos \theta$, $y = Y + a \sin \theta$.

It follows from Eqs. (11), (12) that the susceptibility of the electron beam is a rather complex function of the guiding center coordinates of electrons. This is associated with the fact that in Eq. (4) we took into account not only the terms proportional to F_{0n} (responsible for squared phase bunching of the "O" type) but also the terms proportional to F_{rn}, which lead to orbital drift of the guiding centers (interaction of the "M" type). The situation, however, is significantly simplified in a most important case of weakly relativistic orbital velocities of electrons $\beta_\perp^2 \ll 1$. Here, because the Larmor radius is small compared with the wavelength ($\varkappa_s a \ll 1$), the synchronous harmonic of the rf field has a quasi-stationary structure in the vicinity of the electron trajectory, and expressions (12) reduce to (Petelin and Yulpatov, 1975)

$$O_{ss}^H = \tfrac{1}{2}\beta_\perp^2 M_{ss}^H = 2\pi^2 \beta_\perp^2 [(n\beta_\perp)^{n-1}/2^n (n - 1)!]^2 |L_n|^2. \tag{13}$$

For E modes, $O_{ss}^E = \tfrac{1}{2}\beta_\perp^2 M_{ss}^E = \beta_\parallel^2 O_{ss}^H$ (Ginzburg and Nusinovich, 1979). These expressions are applicable not only for conventional weakly relativistic gyrotrons but also for relativistic ones with a homogeneous magnetic field in which electrons can give the fast waves a rather small part of their initial energy under cyclotron resonance conditions (Ginzburg and Nusinovich, 1979)

$$|(\varepsilon_0 - \varepsilon)/\varepsilon_0| = \tfrac{1}{2}\beta_{\perp 0}^2 |1 - (p_\perp^2/p_{\perp 0}^2)| \lesssim 1/nN \ll 1,$$

where N is the number of electron turns in the interaction space. Corrections of the axial distribution of the external magnetic field make it possible to avoid these limitations. Taking into account Eq. (13), we can define the starting current I_{st} of the gyrotron from the expression [compare with Petelin and Yulpatov (1975)]

$$\frac{2}{\pi} \frac{e I_{st}}{mc^3} Q_s \frac{1}{\beta_\perp^4} \frac{\lambda^3}{N_s} \left[\frac{1}{(n-1)!} \left(\frac{n\beta_\perp}{2} \right)^{n-1} \right]^2 |L_n|^2$$

$$\times \left(-n - \frac{\partial}{\partial \Delta} \right) \left| \frac{1}{2\pi} \int_{-\infty}^{\infty} f_s(\zeta) e^{i\Delta\zeta} d\zeta \right|^2 = 1. \tag{14}$$

This formula is written for an axially symmetric gyrotron in which there is no velocity and guiding center radius spread [the expression taking into account this spread is given by Petelin and Yulpatov (1975)]: $\zeta = \pi(\beta_{\perp 0}^2/\beta_\parallel)(z/\lambda)$ is the normalized axial coordinate, and $\Delta = (2/\beta_{\perp 0}^2)(1 - n\omega_{H0}/\psi)$.

C. NONLINEAR THEORY OF INTERACTION OF A HELICAL ELECTRON BEAM WITH AN ELECTROMAGNETIC WAVE

We restrict our consideration to systems with a small Doppler frequency conversion ($hv_\parallel \ll \omega \simeq n\omega_H$) that is valid for excitation of a wave with a frequency rather close to cutoff. [The theory of CRM with Doppler frequency conversion was developed by Zhurakhovsky (1969), Petelin (1974), Bratman et al. (1979b, 1981), and Ginzburg et al. (1979b, 1981)]. In these systems one can consider the axial momentum of electrons constant and neglect [see, e.g., Bratman et al. (1979b)] the transverse drift of the electron guiding centers.

All the foregoing makes it possible to describe the electron motion by two averaged equations for slowly varying energy ε and phase ϑ of the particles (the phase ϑ is defined as $\vartheta = \omega t - hz - n\varphi$, where φ is given by the relation $p_x + ip_y = p_\perp e^{i\varphi}$):

$$\frac{du}{d\zeta} = (1 - u)^{n/2} \text{Im}(Fe^{i\vartheta}),$$

$$\frac{d\vartheta}{d\zeta} = \Delta - u - \frac{\partial}{\partial u}(1 - u)^{n/2} \text{Re}(Fe^{i\vartheta}). \tag{15}$$

Here, $u = (2/\beta_{\perp 0}^2)[(\varepsilon_0 - \varepsilon)/\varepsilon_0] = 1 - (p_\perp^2/p_{\perp 0}^2)$ is the energy variable and $F = 4\beta_{\perp 0}^{n-4}(eA/mc\omega)(n^n/2^n n!)L_n$ the normalized complex amplitude of the wave. The boundary conditions for Eq. (15) are $u(0) = 0$, $\vartheta(0) = \vartheta_0 \in [0.2\pi]$. V. K. Yulpatov (1967) was the first to obtain these equations when analyzing the interaction between the waveguide mode and weakly relativistic nonisochronously oscillating charged particles with one degree of freedom. Thus, these equations describe a gyro TWT operating close to the cutoff frequency. In a simple gyro TWT with a constant-radius waveguide and a constant magnitude of the external magnetic field H_0, the mismatch $\Delta = (2\beta_{\perp 0}^2)[(\omega - n\omega_{H0} - hv_\parallel)/\omega]$ is constant. For gyro TWTs in which the waveguide radius R and the external magnetic field H_0 are changed in a consistent manner so as to broaden the amplification bandwidth, the mismatch Δ in Eq. (15) should be defined by the function of the axial coordinate $\Delta(\zeta)$. [The theory of such gyro TWTs was developed by Lau et al. (1981) and Lau and Chu (1981).]

To obtain a self-consistent set of equations, one should add to Eq. (15) an equation of excitation (3), which may be rewritten in the form

$$\frac{dF}{d\zeta} = I_0 \frac{i}{2\pi} \int_0^{2\pi} (1-u)^{n/2} e^{-i\vartheta} d\vartheta_0. \tag{16}$$

Here the current parameter is

$$I_0 = 16 \frac{e|I|}{mc^3} \beta_{\perp 0}^{2(n-3)} \frac{k}{h} \frac{J_{m \mp n}^2[2\pi(R_0/\lambda)]}{(v^2 - m^2)J_m^2(v)},$$

where $v = 2\pi R/\lambda$, R_0 is the beam radius, and R the resonator radius. From Eqs. (15) and (16) there follows the law of conservation of power in the beam–wave system:

$$2I_0 \frac{1}{2\pi} \int_0^{2\pi} u(\zeta) \, d\vartheta_0 - |F(\zeta)|^2 = \text{constant}.$$

Other laws of conservation are given by Yulpatov (1967) and Ginzburg, Zarnitsyna, and Nusinovich (1981).

Integrating Eq. (15), we can define the value of the orbital efficiency

$$\eta_\perp = \frac{1}{2\pi} \int_0^{2\pi} u(\zeta_{\text{out}}) \, d\vartheta_0. \tag{17}$$

Preliminary results of integration of Eqs. (15)–(17) were obtained by V. K. Yulpatov in 1965. A more complete investigation of Eqs. (15)–(17) was carried out by Ginzburg et al. (1981) with $\Delta(\zeta) = $ constant. The results are presented in Fig. 4, which shows the equal values of the orbital efficiency in the plane of parameters Δ, I_0 and the optimum values of the normalized length of the system ζ_{out} at given Δ, I_0. These authors also analyzed the dependence of the efficiency $\eta_\perp$ on the mismatch Δ characteristic of band properties of the device.

D. Nonlinear Gyrotron Theory

The nonlinear theory is most easily constructed for a gyrotron in which the diffraction Q of the open resonator

$$Q_{\text{dif}} = 4\pi[(L/\lambda)^2/(1 - |R_1 R_2|)] \tag{18}$$

is much larger than the minimum value $Q_{\text{dif}}^{\text{min}} \simeq 4\pi(L/\lambda)^2$ [in Eq. (18) $R_{1,2}$ are the coefficients of the operating mode reflection at the input and

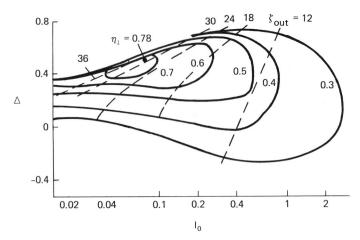

FIG. 4   Lines of isoefficiency (———) and normalized length of interaction space (— —) in the plane of the beam parameter I_0 and frequency mismatch Δ for a gyro TWT.

output cross sections of the resonator]. In this case the effect of the electron beam on the axial structure of the rf field of the open resonator is small. Because of this the function $f(z)$ may be considered fixed and completely defined by the cavity profile [i.e., by the nonuniform string equation (10) with the corresponding boundary conditions]. Stationary self-oscillations in such a gyrotron are described by a self-consistent system that comprises equations of motion (15) [where F is presented in the form $F = F'f(\zeta)$], balance equations (9) for active and reactive power (the second equation may be omitted if we are not interested in electron frequency shift), and an expression defining the electron beam susceptibility χ in the "large signal" regime:

$$\chi = I' \frac{1}{2\pi} \int_0^{2\pi} \left[\int_0^{\zeta_{out}} (1 - u)^{n/2} e^{i\vartheta} f(\zeta)\, dz \right] d\vartheta_0, \tag{19}$$

where

$$I' = \frac{1}{8\pi^4} \frac{e|I|}{mc^3} \frac{\lambda^3}{N_s} |L_n|^2 \left(\frac{n^n}{2^n n!} \right)^2 \beta_\perp^{2(n-3)}.$$

This formula follows from the integral in the right-hand side of Eq. (8), provided that the above designations and the law of charge conservation are taken into account. As follows from Eqs. (16), (19), and the first equation from (15), the imaginary part of susceptibility is related to orbital efficiency by

$$\eta_\perp = F'^2 \chi''(I')^{-1}. \tag{20}$$

Using a self-consistent set of equations (9), (15), (19), (20), one can study the dependence of the gyrotron efficiency on the normalized length of the system ζ_{out}, the cyclotron resonance mismatch Δ, and the self-oscillation amplitude F' (or the beam current). The first results of these investigations were presented by Gaponov et al. (1967), Zhurakhovsky (1972), and Kurayev (1971). These and more recent studies have shown that for a gyrotron operating at the fundamental as well as at gyro-frequency harmonics, the orbital efficiency can be very high. Even in the simplest open-cylinder cavity in which the axial field structure is close to Gaussian, the orbital efficiency of a gyrotron with a constant (along the cavity) magnitude of the external magnetic field exceeds 70% at the first two gyrofrequency harmonics (Nusinovich and Erm, 1972). The results of an investigation of the influence of the axial structure of the rf field and the axial nonuniformity of the external magnetic field in the interaction space on the orbital efficiency, as well as a more detailed calculation of efficiency, can be found in the works by Kurayev (1971) and Zhurakhovsky (1972). Results on the influence of electron-velocity spread on the orbital efficiency were given by Ergakov and Moiseev (1981). All of these problems were reviewed by Gaponov et al. (1981).

If the gyrotron cavity has a diffraction Q close to minimum, i.e., the coefficient of the wave reflection at the output cross section is close to zero [see Eq. (18)] then, the axial structure of the rf field is unfixed, i.e., the electron beam can influence it. In this case a self-consistent set of equations is formed by equations of motion (15) and an equation of excitation, the modified nonuniform string equation (10). In the right-hand side of Eq. (10) a term emerges similar to the one in Eq. (8), which is responsible for the excitation of oscillations by the electron beam (Bratman et al., 1973; Flyagin et al., 1977; Gaponov et al., 1981). The orbital efficiency of a gyrotron with the unfixed axial structure of the resonator field may also be larger than 70% (Bratman et al., 1973). As shown by the calculations (Bratman et al., 1973), the fact that the axial structure of the rf resonator field is unfixed does not essentially affect many characteristics of the gyrotron (in particular, its conditions of self-excitation).

The influence of the space-charge field can be taken into account in a similar way. The term characterizing the influence of the space-charge field on electrons appears in the right-hand side of the equation of motion (15) with the term that describes direct action of the rf resonator field [F in Eq. (15)]. The formula for this term was derived by Bratman and Petelin (1975). As these authors have shown, the influence of the space-charge field on the behavior of the gyrotron under usual conditions is small and can be taken into account when efficiency is estimated using the perturbation method by the space-charge parameter $S = (4/\pi\beta_{\perp 0}^2)(\omega_p^2/\omega_H^2) \ll 1$ (ω_p is the effective plasma frequency of the electrons). In this case, efficiency is defined as

$\eta_\perp = \hat{\eta}_\perp + s(\partial\hat{\eta}_\perp/\partial s)$, where $\hat{\eta}_\perp$ corresponds to the calculation of efficiency neglecting the space-charge field. The derivative $\partial\hat{\eta}_\perp/\partial s$ was calculated by Bratman and Petelin (1975).

In summary, when constructing gyrotrons it usually suffices to use results of computations of efficiency of a gyrotron having an rf field with fixed Gaussian structure. The results of these computations are presented in

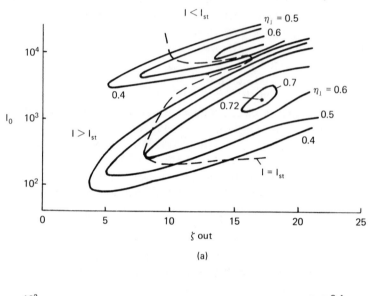

(a)

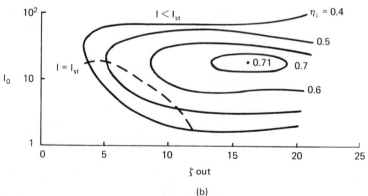

(b)

FIG. 5   Lines of isoefficiency (——) in the plane of the beam parameter I_0 and normalized length of the resonator ζ_{out} for (a) gyrotrons operating at the fundamental ($n = 1$) harmonic of the cyclotron frequency; (b) gyrotrons operating at the second ($n = 2$) harmonic; – –, self-excitation conditions $I_0 = I_0^{st}$; self-excitation takes place for $I_0 > I_0^{st}$.

Fig. 5, taken from the paper by Nusinovich and Erm (1972). The parameter I_0 in this figure is

$$I_0 = (0.24I \cdot Q \times 10^{-3}) \left(\frac{n^n}{2^n \cdot n!} \right)^2 \left(\pi \frac{\beta_{\perp 0}}{\beta_\parallel} \right)^{2(3-n)} \left(\frac{L}{\lambda} \right)^{5-2n} \frac{J_{m \mp n}^2 [2\pi(R_0/\lambda)]}{(v^2 - m^2)J_m^2(v)},$$

where I is the beam current in amperes and R_0 the radius of the electron guiding centers of the annular cylindrical beam. Taking into account the qualities of the electron beam, one can introduce corrections to these results based on the papers by Bratman and Petelin (1975) and Ergakov and Moiseev (1981).

III. Electron Optical Systems

An axially symmetric electron beam is generally used in the gyrotron. The orbital–axial ratio of electron energies $\varepsilon_\perp/\varepsilon_\parallel$ in such a beam should be as large as possible. Its cylindrical shape with a wall thickness usually smaller than $\frac{1}{2}\lambda$ makes it possible to weaken the decrease in efficiency caused by the beam-potential drop associated with the space charge of electrons and by radial nonuniformity of the rf field in the resonator. This effect of potential drop sets an upper limit on $\varepsilon_\perp/\varepsilon_\parallel$. Another and generally stronger limit $\varepsilon_\perp/\varepsilon_\parallel$ is defined by electron velocity spread.

A helical beam in the gyrotron is formed in two stages. First, electrons acquire gyration with cyclotron frequency in a relatively weak magnetic field. Then, moving in a smoothly growing magnetic field, the particles increase their energy of gyration. As a result the required $\varepsilon_\perp/\varepsilon_\parallel$ ratio is attained, and, simultaneously, compression of the beam takes place.

In the "pre-gyrotron" stage of CRMs, several ways were used to impart initial gyration to electrons. These were nonadiabatic beam injection across the magnetic field lines in a shielded gun (Chow and Pantell, 1960), periodic magnetic fields (Hirshfield and Wachtel, 1964), and others. Nonadiabatic methods of initial "twisting" of electrons by beam injection into the region of a sharply changing magnetic field are also used in relativistic gyrotrons (Ginzburg et al., 1979c).

At present all the known weakly relativistic gyrotrons use an adiabatic electron gun of the magnetron type (Gol'denberg and Petelin, 1973), where electrons acquire gyration velocity the very moment they are emitted from the cathode and from the beginning move in weakly nonhomogeneous fields. The adiabatic gun is favored mainly because of its simplicity and the possibility of wide variation in operating conditions on retention of acceptable beam parameters. Exact numerical methods of calculation of an adiabatic gun have been developed (Gol'denberg et al., 1981; Manuilov and

Tsimring, 1981c). Approximate methods (Gol'denberg et al., 1981) are also widely used, because they give a rather good first approximation and a clear description of the basic phenomena in the gun.

A. ELEMENTARY THEORY OF ADIABATIC GUN

1. Orbital and Axial Electron Velocities

Approximate computation of an adiabatic electron gun (Fig. 6) is based on the drift theory of charged-particle motion. This theory is valid if the variations of electric and magnetic fields are rather small at dimensions characteristic of the electron trajectory, and if the electron velocity is small as compared with the velocity of light. Then the electron motion may be presented as gyration with a radius $r_\perp$ around a drifting center whose position is defined by a vector $\mathbf{R}$:

$$\mathbf{r} = \mathbf{R} + \mathbf{r}_\perp. \tag{21}$$

In these fields an adiabatic invariant exists (Landáu and Lifshitz, 1962):

$$I_\perp = \oint \mathbf{p} \cdot d\mathbf{q}, \tag{22}$$

where $\mathbf{q}$ and $\mathbf{p}$ are the generalized coordinate and momentum of the electron. The motion of the drifting center is described by (Northrop, 1963)

$$\dot{\mathbf{R}}_\perp = c \frac{[\mathbf{E} \cdot \mathbf{H}]}{H^2} + \frac{mc}{2e} \frac{v_\perp^2 + 2v_\parallel^2}{H^3} [\mathbf{H} \, \nabla H], \tag{23}$$

$$\dot{v}_\parallel = \frac{e}{m} \frac{(\mathbf{E} \cdot \mathbf{H})}{H} + \tfrac{1}{2} v_\perp^2 \, \mathrm{div} \, \frac{\mathbf{H}}{H}. \tag{24}$$

Here $\dot{\mathbf{R}}_\perp$ and $v_\parallel$ are the drift-velocity components perpendicular and parallel to the magnetic-field direction; overdots denote the time derivative. In weakly nonhomogeneous fields of the adiabatic gun, the second terms in Eqs. (23) and (24) can be neglected and the adiabatic invariant, Eq. (22), can be presented as

$$v_\perp^2 / H(\mathbf{R}) = \text{constant}. \tag{25}$$

At the instant the electron with a zero initial velocity emits, its orbital and transverse drift-velocity components are equal in magnitude and opposite in direction. This makes it possible, using Eqs. (23) and (25), to determine the orbital velocity $v_{\perp c}$ of the electron emitted from a smooth cathode,

$$|v_{\perp c}| = |\dot{\mathbf{R}}_{\perp c}| = c \frac{E_{\perp c}}{H_c}, \tag{26}$$

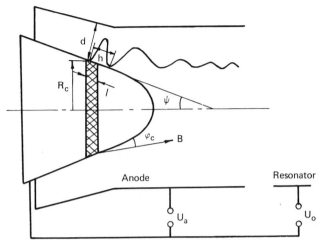

FIG. 6 Schematic diagram of the gyrotron gun. R_c, cathode radius; d, anode–cathode distance; h, step of the cycloid; l, width of the emitter; ψ, angle between the axis and the cathode surface; φ_c, angle between the cathode surface and the magnetic field line; B, magnetic field line.

where the subscript "c" denotes the values referred to the point of electron emission. Electron transition from the near-cathode region to the interaction space with a homogeneous magnetic field H_0 is provided by the axial drift velocity. This is accomplished if the condition

$$U(\mathbf{R}) - (mv_\perp^2/2e) = (mv_\parallel^2/2e) > 0 \tag{27}$$

is fulfilled along all the electron trajectory [$U(\mathbf{R})$ is the beam potential with respect to the cathode]. If condition (27) is violated, the axial velocity becomes zero and changes its sign at the point where $U_\perp = U(\mathbf{R})$. The electrons may be reflected (without any decelerating electric field) in the region of a growing magnetic field ("magnetic mirror").

In the interaction space of the device shown in Fig. 1, an electron moves along a helical path with orbital and axial velocity components that, according to Eqs. (25)–(27), are equal to

$$v_{\perp 0} = \alpha^{1/2}c(E_{\perp c}/H_c), \tag{28}$$

$$v_{\parallel 0} = [2(e/m)U_0 - v_{\perp 0}^2]^{1/2}. \tag{29}$$

Here, $\alpha = H_0/H_c$ and the subscript "0" denotes the parameters referred to the interaction space.

2. Radius of the Electron Orbit Guiding Center

This coordinate may be obtained in an adiabatic approximation by integrating Eqs. (23) and (24). In an axially symmetric system the azimuthal drift does not affect the configuration of the electron beam, and the guiding center radius R_0 can be more easily and exactly determined using Bush's theorem (Pierce, 1954):

$$\dot{\Theta} = (e/m2\pi r^2)(\psi - \psi_c), \tag{30}$$

where Θ is the azimuthal coordinate and ψ the magnetic flux inside the surface formed by the rotation of the magnetic line of force, which contains the point of instant electron location, around the z axis (on the cathode $\dot{\Theta} = 0$).

In a homogeneous magnetic field H_0, the electron moves along a helical line of radius $r_{\perp 0}$ around the guiding center with a radial coordinate R_0. The averaging of Eq. (30) over the period of electron gyration results in (Gol'denberg and Pankratova, 1971)

$$\pi H_0(R_0^2 - r_{\perp 0}^2) = \psi_c. \tag{31}$$

Equation (31) is valid for any continuous distribution of axially symmetric E and H fields in the injector and transient region. If the magnetic field in the injector is paraxial, the beam radius is

$$R_0 = \sqrt{(R_c^2/\alpha) + r_{\perp 0}^2}. \tag{32}$$

3. Electron Velocity Spread in the Absence of Space Charge Effects

Although the total energy is practically equal in all the electrons of the beam, it may be distributed between the orbital and axial components of motion in a different way. This distribution is generally characterized by a relative spread of rotational electron velocities, which is an adiabatic invariant. If the rotational velocity spread exceeds 10–20%, the gyrotron efficiency significantly decreases.

Let us define the maximum pitch factor $g = v_\perp/v_\parallel$, assuming that the rotational velocities of electrons emitted from the injector lie in a rather narrow range $[\bar{v}_{\perp I} - \frac{1}{2}\Delta v_{\perp I}, \bar{v}_{\perp I} + \frac{1}{2}\Delta v_{\perp I}]$. Using Eq. (27) as a condition for complete current transition of the beam, we find that the maximum ratio of the root-mean-square orbital and axial electron velocities is (Gol'denberg and Petelin, 1973)

$$g_{max} = \sqrt{\overline{v_{\perp 0}^2}/\overline{v_{\parallel 0}^2}} = \sqrt{1/\delta v_\perp}, \tag{33}$$

where $\delta v_\perp = \Delta v_{\perp I}/\bar{v}_{\perp I}$. Equation (33) is valid for any gun with an adiabatic magnetic field, irrespective of the injector type.

Within the framework of the adopted limitations, the local velocity spread, caused by regular field nonhomogeneities, is absent in the adiabatic gun if the near-emitter fields satisfy the relation that follows from Eq. (28):

$$E_{\perp c}/H_c^{3/2} = \text{constant}. \tag{34}$$

At the same time, "statistical" spread of initial electron velocities takes place. This includes thermal velocity spread as well as spread caused by the roughness of the emitter surface. In the latter case, the actual rough emitter surface is changed for a closely located smooth surface where initial electron velocities are present. Changing Eq. (26) to take into account the initial gyration velocity spread

$$\Delta v_{\perp c} = c(E_{\perp c}/H_c) \pm \Delta v_\perp, \qquad (35)$$

we obtain, according to Tsimring (1972), the thermal rotational velocity spread at the Boltzman velocity distribution of the emitted electrons,

$$\delta v_{\perp T} \simeq 3.6\sqrt{(kT/eU_a)(d/2r_{\perp c})}, \qquad (36)$$

and the spread caused by the cathode surface roughness,

$$\delta v_{\perp\sim} \simeq 1.6\sqrt{\frac{\rho}{r_{\perp c}}}, \qquad (36b)$$

where ρ is the characteristic size of the emitter roughness. All the other designations are given in Fig. 6.

The contribution of thermal electron velocities and emitter roughness to the total relative velocity spread $\delta v_{\perp 0}$ increases as the operating wavelength shortens, and in the millimeter wavelength range it is usually several percent.

4. Electron Velocity Spread Caused by the Space Charge

The space-charge fields of the electron beam have opposite directions at its boundaries (see Fig. 7). This, in general, violates the conditions of applicability of the drift approximation. However, a regular character of these fields slowly varying from one electron gyration cycle to another enables one to determine their influence on electron velocity distribution within the framework of the elementary theory.

We now consider beam formation in a gun with a narrow emitter width and a small angle φ_c between the magnetic field line and the cathode surface. As seen from Fig. 7a, the electron paths in such a gun intersect. As a result the space-charge fields continuously decelerate gyration of the electrons emitted from the left-hand part of the emitter and accelerate the ones from the right-hand part. This mechanism operates for some gyration cycles until the beam boundaries get diffused. The qualitative estimates together with the experimental measurements of velocity spread caused by the space charge give an approximate formula (Gol'denberg and Petelin, 1973):

$$\delta v_{\perp\rho} \simeq I/I_\rho, \qquad (37)$$

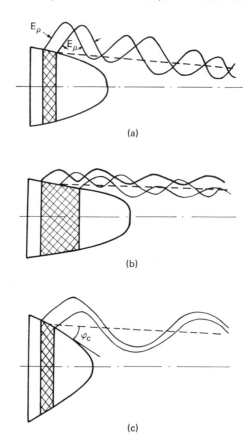

FIG. 7   Types of gyrotron electron guns: (a) gun with a narrow emitter; (b) gun with a wide emitter; (c) gun with a quasilaminar electron beam.

where

$$I_\rho \simeq (0.5)(E_c^2 R_c/H_c) \tag{38}$$

is the current close to the gun current in the regime of a complete space charge [units of measurement: I (A), E_c (kV/mm), R_c (mm), H_c (kOe)].

In a gun with a wide emitter (Fig. 7b), the effect of the forementioned mechanism is much weaker. Instead, importance is given to the screening of the right part of the emitter by the space charge of electrons emitted from the left part. For a wide emitter, Eq. (37) is also valid. Here,

$$I_\rho \simeq 0.5 R_c \sqrt{E_c^3 l \sin \varphi_c}. \tag{39}$$

If the angle of intersection between the magnetic line of force and the cathode surface increases to $\varphi_c \gtrsim 25°$ (Fig. 7c) (Manuilov and Tsimring,

1978), the electron paths in their initial parts do not intersect. After a few gyration cycles, they do intersect, and the above mechanism of increasing velocity spread begins to operate. However, as shown by the numerical computations given in Section III.B, an increase in velocity spread with increasing beam current is slower in such a gun having a quasi-laminar beam than in a gun as presented in Fig. 7a.

5. *Collector*

Relations (25), (29), and (32) enable one to estimate the length of the operating region of electron collection (Gol'denberg *et al.*, 1973):

$$
l_{col} \approx \begin{cases}
\Delta R_0 \dfrac{R_{col}}{R_0 \varphi_{col}} + \dfrac{0.67\sqrt{U_0} R_{col}^2}{H_0 R_0^2 (1 + \varphi_{col}/\gamma)} & \text{for} \quad 0 < \varphi_{col} \leq \gamma, \\[4mm]
\Delta R_0 \dfrac{R_{col}}{R_0 \varphi_{col}} + \dfrac{0.21\sqrt{U_0} R_{col}}{H_0 R_0 \varphi_{col}} & \text{for} \quad \gamma \leq \varphi_{col} \ll \tfrac{1}{2}\pi,
\end{cases}
\tag{40}
$$

where φ_{col} is the angle of intersection between the magnetic field line and the collector surface, $\Delta R_0 = l_c \varphi_c / \sqrt{\alpha}$ the spread of radial coordinates of the electron gyration guiding centers, and $\gamma = 2R_0/\pi R_{col}$. It can be seen from Eq. (40) that for a common case $\Delta R_0 < r_{\perp 0}$ the length of collection l_{col} at a rather small (but not too small) angle φ_{col} [see the condition for φ_{col} in Eq. (40)] is about one step of a helical electron path. In millimeter-wave gyrotrons the condition of smallness for φ_{col} is difficult to fulfill, so l_{col} is usually defined by the lower formula in Eq. (40) as a projection of the magnetic tube of force filled by an electron beam, on the collector surface.

B. NUMERICAL COMPUTATION FOR GYROTRON GUNS

Drift approximation is almost always valid for magnetic fields in magnetron-type gyrotron guns, including the fact that the intrinsic magnetic field of the beam in weakly relativistic gyrotrons is negligibly small (Manuilov and Tsimring, 1981b). For electric fields defined by the electrode configuration, these conditions may be violated, and at a beam current $I \gtrsim 0.1 I_\rho$ they are almost always violated. It was experimentally observed (Avdoshin and Gol'denberg, 1973) that the nonhomogeneities of the electric field in the injector and the fields of the intrinsic space charge dramatically affect the electron velocity distribution. To compute guns with arbitrary currents and electrode configuration, numerical methods were developed that also make it possible to take into account relativistic effects. It is difficult to define maximum possible optimization of such systems (minimum electron velocity spread at a given beam current) in which the number of parameters may be arbitrarily large. The reference spread $\delta v_\perp \simeq 10\%$ is often considered quite satisfactory for a "rough" system, i.e., provided that the value $\delta v_\perp$ changes slightly at small variations of the calculated parameters of the operating conditions and electrode configuration. It is also useful to obtain the initial (zero) approximation from the computation based on the drift theory, which enables one to significantly reduce the number of approximations and the time of computation.

1. *Equations and Methods for Their Solution*

A complete set of equations for a beam that takes into account space charge and relativistic dependence of the electron mass on velocity can be written in dimensionless variables in the form (Manuilov and Tsimring, 1977)

$$d\mathbf{r}/dt = \sqrt{1 - 2y_0 v^2} \left[\tfrac{1}{2}\mathbf{E} + [\mathbf{vB}] - y_0\mathbf{v}(\mathbf{vE})\right],$$
$$\Delta U = -\rho, \quad \text{div } \mathbf{j} = 0, \quad \mathbf{E} = -\nabla U, \quad \mathbf{j} = \rho\mathbf{v}, \tag{41}$$

where $y_0 = \eta U_0/c^2$ (c is the velocity of light). The set of equations (41) is supplemented by the Dirichlet boundary conditions on electrodes and by a given magnetic field. Electrode potentials are presented by means of auxiliary sources (Vashkovsky and Ovcharov, 1971), i.e., by the imaginary charges located behind the electrodes. All the fields in the problem are axially symmetric.

Numerical integration of Eqs. (41) is complicated because the width of the electron beam is much smaller than its length, which makes it difficult to use a common net-point method. An algorithm that enables one to overcome these difficulties is described here (Manuilov and Tsimring, 1981a).

Three supplementary square nets are introduced into the meridian cross section of the electron optical system under consideration: potential net 1 and fine- and large-mesh space-charge nets (2 and 3, respectively) (Fig. 8). The three nets cover only the beam region. Ring charges Q_m equal to the summed charges of all the current tubes crossing the corresponding meshes are placed into the mesh points of net 2. The charges Q_p in the mesh points of net 3 are equal to the sum of charges Q_m in the corresponding mesh points of the large-scale net and are placed in the "mass centers" of the mesh points, i.e., each of them has a radius vector

$$\mathbf{r}_p = \sum_m \mathbf{r}_m Q_m \bigg/ \sum_m Q_m.$$

The large-scale net 3 is introduced to take into account the charges far from a given point A at distances larger than a certain value L. This rough approximation of the distribution of "far" charges provides sufficient accuracy of calculations. Estimates show that the dimensions of nets 1 and 2 can have the same order of magnitude as the radius of electron gyration.

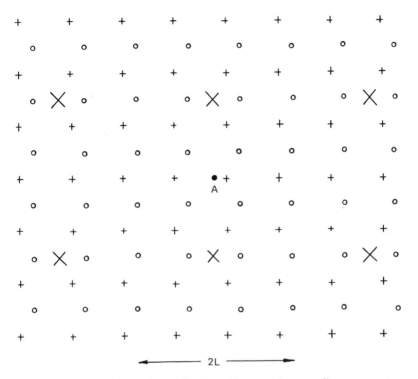

FIG. 8   Mesh points of nets (o, net 1; +, net 2; ×, net 3).

If L is of the same order as the step of the helical electron path, or larger, the one-dimensional net 3 can be used.

The potential in the point A (Fig. 8) is defined by three types of charges: first, charges Q_m in the range $Z_A \pm L$; second, large charges Q_p beyond this range; third, charges located according to the method of supplementary sources behind the electrodes. The electric field in each point of the trajectory is calculated by the values of potentials in the mesh points of net 1 by means of nine-point finite-difference relations. Because of the introduction of a large-scale space-charge net, the time of calculation is a linear function of the length of the beam region $\mathscr{L}$ under study. If only a fine-mesh net is used, this dependence is squared. At $\mathscr{L} \gg L$, a large-scale net saves a lot of computer time. If the beam current is small, i.e., at $I/I_\rho \leq 0.05$, the beam space charge is neglected and the numerical procedure is greatly simplified.

Solution of Eqs. (41), taking into account the space charge, is accomplished by the iteration (Ilyin, 1974). At the initial (zero) iteration, the density of the space charge is assumed zero. Then, all the flow is divided into N current tubes, and central trajectories of each current tube are calculated in the resulting electric field. At every subsequent iteration, the electric field is determined taking into account the distribution of the space charge found in the previous iteration. The process is completed when, at the end of the studied region, the shift of any central trajectory, after the iteration, becomes progressively smaller than the given value.

2. Trajectory Analysis

Figure 9 shows the results of computation (Manuilov and Tsimring, 1977) of electron velocities in a gun with a narrow emitter (Fig. 7a) in which electron paths intersect in the first turn. Figure 9 shows that in comparison with the "cold" local velocity spread, which is 9%, velocity spread increases significantly with growing current. In some guns of this type an increase in velocity spread may be even larger (Lygin and Tsimring, 1978). It is characteristic of gyration velocities of the "left-hand" electrons (tube 1) to increase with the growth of current (see Section III.A.4), although mean gyration velocities of electrons decrease.

In a gun with a quasi-laminar beam (see Fig. 7c), velocity spread under the space-charge effect was determined both in the initial (laminar) and terminal parts of the beam (Lygin and Tsimring, 1978). The calculation demonstrated that the electron velocity spread in the laminar part changes from 6 to 7% as the current increases from 0 to 0.1 I_ρ; in the region where electron paths intersect, the spread $\delta v_\perp$ increases from 7 up to 11%; an in-increase in $\delta v_\perp$ along the z axis ceases when the trajectories of particles emitted from different sections of the cathode intermix. Thus the fact that

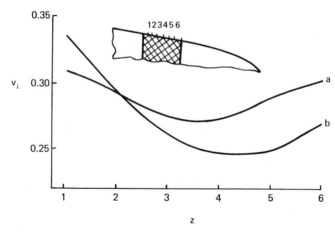

FIG. 9   Averaged velocities of electrons in current tubes that start from different points of the cathode in an electron gun (as shown in Fig. 7a): curve a, $I/I_\rho = 0$; curve b, $I/I_\rho = 0.1$.

the beam is laminar, if only in its initial part, facilitates a significant advance of the operating conditions of a gun to the space-charge region while retaining an acceptable quality of the beam.

Because the irregular intersections of trajectories along all the beam do not result in an increase in velocity spread, Lygin and Tsimring (1978) carried out a trajectory analysis of a gun of the type shown in Fig. 7b. In this gun, the angle φ_c was chosen small enough so that the first step of the electron path was smaller than the width of the emitter. The dependence of the electron velocity spread $\delta v_\perp$ on the beam current (curve b in Fig. 10) in such a gun is relatively weak. As the belt width decreases or the angle φ_c increases, beam boundaries similar to those shown in Fig. 7a appear in the near-cathode region, which results in a sharp increase in the spread $\delta v_\perp$ (curve a in Fig. 10). It should be taken into account, however, that in systems of the type shown in Fig. 7b the velocity of particles escaping from the near-cathode region is relatively small, which enhances the disturbance of the electric field on the cathode by the space charge of the beam. Numerical computations also show that when a certain emission current I_s is exceeded (in the gun for which curve b in Fig. 10 is calculated, $I_s \simeq 0.2I_\rho$), electrons emitted from the left-hand part of the emitter change the direction of axial drift and hit the gun's anode. With an increase in the emitter width, the value I_s increases slightly. Thus, in a gun with mixed electron paths, a beam can be formed with acceptable velocity spread at a current $I \gtrless 0.2I_\rho$.

When comparing the maximum gun currents, one should bear in mind that I_ρ depends on the field amplitude at the cathode E_c [see Eqs. (38) and (39)],

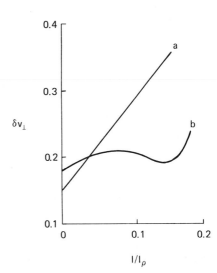

FIG. 10 Dependence of electron velocity spread on beam current in an electron gun (as shown in Fig. 7b): curve a is given for an emitter width two times smaller than that for curve b.

which is chosen close to breakdown. Therefore, to compensate velocity spread in guns intended for operation in the space-charge regime ($I \gtrsim 0.1I_\rho$), the configuration of the anode is complicated, which results in a decrease in the emitter field. This partially reduces the advantages of guns of the types shown in Figs. 7b and 7c over a gun of simpler geometry, shown in Fig. 7a.

3. Method of Synthesis

The computation of gyrotron guns at given parameters of the electron beam is of particular interest, taking into account minimum *a priori* assumptions. Generally speaking, such computations can be carried out by synthesis. This method is applicable to guns in which the initial part of the electron beam is laminar, as in Fig. 7c. A laminar electron beam is described by a set of ordinary differential equations (Tsimring, 1977) whose solutions depend on three dimensionless parameters: angle φ_s, "current parameter" $\gamma = (mj_c/eE_cH_c)$ (j_c is the density of the emission current), and cylindricity parameter $\mu = (mE_c/eH_c^2R_c)$.

The beam is laminar only if φ_c is larger than the critical value φ_{cr} (Manuilov and Tsimring, 1978). The inner problem of synthesis, which calculates the electron motion and the E-field intensity on the electron paths, is solved in a homogeneous magnetic field. The outer problem of synthesis lies in solution of the Laplace equation with Couchy boundary conditions on the outer beam trajectories (Manuilov and Tsimring, 1978; Tsimring, 1977), i.e., in finding equipotentials that define the configuration of the gun electrodes.

Dimensional parameters of an electron-optical system are determined, taking into account the beam parameters in the interaction space. Smooth

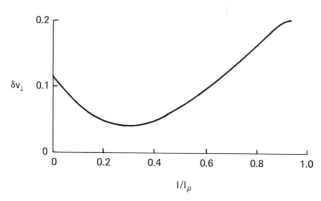

FIG. 11 Dependence of electron velocity spread on beam current for a synthesized electron gun. (From Manuilov and Tsimring, 1981c.)

variation of fields in the transient region permit the use of the drift approximation (Tsimring, 1977) in solving the problem of matching the "synthesized" and the given parts of the beam.

In the final stage of gun calculation, the shape of electrodes is corrected to estimate their intersection with the beam and simplify production of a gun. It is to be noted that at a large cylindricity parameter μ anode equipotentials do not intersect the electron beam (Manuilov and Tsimring, 1981c). The adequacy of correction is checked by trajectory analysis of the gun (see Section III.B.1). Plots of electron velocity spread in one of the types of a synthesized gun presented in Fig. 11 show that this gun can be operated with a much greater space charge than a gun of the type shown in Fig. 7a.

4. Relativistic Effects

The dependence of the electron mass on its velocity is the basic relativistic effect that significantly influences formation of a helical beam in weakly relativistic gyrotrons (Manuilov and Tsimring, 1981b). Neglect of this effect results in overestimation of the pitch factor g, even at voltages lower than 100 kV. For potential guns ($U_a = U_0$) of the conventional adiabatic type shown in Fig. 7a, the correction coefficients of the nonrelativistic Eq. (28) are given in Fig. 12. However, the computed values of electron velocity spread are not strongly affected by taking into account the relativistic effects discussed here.

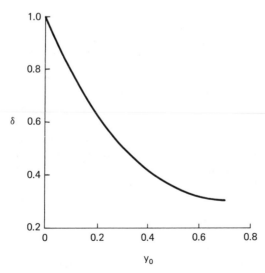

FIG. 12 Correction coefficient taking into account relativistic effects in an adiabatic electron gun: $U_{\perp rel} = \delta U_{\perp nrel}$; $y_0 = eU_0/mc^2$.

IV.   Mode Selection

A.   MODE SELECTION AND THE PROBLEM OF POWER INCREASE

A cardinal way to enhance the electron-tube output power is to increase the operating voltage, allowing for an increase in the beam current as well. The gyrotron is not an exception. At a relativistic voltage of 350 kW, the millimeter-wave gyrotron delivered 25 MW with an efficiency of 6% (Krementsov et al., 1980), and this is not the upper limit. However, for cw (and quasi-cw) gyrotrons of our main interest, relativistic voltages are hardly acceptable because of a very large density of ohmic losses of microwave power in the resonator. An increase in the beam current at a fixed resonator diameter is also restricted either by the ohmic losses or by the space-charge effects in the electron beam (see Section III).

Restrictions associated with ohmic losses in the resonator can be explained in brief by a general balance equation for stationary self-oscillations. This is the relation between the power of the rf field losses in the resonator and the power yielded by the electron beam to the rf field:

$$(\omega/Q)A^2V = \eta_{el}UI, \tag{42}$$

where A is the field amplitude in the resonator with a volume V. The resonator Q is defined by both diffraction emission of microwave radiation into the output waveguide of the gyrotron and ohmic losses of microwave power in the resonator walls:

$$Q = Q_{dif}Q_{ohm}/(Q_{dif} + Q_{ohm}).$$

Generally, the condition $Q_{dif} \ll Q_{ohm}$ is fulfilled in the millimeter-wave gyrotron. Therefore, the power of ohmic losses can be estimated as

$$P_{ohm} = (Q_{dif}/Q_{ohm})P_{out} \simeq (Q_{dif}/Q_{ohm})\eta_{el}UI. \tag{43}$$

As seen from Eqs. (42), (43), reduction of Q_{dif} allows for increase in the limiting output power of the oscillator and simultaneous decrease in the relative portion of ohmic losses. Unfortunately, the diffraction Q of the resonator has a lower limit [see Eq. (18)] defined by the resonator length, which should be sufficiently large to provide high efficiency.

Thus, for gyrotrons whose power is limited by the ohmic-loss density in the resonator, the total power can be increased only by enlarging the resonator diameter, which results in simultaneous growth of the rf field volume and the ohmic Q. This method, however, makes it more difficult to exite only one operating mode, because with an increase in the transverse cross section $S_\perp$ of the resonator (at $S_\perp \gg \lambda^2$) the distance between the eigenfrequencies of its modes decreases on the average as $\Delta\omega/\omega \simeq \lambda^2/S_\perp$. Hence, the limiting power of single-mode oscillations in the gyrotron is defined by successes in mode selection.

Conventionally, there are two sorts of methods of mode selection: electrodynamic and electronic. Electrodynamic methods provide excitation of the operating mode by choosing appropriate configuration of the electrodynamic system; the electronic ones, by choosing the electron-beam parameters. Excitation of the operating gyrotron mode prior to parasitic modes is often sufficient to provide stable single-mode generation (Nusinovich, 1974; Zarnitsyna and Nusinovich, 1974). However, when the distance between the mode frequencies decreases (the resonator diameter increases), it becomes more difficult to excite any definite mode prior to the others. At a rather large density of the eigenfrequency spectrum, several modes with similar field structures may be excited simultaneously. Nonlinear interaction of these modes, which appears in the large-signal regime, may significantly complicate mode selection.

B. ELECTRODYNAMIC MODE SELECTION

Initially, mode selection was developed as applied to traditional resonators (Vlasov et al., 1969), i.e., sections of slightly irregular axially symmetric cylindrical waveguides. Subsequent use of coaxial and slotted (two-mirror) electrodynamic systems has considerably expanded the potentialities of electrodynamic mode selection.

1. Axially Symmetric Cylindrical Cavities

In simple open cavities (Fig. 13), selection of H_{mpq} modes is readily accomplished over the axial index q. This is because, with an increase in q, the group velocity v_{gr} of the waves forming the mode grows, and the coefficient of their reflection at the waveguide irregularity defining the diffraction Q decreases. As a result, the diffraction Q of the resonator $Q_{dif} = \omega L/v_{gr}(1 - |R_1 R_2|)$ [compare Eq. (18)] decreases proportionally to q^{-2} (Vlasov et al., 1969). H_{mp1} are the highest-Q operating modes of such a resonator.

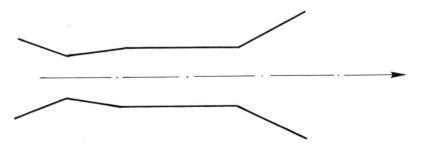

FIG. 13 Profile of a conventional cylindrical resonator.

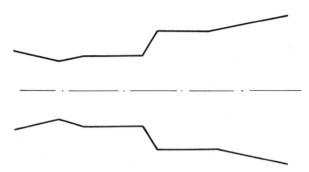

FIG. 14   Profile of an electromagnetic system with coupled resonators.

Changing the waveguide profile (or, to be more precise, the part forming the interaction space) in a more complicated manner, one can obtain an electrodynamic system in the form of coupled resonators [see, e.g., Gaponov *et al.* (1981)]. In this system, the mode with $q > 1$ (e.g., with $q = 2$ in the cavity shown in Fig. 14) has the highest Q factor. Here mode selection over radial indices p is also possible. For example, if the frequency of the H_{mp1} mode in the first of partial resonators coincides with the frequency of the $H_{mp'1}$ mode ($p' > p$) of the second resonator, the Q factor of the mode formed by superposition of this pair of partial modes may be significantly higher than that of other partial modes that cannot be coupled because of the difference in their eigenfrequencies (Gaponov *et al.*, 1981).

Another method of mode selection over the transverse indices m and p can be used in relatively short cavities of large diameter (large as compared with the wavelength). This method is based on the difference in the path lengths of two subsequent reflections at cavity walls $L_{mp} = 2R\sqrt{1 - (m^2/v^2)}$ for the rays that form the fields of different modes (Vainstein, 1966; Vlasov *et al.*, 1981). For example, the ray paths of the whispering gallery modes localized near the cavity wall ($m \gg p$) are less than those of the space-extended modes ($m \lesssim p$). So, it is possible to provide a large Fresnel parameter $C = L^2/4\lambda L_{mp} \gg 1$ for the whispering-gallery modes and a small one $C \lesssim 1$ for the space-extended modes simultaneously. Under these conditions, the space-extended modes will have a low Q factor owing to diffraction diffusion of the waves.

2.   *Coaxial Cavity*

Introduction of an inner conductor into the cylindrical cavity permits one to affect the Q factor of modes with high radial indices p whose fields are disturbed by this rod (Vlasov *et al.*, 1976). If the inner conductor tapers to the output cross section of the cavity (Fig. 15), the rays forming the space-extended mode fields are ejected from the cavity, i.e., the Q factors of these

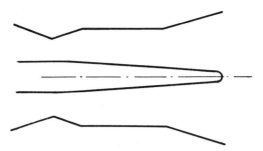

FIG. 15   Profile of a coaxial resonator.

modes decrease. It is also essential for mode selection that the decrease in Q is accompanied by the shortening of the effective field length of the mode. At the same time, the inner conductor weakly affects the Q factor and the effective field length of the whispering gallery modes if its radius is smaller than their caustic radii.

High efficiency of electrodynamic mode selection in a coaxial cavity was proved experimentally (Gaponov et al., 1981). In the gyrotron calculated for operation with high power and efficiency at the $H_{15,1,1}$ mode, parasitic modes with the indices 12, 2, 1; 9, 3, 1 and others in a cylindrical cavity had starting currents lower than $H_{15,1,1}$ and suppressed it. The inner coaxial conductor whose profile was chosen such that the Q factor of the $H_{15,1,1}$ mode was the same as that in the cylindrical cavity and the Q factor of parasitic modes with $p \geq 2$ decreased considerably, made it possible to suppress the excitation of the parasitic modes. In the coaxial-cavity gyrotron, an output power of 1250 kW at the $H_{15,1,1}$ mode was obtained (Gaponov et al., 1981).

3. Slotted Cavities

It is obvious that azimuthal mode selection can be provided only in cavities with a disturbed axial symmetry. For example, the cutting of symmetrically placed slots results in two- or multimirror cavities (Fig. 16) in which modes with the azimuthal index equal to the number of slot pairs have the highest Q. In a slightly elliptical two-mirror cavity, the $H_{1,p,1}$ modes with a transverse field structure similar to $J_1(\varkappa R) \sin \varphi$, for which the slots are out of the caustic, have the highest diffraction Q (Fig. 16). $H_{1,p,1}$ modes are of interest because their ohmic losses in the cavity are relatively small. However, electrons of the axially symmetric beam in such a system interact with an azimuthally nonhomogeneous rf field that lowers the gyrotron efficiency (Luchinin and Nusinovich, 1975). This effect may be weakened by using a beam with "slots" just opposite the resonator slots.

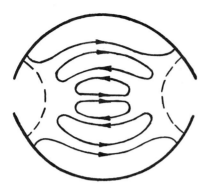

FIG. 16 Slotted resonator with an elliptic cross section.

In gyrotrons with two-mirror cavities, single-mode excitation of the operating modes $H_{1,7,1}$ and $H_{1,10,1}$ was experimentally obtained; at a wavelength of about 3 mm, the power was 300 and 500 kW, respectively (Gaponov *et al.*, 1981). This power was obtained because the radius of the beam was close to that of the cavity. Note that in the gyrotron with a conventional axially symmetric cavity where selective excitation of the $H_{1,7,1}$ and $H_{1,10,1}$ modes requires the location of the electron beam in the inner maximum of the Bessel function $J_{m \mp n}(\varkappa R_0)$, the power at the same modes and a given voltage would be several times smaller. According to the calculations, the efficiency of gyrotrons with two-mirror cavities and axially symmetric beams was rather low, about 17%. The "cutting" of the beam increased the efficiency to 21%.

C. ELECTRON MODE SELECTION

The main electron parameters that define the starting current of modes, as well as their interaction, are the mean radius of electron guiding centers, the beam voltage, and the gyro frequency. The spread of axial electron velocities has considerable influence on the Doppler broadening of the cyclotron resonance band of electrons when the mode fields have a large axial wave number, and also selectively affects modes with the axial index $q > 2$. The appropriate choice of the guiding center radius of electrons in the cavity is the simplest and most effective means of electron-mode selection over radial indices.

1. *The Beam Radius*

For selective excitation of the operating mode, in many cases it is sufficient that the effective coupling impedance of the electron beam with this mode be higher than with other mode fields. For the gyrotron with an axially symmetric resonator, the corresponding coefficient under the self-excitation condition (14) is $|L_n|^2 = J_{m \mp n}^2(\varkappa R_0)$. The resonance mode whose caustic radius is close to the radius of the beam R_0 [i.e., the mode with respect to which the beam enters nearest to the axis maximum of the function $J_{m \mp n}^2(\varkappa R_0)$] has the maximum excitation coefficient. Therefore, when the radius of the beam is close to that of the cavity, i.e., at a maximum possible current, the whispering-gallery modes localized near the cavity wall are advantageous. At these modes, at weakly relativistic voltages 68 kV under the pulse duration 0.1 msec, the output power 1100 kW at the wavelength 3 mm was obtained without any additional means of mode selection. This is maximum power for the short-wave part of the millimeter-wavelength range (Flyagin et al., 1978). Note that when the whispering-gallery modes with a radial index $p = 1$ are excited, alignment of the tube should be very precise, because the maximum of the function $J_{m-1}^2(\varkappa R_0)$ corresponds to $R - R_0 \simeq \lambda/2\pi$ and the Larmor radius of electrons $a = \beta_\perp(\lambda/2\pi)$; hence, the outer boundary of the electron beam is at a distance $(\lambda/2\pi)(1 - \beta_\perp)$ from the cavity wall.

If modes with close eigenfrequencies and similar transverse field structures compete, the choice of the beam radius may be insufficient for the excitation of a given mode. In this case, the mode that is excited first when the gyrotron starts operating is advantageous.

2. Start-Up Operation

Different ways of gyrotron feeding provide different trajectories in the plane of electron velocity components (Fig. 17) (Nusinovich, 1974). These trajectories pass the self-excitation region of the modes whose resonance frequencies are close to the cyclotron frequency (or its harmonic) and stop when the required beam voltage is attained, as the final point, which corresponds to the high-efficiency regime. An example of the analogous trajectory in the plane of gyrotron parameters I_0, ζ_{out} given in Section II (Fig. 5) for the case of $U_a = U_0$ is shown in Fig. 17c ($I_0^{fin} = 2 \times 10^3$, $\zeta_{out}^{fin} = 17$). Varying these trajectories, i.e., choosing the values and order of switching on the anode voltage U_a and the cavity voltage U_0, one can provide a definite mode excitation. Usually, the time of switching on (the duration of the front at pulse operation) is much larger than the time of oscillation growth ($\simeq Q/\omega$), and the transit time of electrons. This allows one to consider the voltage increase as a quasi-stationary process. Figures 17a,b show the self-excitation regions of the operating and parasitic modes. The minimum starting currents of all the competing modes are supposed to be equal, and all the gyrotron parameters at full voltages are chosen to provide the maximum efficiency of the operating mode (its self-excitation region is shaded). Note that the

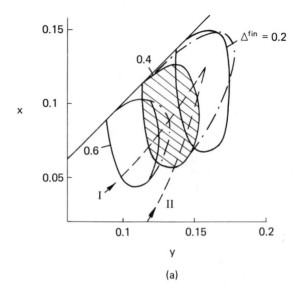

(a)

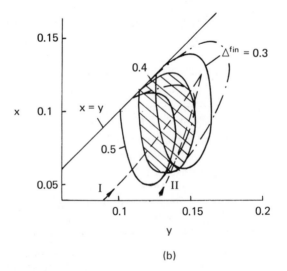

(b)

FIG. 17 Start-up operation: (a), self-excitation zones (———) and voltage trajectories (— —) for a "rare" eigenfrequency spectrum ($x = \frac{1}{2}\beta_\perp^2$, $y = \frac{1}{2}\beta^2$); (b), the same for a gyrotron with a "dense" spectrum; (c), example of start-up trajectories in the plane of beam current I_0 and frequency mismatch Δ parameters. ———, $I = I_{st}$; — —, $U_0 = U_a$.

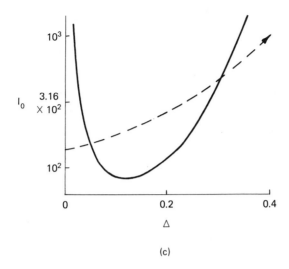

(c)

point of maximum efficiency lies in the region of hard self-excitation of the mode and must be achieved by transition from the soft self-excitation region. To the left of the shaded self-excitation region of the operating mode there lies a region of a higher-frequency mode, and to the right a region of a lower-frequency mode. When the difference between the mode frequencies is large enough and the anode and cavity voltages are equal (trajectory I in Fig. 17a), a higher-frequency parasitic mode is excited first, then its oscillation breaks (dashed line) and the operating mode is excited. If in the same oscillator the voltages are fed with the potential difference $U_0 - U_a = $ constant (trajectory II in Fig. 17a), only one operating mode is excited. The operating mode can also be excited when the resonator voltage grows under the fixed anode voltage (in Fig. 17a y grows at fixed x). Some examples of the gyrotron start-up are also analyzed by Kreischer and Temkin (1981).

At a denser spectrum of the competing modes (Fig. 17b), the voltage feeding by the first method ($U_a = U_0$) results in initial excitation of a higher-frequency parasitic mode whose oscillations break in the self-excitation zone of a lower-frequency parasitic mode. Here, the second method of voltage feeding provides practically simultaneous excitation of the operating and lower-frequency parasitic modes. To determine the character of stationary oscillations in such a generator, it seems necessary to analyze nonlinear interaction of the excited modes.

D. Nonlinear Mode Interaction

When the gyrotron start-up provides self-excitation of several non-degenerate cavity modes, the oscillations of these modes grow independently until the oscillator acquires the large-signal regime. (Degenerate modes are coupled even in the small-signal regime.) Here nonlinear effects (effects of saturation) become significant; one of them is mode interaction.

The gyrotron behavior depends on the character of the excited modes spectrum. As the spectrum becomes more dense, first the neighboring modes differing both in azimuthal and radial indices appear in the cyclotron resonance band. These modes have a nonequidistant spectrum of eigenfrequencies and their interaction is in amplitude, i.e., only intensities of all the modes affect the oscillations on each of them. As the resonant cavity diameter increases, several modes differing only in one of the transverse indices can appear in the cyclotron resonance band. The spectrum of these modes is close to the equidistant one, which, under conditions of time $|2\omega_s - \omega_{s-1} - \omega_{s+1}| \lesssim \omega/Q$ and space $2\mathbf{k}_s = \mathbf{k}_{s-1} + \mathbf{k}_{s+1}$ synchronism [see Nusinovich (1981a)], the mode interaction acquires the amplitude-phase character. This means that the oscillations of the Sth mode depend not only on the intensities of the other modes but also on slowly varying difference phases $\psi_{si} \simeq (2\omega_s - \omega_{s-i} - \omega_{s+i})t$.

The basic features of a purely amplitude-mode interaction can be illustrated by a two-mode oscillator. A quasi-linear model of a two-mode gyrotron discussed by Nusinovich (1981a) is described by the following equations for the mode amplitudes:

$$dF_1/d\tau = F_1(\sigma_1 - \beta_1 F_1^2 - \gamma_{12} F_2^2),$$
$$dF_2/d\tau = qF_2(\sigma_2 - \beta_2 F_2^2 - \gamma_{21} F_1^2).$$

$$(44)$$

Here σ_s characterize the excess of the beam current over the threshold values; the coefficient $q = |L_{n,2}|^2/|L_{n,1}|^2$ defines the difference in the effective coupling impedances between the electrons and the mode fields; the coefficients β_s and $\gamma_{ss'}$ are the functions of the transit angles $\theta_s = (\omega_s - \omega_H)L/v_\parallel$, $\gamma_{ss'}$ depends on the relative location of the mode frequencies (see Fig. 18), where $K = (\omega_{s'} - \omega_H)/(\omega_s - \omega_H)$.

As shown by Lamb (1964) in the oscillator described by Eqs. (44), single-mode oscillations are stable if the condition of "strong" coupling of modes is fulfilled:

$$\gamma_{12}\gamma_{21} > \beta_1\beta_2.$$

$$(45)$$

If the coupling is "weak,"

$$\gamma_{12}\gamma_{21} < \beta_1\beta_2,$$

$$(45)$$

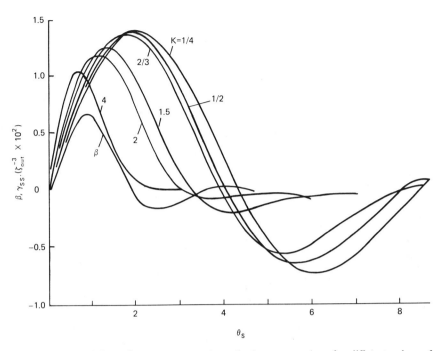

FIG. 18   Coefficients β, $\gamma_{ss'}$ versus transit angle θ_s; $\gamma_{ss'}$ are given for different values of ratio $k = (\omega_{s'} - \omega_{\mathrm{H}})/(\omega_s - \omega_{\mathrm{H}})$. (From Nusinovich, 1981a.)

a biharmonic regime (oscillations of two modes) is stable (see the phase portraits in Figs. 19a,b). Figure 18 shows [see also Nusinovich (1981a)] that condition (45a) is fulfilled all through the region of soft self-excitation ($\beta > 0$). From this it follows that, in the case considered, initial excitation of the operating mode provides stable single-mode oscillation. For example, in trajectory II in Fig. 17b one can expect to obtain steady single-mode oscillations of the operating mode.

A more complex nonlinearity complicates the phase portrait of the oscillator with strong mode coupling. Figure 19c gives a phase portrait of an oscillator with hard self-excitation (nonlinearity of the fifth degree) of two modes with close frequencies ($K \approx 1$) (Moiseev and Nusinovich, 1974). One can see that, together with single-mode oscillations, stable two-mode oscillations are also possible in such a system. Analysis of the ways to switch on the oscillator shows, however, that when passing from the region of soft self-excitation where oscillations increase to that of hard self-excitation, the system acquires one of the stable single-mode equilibrium states (see Fig. 19c) before a stable two-mode equilibrium state appears in the phase space. Thus oscillations of one mode are established in such an oscillator.

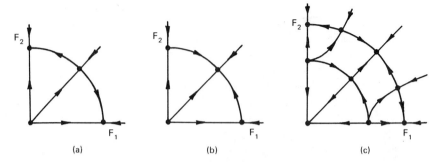

FIG. 19   Phase plane for a "soft"-excited gyrotron with (a) "strong" or (b) "weak" coupling between modes; (c) phase portrait of a "rigid"-excited gyrotron with "strong" coupling.

At a definite distance between mode frequencies (when the first mode lies in the zone of soft self-excitation and the second lies in the hard self-excitation zone), mode interaction may change qualitatively. This is illustrated in Fig. 18, where it is evident that the coefficient $\gamma_{ss'}$ is negative at certain values of the transit angle θ_s and the ratio $K = (\omega_{s'} - \omega_H)/(\omega_s - \omega_H)$. In this case, as seen from Eq. (44), an increase in the intensity of the Sth mode improves the conditions of excitation of the Sth mode. In particular, even when the conditions of self-excitation of the Sth mode are not fulfilled in the unexcited oscillator, this mode may be excited following the Sth mode. Excitation of such a mode in the region of hard self-excitation is usually possible owing to a lower-frequency mode with a small starting current. This effect of non-linear mode excitation caused by dispersion of the electron beam non-linearity was discussed in detail by Nusinovich (1981a).

Many essential features of amplitude–phase mode interaction can be studied in a three-mode model of a gyrotron, which is described by a set of four equations for the amplitudes of three modes and the difference phase $\psi \simeq (2\omega_2 - \omega_1 - \omega_3)t$. Study of these equations shows that in the interaction of three modes with equal Q factors and close eigenfrequencies ($|\omega_s - \omega_{s'}| \ll \Delta\omega_{res}$), it is the mode competition that causes oscillations of of the central mode to be stable (Nusinovich, 1981a,b). Deterioration of the Q factors of side satellites improves the stability of the central mode. Also, a high Q mode can sustain the oscillations of lower Q satellites for which conditions of time and space synchronism are fulfilled in a gyrotron with small active nonlinearity, as described in equations of the type (44) with complex coefficients $\beta = \beta' + i\beta''$ that characterize the nonlinearity $|\beta'| \ll |\beta''|$, which occurs near the boundary between the regions of soft and hard self-excitation. In this situation, regimes with static or periodic radiation automodulation, as well as stochasticity, are possible (Rabinovich and Fabricant, 1979). Mode locking corresponding to the automodulation of radiation is also possible in the gyrotron with a quasi-equidistant spectrum of frequencies if the distance between the frequencies and the beam current are large enough. Here the mode with the minimum starting current, as mentioned previously, may excite higher-frequency modes. Gyrotrons in which effects of delay caused by the finiteness of the electron transit time are significant behave in a similar way (Petelin, 1981).

Thus, when two modes are excited simultaneously, the main role generally belongs to the effects of mode competition, which results in the establishment of single-mode oscillations. In this situation, using the aforementioned methods of switching on the oscillator, it is possible to achieve stable oscillation of one operating mode with high efficiency. If several modes whose frequencies form a spectrum close to the equidistant one are present in the cyclotron resonance band, stable single-mode oscillations of the operating mode can be obtained only at a relatively small beam current. Analysis of similar processes in other electron microwave oscillators (Bogomolov et al., 1981; Ginzburg et al., 1979a) shows that, as a rule, single-mode oscillations become unstable when the current exceeds the value optimum for efficiency.

V. Prospectus

Immediately after its appearance, the gyrotron was considered (and soon acknowledged) as a powerful device in the shortest microwave range. An increase in power and a shortening of the wavelength are still the main directions of its development. There are two ways to increase the power, depending on thermal problems. In short-pulse gyrotrons without any heating restrictions, voltage escalation is basic. This leads to the relativistic electronics (Ginzburg et al., 1979c; Krementsov et al., 1980) whose prospects are not considered here. But we will discuss the role of thermal restrictions of the output power in cw and long-pulse ($\tau \geq 10$ msec) gyrotrons. The possibilities of shortening the wavelength in gyrotrons will be analyzed in a wider voltage range.

A. POTENTIALITIES OF POWER INCREASE

The power of order 1 MW obtained in weakly relativistic pulsed gyrotrons at a wavelength of 3–7 mm (Andronov et al., 1978; Flyagin et al., 1978) can be reproduced at cw or long-pulse operation only after essential changes in the structure and operating conditions of the tube. These changes are aimed mainly at mitigating thermal loads in the resonator and collector walls and in the output window.

To avoid large thermal-loss densities in the resonator, gyrotrons with pulse duration over 1 msec should not be operated at voltages higher than 100 kV. The reason for this restriction is that the growth of ohmic losses outstrips the growth of the output power (see Section IV). Also, high voltages result in growth of the axial electron velocity and, hence, in the broadening of the cyclotron resonance band. This fact aggravates the problem of mode competition in gyrotrons with space-extended resonators.

Therefore, further increase in transverse dimensions of the interaction space can be considered the main way to increase gyrotron power. Analysis of rf ohmic losses in the cavity, estimation of potentialities of mode selection, and preliminary results of experimental studies of selective properties of gyrotrons with large-diameter cavities show that output power up to 3 MW at wavelengths of about 3 mm can be obtained at stable single-mode operation. With the shortening of the wavelength, the power limit decreases because of increasing density of ohmic losses in the cavity.

The problem of collector heating in gyrotrons is sometimes solved by enlarging the electron-collecting section of the output waveguide (Jory et al., 1982). This can, however, result in parasitic mode conversion in the output waveguide and, therefore, in deterioration of spatial distribution of the output microwave radiation.

The collector problem may be radically solved if the collector is separated from the waveguiding system. This eliminates restrictions on the collector diameter. In gyrotrons intended for electron–cyclotron plasma heating in T-10, such separation was accomplished by means of a quasi-optical converter of the operating mode to a linearly polarized wave beam (Gaponov et al., 1981). Figure 20 is a photo of such a gyrotron. The wavelength of this gyrotron is 3.6 mm, and the output power exceeds 200 kW with a pulse duration of 150 msec. Microwave radiation of this gyrotron effectively excites the lowest $H_{1,1}$ mode in an oversize waveguide through which the microwave power is transmitted from the gyrotron to the tokamak with relatively small ohmic losses (Flyagin et al., 1982a).

A dielectric output window absorbs part of the microwave energy transmitted through it and, as a result, is heated. The estimates show that, for conventional flat-disk windows cooled over their perimeter, the maximum microwave power of the short-wave millimeter range with pulse duration $\simeq 0.1$–1.0 sec is up to 300–500 kW. Double-disk windows with fluorocarbon (or other) coolant circulated between the disks (Jory et al., 1982) can withstand several times larger power. When developing large-power microwave complexes for electron-cyclotron plasma heating, one can also use "windowless" coupling of a gyrotron having an output power on the order of some megawatts directly with a vacuum vessel of the tokamak. This idea was put forward by Granatstein et al. (1979) and Gaponov et al. (1980).

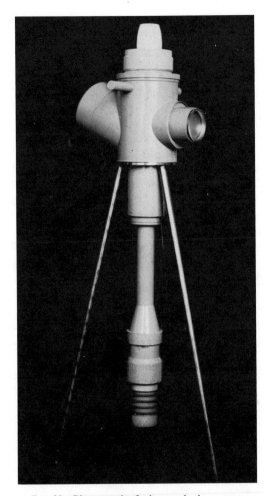

FIG. 20   Photograph of a long-pulsed gyrotron.

B.   POSSIBILITIES OF WAVELENGTH SHORTENING

Creation of strong magnetic fields in rather large volumes is the main problem that inhibits the advance of gyrotrons to the submillimeter-wavelength range. At present, superconducting magnetic fields allow one to obtain up to 10–15 T in the required field volumes. This serves as a basis for obtaining cw (and quasi-cw) generation in the wavelength range 0.8–1.1 mm at the fundamental gyrofrequency ω_H.

The doubling of the operating frequency is possible at the second harmonic ω_H. This allows one to obtain cw generation of 1.5 kW at a wavelength

0.92 mm (Zaytsev et al., 1974). Generally speaking, in gyrotrons operating at the second harmonic it is difficult to suppress parasitic mode excitation at the fundamental (Zapevalov et al., 1977). This can be partially overcome by some methods of selection (Gaponov et al., 1981; Zapevalov et al., 1977). Note that the competing abilities of modes excited at the second harmonic are somewhat improved with an increase in voltage, because the electron radiation intensity at gyro-frequency harmonics grows with the orbital velocity of particles. The effective impedance of electron coupling with the rf field is proportional to $\beta_{\perp}^{2n}$ (n is the harmonic number). Thus, to obtain optimum efficiency at high harmonics in a weakly relativistic gyrotron, one has to increase the diffraction Q of the resonator. Therefore, as the harmonic numbers grow, the ohmic losses of microwave power in the resonator increase, $P_{\text{ohm}} = (Q_{\text{dif}}/Q_{\text{ohm}})P_{\text{out}}$, Hence, cw gyrotrons at harmonics (even at the second one) are not comparable in power and efficiency with gyrotrons at the fundamental resonance.

For pulsed gyrotrons, strong magnetic fields can be obtained using pulsed solenoids (Antakov et al., 1965; Bott, 1964; Nikolayev and Ofitserov, 1974). A recently constructed gyrotron with a pulsed magnetic field up to 25 T gave an output power 100–120 kW with an efficiency of 8–15% at 0.6–0.8-mm wavelengths (Flyagin et al., 1982b; Luchinin et al., 1982; Luchinin et al., 1983). A photo of a gyrotron with a pulsed solenoid is shown in Fig. 21.

The Doppler frequency shift can also be used to increase the electron radiation frequency in CRMs. This effect is used in all free-electron lasers [see, e.g., Deacon et al. (1977) and Bratman et al. (1979a)]. As applied to CRMs with waves propagating along the external magnetic field, it allows for an increase in the frequency by a factor of $2\gamma^2$ [$\gamma = (1 - v^2/c^2)^{-1/2}$ is the Lorentz factor] when using the resonance between relativistic electrons and fast waves. Owing to the autoresonance effect (Davydovsky, 1962; Kolomensky and Lebedev, 1962), high-efficiency can, in principle, be obtained up to $\omega \simeq \gamma^2 \omega_H$ (Ginzburg et al., 1981; Petelin, 1974). In CRMs with

FIG. 21   Photograph of a submillimeter-wave gyrotron with a pulsed solenoid. (From Luchinin et al., 1983.)

slow waves, the Doppler frequency upconversion can be substantial even at weakly relativistic beam voltages. The potentialities of these CRMs depend on dispersion characteristics of resonance structures.

For cyclotron autoresonance masers (CARMs), high-selectivity electrodynamic systems are needed in which, contrary to resonators used in conventional gyrotrons, the Q factor of the mode formed by the waves propagating at a small angle to the resonator axis should be higher than for the quasi-transverse modes generally used in gyrotrons. Such systems in the form of open cylindrical cavities with a corrugated wall were analyzed by Denisov and Reznikov (1982). These systems made it possible to construct the high-efficiency CARM that at steady single-mode operation at wavelength 4 mm (frequency conversion $\omega/\omega_H = 3\text{-}4$) delivered 6 MW output power with 4% efficiency (Botvinnik et al., 1982).

In summary, we can expect that soon cw or long-pulsed gyrotrons will be constructed with an output power of about 1 MW at frequencies up to 100–200 GHz. At 300 GHz and somewhat higher, an output power of long-pulsed gyrotrons can achieve a 100-kW level, and at short-pulse operation ($\tau \lesssim 1$ msec) with pulsed magnetic fields the gyrotron output power can be several times higher.

ACKNOWLEDGMENTS

The authors are thankful to A. V. Gaponov for useful comments on the manuscript.

REFERENCES

Alikaev, V. V. (1981). *Proc. 10th European Conf. Controlled Fusion Plasma Phys.*, Report H-3, Moscow, II, pp. 11–13.

Alikaev, V. V., Bobrovsky, G. A., Ofitserov, M. M., Poznyak, V. I., and Razumova, K. A. (1972). *JETP Lett.* **15**(1), 27–30.

Andronov, A. A., et al. (1978). *Infrared Phys.* **18**(6), 385–393.

Antakov, I. I., Klymov, V. G., and Lin'kov, R. V. (1965). *Izv. VUZov Radiofizika* **8**(5), 948–954.

Antakov, I. I., Gaponov, A. V., Malygin, O. V., and Flyagin, V. A. (1966). *Radiotekh. Elektron.* **11**(12), 2254–2257.

Avdoshin, E. G., and Gol'denberg, A. L. (1973). *Radiophys. Quantum Electron.* **16**(10), 1241–1246.

Bogomolov, Ya. L., Bratman, V. L., Ginzburg, N. S., Petelin, M. I., and Yunakovsky, A. D. (1981). *Opt. Commun.* **36**(3), 209–213.

Bott, J. B. (1964). *Proc. IEEE* **52**(3), 330–332.

Botvinnik, I. E., et al. (1982). *Pis'ma Zh. Eksp. Teor. Fiz.* **35**(10), 418–420.

Bratman, V. L., and Petelin, M. I. (1975). *Radiophys. Quantum Electron.* **18**(10), 1136–1140.

Bratman, V. L., Moiseev, M. A., Petelin, M. I., and Erm, R. E. (1973). *Radiophys. Quantum Electron.* **16**(4), 474–479.

Bratman, V. L., Ginzburg, N. S., and Petelin, M. I. (1979a). *Opt. Commun.* **30**(3), 409–412.

Bratman, V. L., Ginzburg, N. S., Nusinovich, G. S., Petelin, M. I., and Yulpatov, V. K. (1979b). *In* "High-Frequency Relativistic Electronics," pp. 157–216. Institute of Applied Physics, Academy of Sciences of the USSR, Gorky, USSR.

224   V. A. FLYAGIN, A. L. GOLDENBERG, AND G. S. NUSINOVICH

Bratman, V. L., Ginzburg, N. S., Nusinovich, G. S., Petelin, M. I., and Strelkov, P. S. (1981). *Int. J. Electron.* **51**(4), 541–568.
Bykov, Yu. V., Gaponov, A. V., and Petelin, M. I. (1974). *Radiophys. Quantum Electron.* **17**(8), 928–931.
Carmel, Y., et al. (1982), *Int. J. Infrared and Millimeter Waves* **3**(5), 645–666.
Chow, K. K., and Pantell, R. H. (1960). *Proc. IRE* **48**(11), 1864–1870.
Chu, K. R., and Hirshfield, J. L. (1978). *Phys. Fluids* **21**(3), 461–466.
Davydovsky, V. Ya. (1962). *Zh. Eksp. Teor. Fiz.* **43**[3(9)], 886–889.
Deacon, D. A., et al. (1977). *Phys. Rev. Lett.* **38**(16), 892–894.
Denisov, G. G., and Reznikov, M. G. (1982). *Izv. VUZov Radiofizika* **25**(5), 562–569.
Ergakov, V. S., and Moiseev, M. A. (1981). *In* "Gyrotron," p. 53–61. Institute of Applied Physics, Academy of Sciences of the USSR, Gorky, USSR.
Faillon, G., and Mourier, G. (1982). *Proc. 3rd Int. Symp. Heating Toroidal Plasmas* **III**, 1051–1058. CEN, Grenoble, France.
Flyagin, V. A., Gaponov, A. V., Petelin, M. I., and Yulpatov, V. K. (1977). *IEEE Trans. Microwave Theory Tech.* **MTT-25**(6), 514–521.
Flyagin, V. A., et al. (1978). *Proc. 1st Int. Symp. Heating Toroidal Plasmas* **II**, 339–349. CEN, Grenoble, France.
Flyagin, V. A., et al. (1982a). *Proc. 3rd Int. Symp. Heating Toroidal Plasmas* **III**, 1059–1066. CEN, Grenoble, France.
Flyagin, V. A., Luchinin, A. G., and Nusinovich, G. S. (1982b). *Int. J. Infrared and Millimeter Waves* **3**(6), 765–770.
Gaponov, A. V. (1959a). *Izv. VUZov, Radiofizika* **2**, 450–462, 836–837.
Gaponov, A. V. (1959b). *Report Presented at the All-Union Session of Popov's Society,* Moscow, USSR.
Gaponov, A. V. (1960). *JETP* **39**[2(8)], 326–331.
Gaponov, A. V. (1961). *Izv. VUZov, Radiofizika* **4**(3), 547–560.
Gaponov, A. V., and Yulpatov, V. K. (1967). *Radiotekh. Elektron.* **12**(4), 627–633.
Gaponov, A. V., et al. (1965). *JETP Lett.* **2**(9), 430–435.
Gaponov, A. V., Petelin, M. I., and Yulpatov, V. K. (1967). *Radio Phys. Quantum Electron.* **10**(9–10), 794–813.
Gaponov, A. V., Gol'denberg, A. L., Petelin, M. I., and Yulpatov, V. K. (1976). Copyright No. 223931 with priority of March 24, 1967, Official Bulletin KDIO of SM USSR, No. 11, p. 200.
Gaponov, A. V., et al. (1980). *Int. J. Infrared and Millimeter Waves* **1**(3), 351–372.
Gaponov, A. V., et al. (1981). *Int. J. Electron.* **51**(4), 277–302.
Ginzburg, N. S., and Nusinovich, G. S. (1979). *Radio Phys. Quantum Electron.* **22**(6), 522–528.
Ginzburg, N. S., Kuznetsov, S. P., and Fedoseeva, T. N. (1979a). *Radio Phys. Quantum Electron.* **22**, 728–736.
Ginzburg, N. S., Zarnitsyna, I. G., and Nusinovich, G. S. (1979b). *Radiotekh. Elektron.* **24**(6), 1146–1152.
Ginzburg, N. S., Krementsov, V. I., Petelin, M. I., Strelkov, P. S., and Shkvarunets, A. G. (1979c). *Zh. Tekh. Fiz.* **49**(2), 378–386.
Ginzburg, N. S., Zarnitsyna, I. G., and Nusinovich, G. S. (1981). *Radio Phys. Quantum Electron.* **24**(4), 331–339.
Gol'denberg, A. L., and Pankratova, T. B. (1971). *Elektron. Tekh., Ser. I, SVCh-Elektronika* (9), 81–89.
Gol'denberg, A. L., and Petelin, M. I. (1973). *Radio Phys. Quantum. Electron.* **16**(1), 106–112.
Gol'denberg, A. L., Petelin, M. I., and Shestakov, D. I. (1973). *Elektron. Tekh., Ser. I, SVCh-Elektronika* (5), 73–80.

Gol'denberg, A. L., Lygin, V. K., Manuilov, V. N., Petelin, M. I., and Tsimring, Sh. E. (1981). *In* "Gyrotron," p. 86–106. Institute of Applied Physics, Academy of Sciences of the USSR, Gorky, USSR.

Granatstein, V. L., Herndon, M., Sprangle, P., Carmel, Y., and Nation, J. A. (1975). *Plasma Phys.* **17**(1), 23–28.

Granatstein, V. L., *et al.* (1979). *Fourth Int. Conf. Infrared and Millimeter Waves Digest,* Miami Beach, Florida, pp. 122–124.

Hirshfield, J. L., and Wachtel, J. M. (1964). *Phys. Rev. Lett.* **12**(19), 533–536.

Ilyin, V. P. (1974). "Numerical Methods of Solving Electrostatic Problems." Nauka, Novosibirsk, USSR.

Jory, H., Hegji, S., Shively, J., and Symons, R. (1978). *Microwave J.* **21**(8), 30–32.

Jory, H., Evans, S., Felch, K., Shively, J., and Spang, S. (1982). *Proc. 3rd Int. Symp. Heating Toroidal Plasmas* **III**, 1073–1078. CEN, Grenoble, France.

Jory, H. (1984). *Proc. 4th Int. Symp. Heating Toroidal Plasmas,* Rome, Italy.

Kolomensky, A. A., and Lebedev, A. N. (1962). *Doklady Akademii Nauk SSSR* **145**, 1259–1261.

Kreischer, K. E., and Temkin, R. J. (1981). *Int. J. Infrared and Millimeter Waves* **2**(2), 174–196.

Krementsov, V. I., Strelkov, P. S., and Shkvarunets, A. G. (1980). *Zh. Tekh. Fiz.* **50**, 2469–2472.

Kurayev, A. A. (1971). "Microwave Devices with Curved Electron Beams." Nauka i Tekhnika, Minsk, USSR.

Lamb, W. E. (1964). *Phys. Rev. A* **134**(6), 1429–1450.

Landau, L. D., and Lifshitz, E. M. (1962). "The Field Theory." Fizmatgiz, Moscow, USSR.

Lau, Y. Y., and Chu, K. R. (1981). *Int. J. Infrared and Millimeter Waves* **2**(3), 415–426.

Lau, Y. Y., Chu, K. R., Barnett, L. R., and Granatstein, V. L. (1981). *Int. J. Infrared and Millimeter Waves* **2**(3), 373–393.

Luchinin, A. G., and Nusinovich, G. S. (1975). *Elektron. Tekh., Ser. I, SVCh-Elektronika* (11), 26–36.

Luchinin, A. G., Malygin, O. V., Nusinovich, G. S., Fix, A. Sh., and Flyagin, V. A. (1982). *Pis'ma Zh. Tekh. Fiz.* **8**(18), 1147–1149.

Luchinin, A. G., Malygin, O. V., Nusinovich, G. S., and Flyagin, V. A. (1983). *Zh. Tekh. Fiz.* **53**(8), 1629–1632.

Lygin, V. K., and Tsimring, Sh. E. (1978). *Radiophys. Quantum Electron.* **21**(9), 948–953.

Manuilov, V. N., and Tsimring, Sh. E. (1977). *Elektron. Tekh., Ser. I, SVCh-Elektronika* (4), 67–76.

Manuilov, V. N., and Tsimring, Sh. E. (1978). *Radiotekh. Elektron.* **23**(7), 1486–1496.

Manuilov, V. N., and Tsimring, Sh. E. (1981a). *Radiophys. Quantum Electron.* **24**(4), 338–343.

Manuilov, V. N., and Tsimring, Sh. E. (1981b). *Zh. Tekh. Fiz.* **51**(12), 2483–2488.

Manuilov, V. N., and Tsimring, Sh. E. (1981c). *In* "Gyrotron," pp. 107–121. Institute of Applied Physics, Academy of Sciences of the USSR, Gorky, USSR.

Moiseev, M. A. (1977). *Radiophys. Quantum Electron.* **20**(8), 846–849.

Moiseev, M. A., and Nusinovich, G. S. (1974). *Radiophys. Quantum Electron.* **17**(11), 1305–1311.

Nikolayev, L. V., and Ofitserov, M. M. (1974). *Radio Eng. Electron. Phys.* **19**(3), 139–140.

Northrop, T. (1963). "The Adiabatic Motion of Charged Particles." Wiley (Interscience), New York.

Nusinovich, G. S. (1974). *Elektron. Tekh., Ser. I, SVCh-Elektronika* (3), 44–49.

Nusinovich, G. S. (1981a). *Int. J. Electron.* **51**(4), 457–474.

Nusinovich, G. S. (1981b). *In* "Gyrotron," p. 146–168. Institute of Applied Physics, Academy of Sciences of the USSR, Gorky, USSR.

Nusinovich, G. S., and Erm, R. E. (1972). *Elektron. Tekh., Ser. I, SVCh-Electronika* (8), 55–59.

Pantell, R. H. (1959). *Proc. IRE* **47**(6), 1146–1147.

Petelin, M. I. (1974). *Radiophys. Quantum Electron.* **17**(6), 686–690.

Petelin, M. I. (1981). *In* "Gyrotron," p. 77–85. Institute of Applied Physics, Academy of Sciences of the USSR, Gorky, USSR.

Petelin, M. I., and Yulpatov, V. K. (1975). *Radiophys. Quantum Electron.* **18**(2), 212–218.

Pierce, J. R. (1954). "Theory and Design of Electron Beams." D. Van Nostrand Co., Princeton, New Jersey.

Rabinovich, M. I., and Fabricant, A. L. (1979). *Zh. Eksp. Teor. Fiz.* **77**(2), 617–624.

Schneider, J. (1959). *Phys. Rev. Lett.* **2**, 504–505.

Sprangle, P., Vomvoridis, J. L., and Manheimer, W. M. (1981). *Phys. Rev. A* **23**(6), 3127–3138.

Temkin, R. J., Kreischer, K. E., Mulligan, W. J., MacCabe, S., and Fetterman, H. R. (1982). *Int. J. Infrared and Millimeter Waves* **3**(4), 427–438.

Tsimring, Sh. E. (1972). *Radiophys. Quantum Electron.* **15**(8), 952–961.

Tsimring, Sh. E. (1977). *Radiophys. Quantum Electron.* **20**(10), 1068–1075.

Twiss, R. Q. (1958). *Australian J. Phys.* **11**, 564–579.

Uckan, N. A. (1981). *Proc. 10th European Conf. Controlled Fusion Plasma Phys.*, **II**, p. 101–103. Moscow, USSR.

Vainstein, L. A. (1966). "Open Resonators and Open Waveguides." Soviet Radio, Moscow, USSR. (Translated into English by P. Beckmann, Golem Press, Boulder, Colorado, 1969).

Vainstein, L. A., and Solntsev, V. A. (1972). "Lectures on Microwave Electronics." Soviet Radio, Moscow, USSR.

Vashkovsky, A. V., and Ovcharov, V. T. (1971). *Elektron. Tekh., Ser. I, SVCh-Elektronika* (9), 34–42.

Vlasov, S. N., Zhislin, G. M., Orlova, I. M., Petelin, M. I., and Rogacheva, G. G. (1969). *Radiophys. Quantum Electron.* **12**(8), 972–978.

Vlasov, S. N., Zagryadskaya, L. I., and Orlova, I. M. (1976). *Radiotekh. Elektron.* **21**(7), 1485–1492.

Vlasov, S. N., Orlova, I. M., and Petelin, M. I. (1981). *In* "Gyrotron," p. 62–76. Institute of Applied Physics, Academy of Sciences of the USSR, Gorky, USSR.

Yulpatov, V. K. (1967). *Radiophys. Quantum Electron.* **10**(6), 471–476.

Zapevalov, V. E., Korablev, G. S., and Tsimring, Sh. E. (1977). *Radiotekh. Elektron.* **22**(8), 1661–1669.

Zarnitsyna, I. G., and Nusinovich, G. S. (1974). *Radiophys. Quantum Electron.* **17**(12), 1418–1424.

Zarnitsyna, I. G., and Nusinovich, G. S. (1978). *Radio Eng. Electron. Phys.* (*USSR*), **23**(6), 74–78.

Zaytsev, N. I., Pankratova, T. B., Petelin, M. I., and Flyagin, V. A. (1974). *Radio Eng. Electron. Phys.* **19**(5), 103–107.

Zheleznyakov, V. V. (1960). *Izv. VUZov, Radiofizika* **3**(1), 57–67.

Zhurakhovsky, V. A. (1969). *Radiotekh. Elektron.* **14**(1), 128–136.

Zhurakhovsky, V. A. (1972). "Nonlinear Oscillations of Electrons of Magneto-Directed Flows." Naukova Dumka, Kiev, USSR.

CHAPTER 6

Some Perspectives on Operating Frequency Increase in Gyrotrons

G. S. Nusinovich

Institute of Applied Physics
Academy of Sciences of the USSR
Gorky, USSR

I. Introduction

Cyclotron resonance masers (CRMs), as well as other *Bremsstrahlung* generators, combine the virtues of both classical microwave devices and quantum optical devices. As in lasers, electrons in CRMs can resonate with fast waves. Because of this, CRM electrodynamic systems can represent open resonators and waveguides without any small-scale elements, such as slow-wave structures in Cherenkov devices (traveling-wave tubes, backward-wave oscillators, magnetrons) or cavities with lengths on the order of the wavelength in conventional klystrons based on transit radiation. Thus, the problems of diminishing the interaction space and heat transfer from the electrodynamic system (at cw or long-pulse operation) that arise with the shorter wavelengths are not so dangerous in CRMs as in conventional microwave devices.

At the same time, if the active medium of lasers has an essentially non-equidistant spectrum of energy levels, and correspondingly only one transition between two given levels is usually used, in CRMs electrons rotating in the external homogeneous magnetic field can be presented as an active medium with the energy spectrum close to the equidistant one. This fact permits multiphoton radiation of each electron into the given mode of the resonator. From this point of view, CRMs as well as other conventional microwave devices demonstrate the absence of a very important defect, single-photon radiation of each active particle, which leads to a drop in laser efficiency for the infrared and submillimeter wavelengths.

The combination of the virtues of classical microwave devices with the advantages of quantum optical generators makes CRMs attractive for mastering the intermediate millimeter and submillimeter wavelengths.

Induced coherent radiation of electrons takes place under the cyclotron resonance condition

$$\omega \simeq k_{\parallel} v_{\parallel} + n\omega_{\mathbf{H}}, \tag{1}$$

where ω is the radiation frequency, $k_{\parallel}$ the axial wave number (in the direction of the external magnetic field $\mathbf{H}_0$), $v_{\parallel}$ the axial velocity of the electrons, and n the resonant harmonic number of the cyclotron frequency $\omega_{\mathbf{H}}$. Coherent radiation of electrons in CRMs is possible owing to phase (orbital) bunching of electrons, as well as to axial bunching (Gaponov, 1960; Gaponov et al., 1967). Orbital bunching of electrons is caused by the relativistic dependence of electron mass (and cyclotron frequency) on electron energy. Axial bunching of electrons occurs if the wave field acts on the axial momentum of particles (the "recoil" effect under quantum radiation). Both mechanisms of electron bunching are inertial, i.e., the bunching develops under the initial modulation of electron energy by the field, even outside the interaction space. [The effects caused by the force bunching are discussed elsewhere; see, e.g., Gaponov et al. (1967).] Because an increase in electron energy leads to an increase in axial velocity and a decrease in cyclotron frequency, both of these changes have opposite signs in the cyclotron resonance condition (1). When the axial phase velocity of the wave is larger than the velocity of light (fast waves), the shift in cyclotron frequency exceeds the change in the Doppler term $|\delta(\omega_{\mathbf{H}})| > |k_{\parallel} \delta(v_{\parallel})|$. For slow waves the change in the Doppler term is dominant. When the axial phase velocity equals the velocity of light (luminous waves), both the shifts are the same. Therefore, unlimited change in electron energy can take place under the cyclotron resonance condition. This effect is known as autoresonance (Davidovsky, 1962; Kolomensky and Lebedev, 1962). In this case the two mechanisms of inertial bunching compensate each other, and the electron flow behaves as an ensemble of charged linear oscillators (Gaponov, 1960).

In fact, the cyclotron resonance condition (1) shows all the possibilities of the operating frequency increase in CRMs:

(1)   an increase in the external magnetic field (cyclotron frequency);
(2)   operation at cyclotron harmonics $n > 1$;
(3)   "gambling" with the Doppler frequency upconversion.

Let us analyze all the potentialities of not only the most popular device of the CRM family, the gyrotron, but other possible varieties of CRMs as well.

II.   The Gyrotron

The gyrotron (Fig. 1 in Chapter 5, page 181) is a device in which an electron gun of the magnetron type produces an electron flow that excites one of eigenmodes of the open resonator. In gyrotron open resonators, the modes with a small axial wave number have the highest diffraction Q factors and, correspondingly, the lowest starting currents. These are operating modes. The smallness of the axial wave number diminishes the influence of the axial-velocity spread on the cyclotron resonance condition and, hence, on gyrotron efficiency. At the same time, this smallness makes the axial bunching of electrons insignificant, and the Doppler frequency upconversion is impossible in gyrotrons.

A.   How High Can Magnetic Fields Be Increased in Gyrotrons?

In millimeter-wave gyrotrons, superconducting solenoids are usually used for magnetic field formation. These solenoids can produce magnetic fields up to 10–15 T in sufficiently large volumes necessary for gyrotrons. For gyrotrons operating at the fundamental harmonic, these fields yield gyrotron radiation at wavelengths greater than 0.8–0.9 mm. However, until the present gyrotrons with superconducting solenoids were operated with magnetic fields smaller than 7 T. Because of this, cw operation at the fundamental took place in gyrotrons where the shortest wavelength was approximately 2 mm; output power reached a 22-kW level (Andronov et al., 1978). In cw gyrotrons operating at the second harmonic, the shortest wavelength was 0.92 mm with output of 1.5 kW (Zaytsev et al., 1974).

Pulse solenoids provide some reserves in producing high magnetic fields (Antakov et al., 1965; Bott, 1964; Nikolaev and Ofitserov, 1974). Estimations show that in volumes sufficiently large for gyrotrons, pulse solenoids can produce magnetic fields up to 30 T, and correspondingly the wavelength range of gyrotrons with pulse solenoids can be increased to 0.4 mm. For the pulsed magnetic field of duration over 1 msec, the skinning of the alternating magnetic field in a sufficiently thin body of a gyrotron is negligibly small.

The gyrotron with a pulse solenoid producing 100–120 kW peak power (microwave pulse duration 100 μsec) with efficiency 10–15% at a wavelength of 0.6–0.8 mm (Flyagin et al., 1982, 1983) can serve as a convincing example of the potentials of these systems.

B.   Can the Gyrotron Be Operated at High Harmonics?

The affirmative answer to this question is obvious; cw operation at the second harmonic in the submillimeter-wave region has been discussed. However, one should bear in mind that operation at harmonics, especially for high-resonator modes, is often complicated owing to the excitation of parasitic rf oscillations at the fundamental. To avoid this excitation, a thorough analysis of gyrotron "large-signal" operation and different methods of mode selection are necessary. With an increase in the operating voltage, the orbital velocity of electrons grows, and correspondingly the possibility of obtaining stable operation at harmonics becomes more likely.

The recent experiment of S. A. Malygin and Sh. E. Tsimring demonstrates the possibility of successful operation of the gyrotron at harmonics. Using various methods of mode selection, these authors obtained stable single-mode operation at the third harmonic of cyclotron frequency (wavelength 5.6 mm, output power more than 100 kW, electron efficiency 10–12%).

Another interesting example of potentialities of operation at high cyclotron harmonics is a CRM with frequency multiplication. The principal idea of such a device was discussed by Gaponov et $al.$ (1967). In the device described by Kupiszewski et $al.$ (1981), a thin electron beam with a current of about several tens of milliamperes was accelerated in the input resonator by the rf field, resonant at the fundamental cyclotron harmonic, up to energies 300–500 keV. The transverse electric field of the H_{111} mode of this resonator provoked the orbital velocity of electrons. As a result of this modulation, the rotating electron beam in the output resonator excited the H_{n11} mode at the nth cyclotron harmonic ($n = 5$–10). Wavelength shortening in such devices seems to depend on the possibilities of the beam focusing on the filament with a cross section smaller than λ^2 and on the axial-velocity spread that significantly affects the efficiency of CRMs operating at high cyclotron harmonics.

III.   CRMs with Doppler Frequency Upconversion

In recent years the possibility of Doppler frequency upconversion has been actively investigated in CRMs with relativistic electron beams (several varieties of free-electron lasers and masers). Indeed, it follows from the cyclotron resonance condition (1), which can be rewritten in the form $\omega \simeq n\omega_H/(1 - v_{\parallel}/v_{ph})$, that significant Doppler frequency upconversion of fast waves ($v_{ph} > c$) can be realized only with relativistic electrons ($v_{\parallel} \simeq c$). The upper limit of the Doppler frequency upconversion in CRM with fast waves is $\omega/n\omega_H \simeq 2\gamma_0^2$ (γ_0 is the Lorentz factor of the electrons), and the developed theory predicts feasibility of high efficiency in CRMs with frequency upconversion from 0 to γ_0^2 (Bratman et $al.$, 1979a, 1981; Ginzburg et $al.$, 1981).

It should be noted that the developed theory is also valid for CRMs with slow waves. In these devices the frequency upconversion can be arbitrary for arbitrary energy of electrons. The main restrictions on the wavelength shortening in this kind of CRM are believed to be associated with the

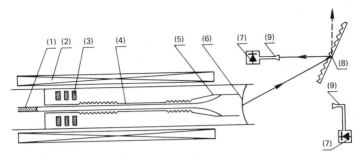

FIG. 1 Schematic diagram of the cyclotron resonance maser: (1) cathode, (2) pulse solenoid, (3) rings for electron twisting, (4) resonator with Bragg's mirrors, (5) collector, (6) vacuum window, (7) microwave detector, (8) echelette grating, (9) receiver. (From Botvinnik *et al.*, 1982a, b.)

problems of slow-wave structures. (The same problem is cardinal for TWT and other Cherenkov devices.)

A cyclotron autoresonance maser with an open resonator (Botvinnik *et al.*, 1982a) can serve as an interesting example of CRMs with Doppler frequency upconversion. In such a device, the resonator [see Denisov and Reznikov (1982) and the cited literature] represents a section of a smooth cylindrical waveguide bounded by two sections of a waveguide with corrugated walls (Fig. 1). In this system small diffraction losses exist in the mode formed by two fast waves,

$$A_1 \simeq \exp\{i(\omega t - m_1 \varphi - k_{\parallel,1} z)\}, \qquad A_2 \simeq \exp\{i(\omega t - m_2 \varphi + k_{\parallel,2} z)\},$$

for which Bragg's scattering conditions $\bar{k}_{\parallel} \simeq k_{\parallel,1} + k_{\parallel,2}, \bar{m} = \pm(m_1 - m_2)$ are valid at the corrugated surface of the waveguide,

$$R = R_0 + R_1 \cos(\bar{m}\varphi + \bar{k}_{\parallel} z).$$

The waves with other Brillouin angles are scattered at the corrugated walls. In recent experiments on these devices (beam voltage $U = 0.7$ MV) stable single-mode oscillations were observed at wavelength 4 mm (output power 6 MW, electron efficiency 4%) and at 2.4 mm (output power 10 MW, electron efficiency 2%) (Botvinnik *et al.*, 1982a, b). A rather high efficiency in this type of free-electron lasers may be explained by a relatively good quality of the electron beam and also by the fact that in CRMs with quasi-luminous waves ($v_{ph} \simeq c$), the aforementioned autoresonant compensation of the change in cyclotron frequency by the Doppler shift mitigates the influence of electron-energy spread on the electron efficiency (Bratman *et al.*, 1979a; Ginzburg *et al.*, 1981).

IV.   Submillimeter-Wave Gyrotron Efficiency

Two effects become significant with the wavelength shortening in gyrotrons as well as in other microwave devices, and both effects lead to a decrease in efficiency. The first effect is the growth of ohmic losses: when the ohmic Q factor is of the order of the diffraction Q, the output efficiency

$$\eta_{\text{out}} = [Q_{\text{ohm}}/(Q_{\text{dif}} + Q_{\text{ohm}})]\eta_{\text{el}} \tag{2}$$

is significantly smaller than the electron efficiency η_{el}. The second effect is based on contradictory tendencies. On the one hand, as the wavelength shortens, the beam current density in the elementary interaction cell (with a cross section equal to λ^2) diminishes. On the other hand, the value of the rf field that is optimum for electron efficiency grows proportionally to the magnetic field. As a result, at very short wavelengths the beam current density becomes smaller than the value optimum for efficiency.

Depending on the possibilities of electron-optical systems and the quality of the resonator surfaces, each of these two effects enters the game sooner or later. If the current density is sufficiently large, ohmic losses are the first to play their role. In such a situation it is necessary to determine the optimum gyrotron parameters using nonlinear equations of the gyrotron and the expression for the output efficiency, Eq. (2). The self-consistent set of gyrotron equations (Gaponov et al., 1967; Petelin and Yulpatov, 1974) obtained by Yulpatov in the beginning of the 1960s contains the averaged equations for energy ε and phase $\vartheta = n \int_0^t \omega_{\text{H}} \, dt' - \omega t + \vartheta_0$ of electron gyration disturbed by the rf field:

$$d\varepsilon/dz = e \, \text{Re}[Af(z) \, \partial H/\partial \vartheta], \tag{3}$$

$$(d\vartheta/dz) + (\omega/v_{\parallel})[1 - (n\omega_{\text{H}}/\omega)] = -e \, \text{Re}[Af(z) \, \partial H/\partial \varepsilon], \tag{4}$$

the expression for electron efficiency,

$$\eta_{\text{el}} = \frac{1}{(1 - \gamma_0^{-1})} \left[1 - \frac{1}{2\pi} \int_0^{2\pi} \frac{\varepsilon(z_{\text{out}})}{\varepsilon_0} \, d\vartheta_0 \right], \tag{5}$$

and the balance equation that defines the dependence of the rf field amplitude A on the beam current I,

$$\eta_{\text{el}} UI = (\omega/Q)A^2 V \tag{6}$$

[V is the resonator volume, $Q = Q_{\text{dif}} Q_{\text{ohm}}/(Q_{\text{dif}} + Q_{\text{ohm}})$]. In Eqs. (3)–(4) z is the axial coordinate of the interaction space, the function $f(z)$ describes the axial structure of the rf field, and the function H, which plays the role of a Hamiltonian in the "canonically conjugated" set of equations (3)–(4) for H modes, is defined as

$$H = i(p_\perp/p_\parallel) J_n'(\xi) J_{m \mp n}(k_\perp R) e^{-i\vartheta}.$$

Here $p_{\perp,\parallel}$ are the components of the electron momentum, $\xi = n\beta_\perp$; the Bessel function $J_{m\mp n}(k_\perp R)$ describes the radial structure of the rf Lorentz force acting on the electron with a guiding center radius R.[†] The results of numerical study of the set of equations (3)–(6) were presented by Gaponov *et al.* (1967) and in many other papers. The optimization of gyrotron parameters for maximum output efficiency (2) were given by Temkin *et al.* (1979) and Nusinovich and Pankratova (1981).

If the current density is smaller than the value optimum for efficiency, it seems expedient, according to Petelin's idea (Bratman *et al.*, 1979b), to increase the resonator length so as to obtain the effective orbital bunching of electrons in the weak rf field. In such a case, Eqs. (3)–(4) can be simplified, because for a small amplitude A of the field the change in electron energy is small. Consequently, the function H in the right-hand side of Eq. (3) can be assumed equal to the initial value $H_0(p_{\perp 0}, p_{\parallel 0})$; in the right-hand side of Eq. (4), the term responsible for the direct action of the rf field on the electron phase can be omitted and the change in the cyclotron frequency [the term $n\omega_H/\omega$ in the left-hand side of Eq. (4)] caused by the change in energy can be linearized ($\omega_H \simeq \omega_{H_0} + (d\omega_H/d\varepsilon)\Delta\varepsilon$). As a result, introducing a new axial variable

$$\zeta = \sqrt{(n/v_\parallel)|d\omega_H/d\varepsilon|eA|H_0|}\, z,$$

one can reduce Eqs. (3)–(4) to one equation,

$$(d^2\vartheta/d\zeta^2) + f(\zeta)\sin\vartheta = 0, \tag{7}$$

similar to the one known in the theory of resonant TWT. This equation is valid for several varieties of free-electron lasers (Bratman *et al.*, 1979c). For the homogeneous structure of the rf field $f(\zeta) = 1$, Eq. (7) represents a pendulum nonlinear equation. Results of the investigations of Eqs. (2), (5)–(7) were presented by Nusinovich (1981a). It is significant that owing to the increase in the resonator length the electron efficiency decreases proportionally to $\sqrt{A}$, whereas with the constant resonator length the decrease in the current density leads to the dependence $\eta_{el} \propto A^2$ (Bratman *et al.*, 1979b). It is also interesting to note that Eq. (7) does not contain the number of the resonance cyclotron harmonic and, therefore, the results obtained are valid for gyrotrons operating at arbitrary harmonics.

[†] The azimuthal drift of the guiding centers can be easily taken into account by the introduction of the phase $\vartheta = \vartheta + (m - n)\psi$ (ψ is the azimuthal coordinate of the guiding center). The radial drift of the guiding centers in gyrotrons is small owing to the smallness of the change in the total electron energy under the cyclotron resonance condition $|\delta\varepsilon|/\varepsilon_0 = |\delta\omega_H|/\omega_H \lesssim 1/nN$, where N is the number of electron turns in the interaction space.

The foregoing consideration is significant mainly for pulsed gyrotrons, because for submillimeter-wave and short millimeter-wave cw gyrotrons the heating of a resonator by the rf ohmic losses plays an important role even when the ohmic Q is sufficiently greater than the diffraction Q. This fact modifies the methods of determining the gyrotron parameters corresponding to the maximum output power for cw operation.

V. Nonstationary Processes in CRMs with Extended Interaction Space

To avoid miniaturization of the interaction space and the corresponding drop in the microwave power with the wavelength shortening, it is necessary to increase the cross section of the resonator (in the wavelength scale). This leads to operation at high modes of the resonator in the region of a dense eigenfrequency spectrum and, hence, the electron beam can excite several neighboring resonator modes simultaneously. To understand the behavior of such an oscillator, it is necessary to analyze equations describing non-stationary processes of multimode CRMs.

At large-signal operation, the excited modes interact because of the non-linear features of the oscillator. The main effect of this interaction is mode competition inherent to CRMs as well as to other microwave electron devices (Moiseev and Nusinovich, 1974; Nusinovich, 1981b). This effect is stipulated by the fact that the azimuthally rotating modes of the resonator envelop the annular electron beam and interact with electrons having different azimuthal coordinates of the guiding centers with the same efficacy. As a result, strong coupling between such modes takes place, and the mode with the greatest initial increment suppresses the others. This effect makes us expect to obtain stable single-mode operation even if conditions of self-excitation are fulfilled for several modes at the initial moment of the gyrotron switching. The main problem for gyrotrons with mode competition is to provide at the initial moment the largest increment for the mode capable of delivering large microwave power with high efficiency.

Besides mode competition, other effects of nonlinear mode interaction are possible. In some cases a mode with a large increment can support oscil-lations of another mode having a larger eigenfrequency. This effect appears at a definite location of mode frequencies in the cyclotron resonance band and at a rather large beam current. The results of numerical investigations of two-mode excitation in the gyrotron are shown in Fig. 2.

Another example demonstrating the possibilities of apparently compli-cated oscillations is a CRM with a Fabry–Perot resonator (Fig. 3). The theory of stationary oscillations in these and similar devices was developed by Sprangle et al. (1981) and Zarnitsyna and Nusinovich (1978). Here we

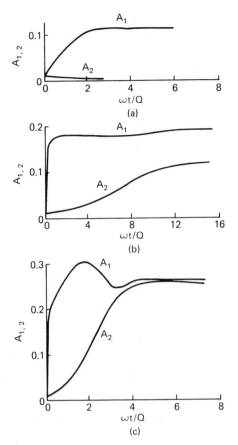

FIG. 2   Nonstationary processes in the two-mode gyrotron (in the notation adopted by Nusinovich (1981b), $f_{1,2}(\zeta) = \exp[-(2\zeta/\zeta_{out})^2]$, $\zeta_{out} = 10$, $\Delta_1 = 0.3$, $\Delta_2 = 0.8$). (a) $I/I_{st} \simeq 7$; (b) $I/I_{st} \simeq 32$; (c) $I/I_{st} \simeq 110$.

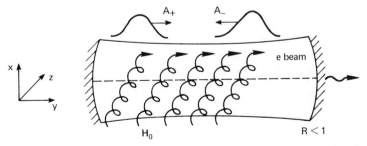

FIG. 3   Schematic representation of a quasi-optical cyclotron maser. (After Sprangle et al., 1981.)

will discuss some results of the theory of nonstationary processes (Nusinovich and Sher, 1984). If the distance between the resonator mirrors is large (this case is attractive for high power generation), the frequencies of many axial modes can be in the cyclotron resonance band. It is simpler to investigate the dynamics of such an oscillator without representation of the rf field as a sum of eigenmodes. We consider this field as a sum of two waves propagating in opposite directions between the mirrors:

$$\mathbf{E} = \text{Re}\{\mathbf{x}_0 E(x, z)[A_+(y, t) \exp[i(\omega t - ky)] + A_-(y, t) \exp[i(\omega t + ky)]]\}.$$

Equations for the envelope amplitudes of these waves $A_\pm(y, t)$ have the form

$$\frac{1}{c}\frac{\partial A_+}{\partial t} + \frac{\partial A_+}{\partial y} = \frac{1}{N}\left\langle \int_0^{x_\text{out}} \int_0^{z_\text{out}} j_{\omega_x} e^{iky} E(x, z)\, dz\, dx \right\rangle,$$

$$\frac{1}{c}\frac{\partial A_-}{\partial t} - \frac{\partial A_-}{\partial y} = \frac{1}{N}\left\langle \int_0^{x_\text{out}} \int_0^{z_\text{out}} j_{\omega_x} e^{-iky} E(x, z)\, dz\, dx \right\rangle,$$

$$(8)$$

where N is the norm of the wave, the brackets $\langle \cdots \rangle$ denote the averaging over the wavelength in the y direction, and the current density $\mathbf{j}$ is presented as $\mathbf{j} = \text{Re}\{\mathbf{j}_\omega e^{i\omega t}\}$. It is significant that the right-hand terms in Eq. (8), describing the rf field excitation by the electron beam, are nonlinear functions of the rf field amplitude with the delay argument $t - \tau_\text{int}$ (here τ_int is the time of electron–rf field interaction). The presence of the delay argument is inherent in electron microwave oscillators characterized by the space–time dispersion of the active medium. As is known [see, e.g., Ginzburg et al. (1978); Bogomolov et al. (1981)], this fact can lead to instability of stationary single-mode oscillations.

Numerical study of the self-consistent set of equations comprising modified Eqs. (8), supplemented by the equation of electron motion in the field of two opposite waves [given by Zarnitsyna and Nusinovich (1978)], demonstrates the change in the oscillator dynamics with the increase in the beam current. The results presented in Fig. 4 show that when the total beam current proportional to the distance between the mirrors grows, the stable single-mode oscillations with a constant amplitude give place to oscillations with automodulation of radiation. These oscillations are complicated with an increase in the beam current and eventually stochastic oscillations appear. Correspondingly, the space structure of the rf field also becomes more complicated (Fig. 4b).

It should be emphasized that in this example, as well as in the previous one, stationary single-mode oscillations are established in the multimode systems at the value of the beam current optimum for efficiency.

In conclusion, it seems expedient to note that the developed theory

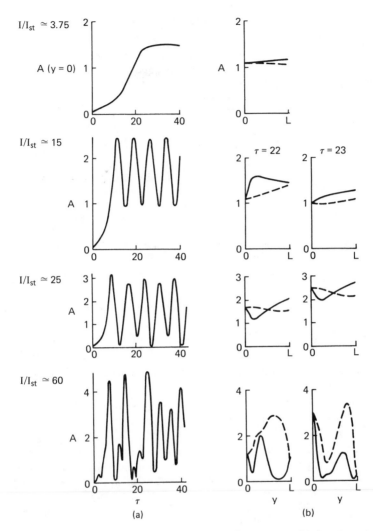

FIG. 4 Nonstationary processes in a quasi-optical cyclotron maser: (a) slow-time evolution of the rf-field amplitude on the left mirror; (b) space structure of both opposite waves at fixed moments in time $\tau \simeq I\omega t$.

predicts rather high values of the output efficiency in spite of the ohmic losses for all the varieties of CRMs discussed, and the results achieved experimentally permit one to suppose that soon we shall see convincing coverage of the submillimeter-wave range by different pulsed cyclotron resonance masers.

REFERENCES

Andronov, A. A., et al. (1978). Infrared Phys. **18**(6), 385–393.

Antakov, I. I., Clymov, V. G., and Lin'kov, R. V. (1965). Izv. VUZov, Radiofiz. **8**(5), 948–951.

Bogomolov, Ya. L., Bratman, V. L., Ginzburg, N. S., Petelin, M. I., and Yunakovsky, A. D. (1981). Opt. Commun. **36**(3), 209–212.

Bott, I. B. (1964). Proc. IEEE **52**, 330–332.

Botvinnik, I. E., et al. (1982a). Pis'ma Zh. Eksp. teor. Fiz. **35**(10), 418–420.

Botvinnik, I. E., et al. (1982b). Pis'ma Zh. Tech. Fiz. **8**(22), 1386–1389.

Bratman, V. L., Ginzburg, N. S., Nusinovich, G. S., Petelin, M. I., and Yulpatov, V. K. (1979a). In "High-Frequency Relativistic Electronics," pp. 157–216. Institute of Applied Physics, Academy of Sciences of the USSR, Gorky, USSR.

Bratman, V. L., Ginzburg, N. S., Kovalev, N. F., Nusinovich, G. S., and Petelin, M. I. (1979b). In "High-Frequency Relativistic Electronics," pp. 249–274. Institute of Applied Physics, Academy of Sciences of the USSR, Gorky, USSR.

Bratman, V. L., Ginzburg, N. S., and Petelin, M. I. (1979c). Opt. Commun. **30**(3), 409–412.

Bratman, V. L., Ginzburg, N. S., Nusinovich, G. S., and Petelin, M. I. (1981). Proc. 4th Int. Conf. High-Power Electron and Ion Beams, Vol. II, pp. 853–859. Palaiseau, France.

Davydovsky, V. Ya. (1962). Zh. Eksp. Teor. Fiz. **43**, 886–889.

Denisov, G. G., and Reznikov, M. G. (1982). Izv. VUZov, Radiofizika **25**(5), 562–567.

Flyagin, V. A., Luchinin, A. G., and Nusinovich, G. S. (1982). Int. J. Infrared and Millimeter Waves **3**(6), 765–770.

Flyagin, V. A., Luchinin, A. G., and Nusinovich, G. S. (1983). Int. J. Infrared and Millimeter Waves **4**(4), 629–637.

Gaponov, A. V. (1960). JETP **39** [2(8)], 326–331.

Gaponov, A. V., Petelin, M. I., and Yulpatov, V. K. (1967). Radio Phys. Quantum Electron. **10**, 794–833.

Ginzburg, N. S., Kuznetsov, S. P., and Fedoseeva, T. N. (1978). Izv. VUZov, Radiofizika **21**(7), 1037–1044.

Ginzburg, N. S., Zarnitsyna, I. G., and Nusinovich, G. S. (1981). Radio Phys. Quantum Electron. **24**(4), 331–340.

Kolomensky, A. A., and Lebedev, A. N. (1962). Doklady Akademii Nauk, SSSR **145**, 1259–1262.

Kupiszewski, A., Luhmann, N. C., and Jory, H. (1981). Proc. 6th Int. Conf. Infrared and Millimeter Waves, Miami Beach, Florida, Report W-2-5.

Moiseev, M. A., and Nusinovich, G. S. (1974). Radio Phys. Quantum Electron. **17**, 1305–1313.

Nikolaev, L. V., and Ofitserov, M. M. (1974). Radio Eng. Electron. Phys. **19**, 139–140.

Nusinovich, G. S. (1981a). Elektron. Tekh., Ser. I, SVCh Elektronika (1), 16–19.

Nusinovich, G. S. (1981b). Int. J. Electron. **51**(4), 457–474.

Nusinovich, G. S., and Pankratova, T. B. (1981). In "Gyrotron," pp. 169–184. Institute of Applied Physics, Academy of Sciences, Gorky, USSR.

Nusinovich, G. S., and Sher, E. M. (1984). Int. J. Electron. **56**(3), 275–286.

Petelin, M. I., and Yulpatov, V. K. (1974). Lectures on Microwave Electronics IV, 95–178. Saratov University, Saratov, USSR.

Sprangle, P., Vomvoridis, J. L., and Manheimer, W. M. (1981). Phys. Rev. A **23**(6), 3127–3138.

Temkin, R. J., Kreischer, K., Wolfe, S. M., Cohn, D. R., and Lax, B. (1979). J. Magn. Magn. Mater. **11**, 368–371.

Zarnitsyna, I. G., and Nusinovich, G. S. (1978). Radiotekh. Elektron. **23**(6), 1212–1216.

Zaytsev, N. I., Pankratova, T. B., Petelin, M. I., and Flyagin, V. A. (1974). Radio Eng. Electron. Phys. **19**(5), 103–107.

CHAPTER 7

Phase Noise and AM Noise Measurements in the Frequency Domain

Algie L. Lance, Wendell D. Seal, and Frederik Labaar

TRW Operations and Support Group
One Space Park
Redondo Beach, California

I. Introduction

Frequency sources contain noise that appears to be a superposition of causally generated signals and random, nondeterministic noises. The random noises include thermal noise, shot noise, and noises of undetermined origin (such as flicker noise). The end result is time-dependent phase and amplitude fluctuations. Measurements of these fluctuations characterize the frequency source in terms of amplitude modulation (AM) and phase modulation (PM) noise (frequency stability).

239

The term *frequency stability* encompasses the concepts of random noise, intended and incidental modulation, and any other fluctuations of the output frequency of a device. In general, frequency stability is the degree to which an oscillating source produces the same frequency value throughout a specified period of time. It is implicit in this general definition of frequency stability that the stability of a given frequency decreases if anything except a perfect sine function is the signal wave shape.

Phase noise is the term most widely used to describe the characteristic randomness of frequency stability. The term *spectral purity* refers to the ratio of signal power to phase-noise sideband power. Measurements of phase noise and AM noise are performed in the *frequency domain* using a spectrum analyzer that provides a *frequency window* following the detector (double-balanced mixer). Frequency stability can also be measured in the *time domain* with a gated counter that provides a *time window* following the detector. [ ]

Long-term stability is usually expressed in terms of parts per million per hour, day, week, month, or year. This stability represents phenomena caused by the aging process of circuit elements and of the material used in the frequency-determining element. Short-term stability relates to frequency changes of less than a few seconds duration about the nominal frequency.

Automated measurement systems have been developed for measuring the combined phase noise of two signal sources (the two-oscillator technique) and a single signal source (the single-oscillator technique), as reported by Lance *et al.* (1977) and Seal and Lance (1981). When two source signals are applied in quadrature to a phase-sensitive detector (double-balanced mixer), the voltage fluctuations analogous to *phase fluctuations* are measured at the detector output. The single-oscillator measurement system is usually designed using a frequency cavity or a delay line as an FM discriminator. Voltage fluctuations analogous to *frequency fluctuations* are measured at the detector output.

The integrated phase noise can be calculated for any selected range of Fourier frequencies. A representation of fluctuations in the frequency domain is called *spectral density* graph. This graph is the distribution of power variance versus frequency.

II. Fundamental Concepts

In this presentation we shall attempt to conform to the definitions, symbols, and terminology set forth by Barnes *et al.* (1970). The Greek letter v represents frequency for carrier-related measures. Modulation-related frequencies are designated f. If the carrier is considered as dc, the frequencies measured with respect to the carrier are referred to as baseband, offset from the carrier, modulation, noise, or Fourier frequencies.

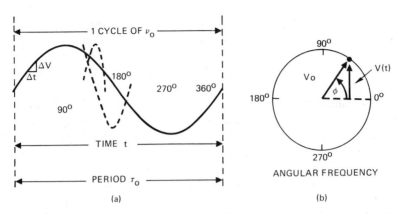

FIG. 1 Sine wave characteristics: (a) voltage V changes with time τ as (b) amplitude changes with phase angle ϕ.

A sine wave generator produces a voltage that changes in time t as the amplitude V changes with the phase angle ϕ, shown in Fig. 1. Phase is measured from a zero crossing, as illustrated by plotting the phase angle as the radius vector rotates at a constant angular rate determined by the frequency. The ideal (perfect) sine-wave-related parameters are as follows: v_0, average (nominal) frequency of the signal; $v(t)$, instantaneous frequency of a signal

$$v(t) = \frac{1}{2\pi} \frac{d\phi}{dt}(t);$$ (1)

V_0, nominal peak amplitude of a signal source output; τ, period of an oscillation $(1/v_0)$; Ω, signal (carrier) angular frequency (rate of change of phase with time) in radians

$$\Omega = 2\pi v_0 ;$$ (2)

Ωt, instantaneous angular frequency; $V(t)$, instantaneous output voltage of a signal. For the ideal sine wave signal of Fig. 1, in volts,

$$V(t) = V_0 \sin(2\pi v_0 t).$$ (3)

The basic relationship between phase ϕ, frequency v_0, and time interval τ of the ideal sine wave is given in radians by the following:

$$\phi = 2\pi v_0 \tau,$$ (4)

where $\phi(t)$ is the instantaneous phase of the signal voltage, $V(t)$, defined for the ideal sine wave in radians as

$$\phi(t) = 2\pi v_0 t.$$ (5)

The instantaneous phase $\phi(t)$ of $V(t)$ for the noisy signal is

$$\phi(t) = 2\pi v_0 t + \phi(t), \tag{6}$$

where $\phi(t)$ is the instantaneous phase fluctuation about the ideal phase $2\pi v_0 \tau$ of Eq. (4).

The simplified illustration in Fig. 1 shows the sine-wave signal perturbed for a short instant by noise. In the perturbed area, the Δv and Δt relationships correspond to other frequencies, as shown by the dashed-line waveforms. In this sense, frequency variations (phase noise) occur for a given instant within the cycle.

The instantaneous output voltage $V(t)$ of a signal generator or oscillator is now

$$V(t) = [V_0 + \varepsilon(t)] \sin[2\pi v_0 t + \phi(t)], \tag{7}$$

where V_0 and v_0 are the nominal amplitude and frequency, respectively, and $\varepsilon(t)$ and $\phi(t)$ are the instantaneous amplitude and phase fluctuations of the signal.

It is assumed in Eq. (7) that

$$\varepsilon(t)/V_0 \ll 1 \quad \text{and} \quad \frac{\dot\phi(t)}{v_0} \ll 1 \qquad \text{for all} \quad (t), \dot\phi(t) = d\phi/dt. \tag{8}$$

Equation (7) can also be expressed as

$$V(T) = [V_0 + \delta\varepsilon(t)] \sin[2\pi v_0 t + \phi_0 + \delta\phi(t)], \tag{9}$$

where ϕ_0 is a constant, δ is the fluctuations operator, and $\delta\varepsilon(t)$ and $\delta\phi(t)$ represent the fluctuations of signal amplitude and phase, respectively.

Frequency fluctuations δv are related to phase fluctuations $\delta\phi$, in hertz, by

$$\delta v = \frac{\delta\Omega}{2\pi} = \frac{1}{2\pi}\frac{d(\delta\phi)}{dt}, \tag{10}$$

i.e., radian frequency deviation is equal to the rate of change of phase deviation (the first-time derivative of the instantaneous phase deviation).

The fluctuations of time interval $\delta\tau$ are related to fluctuations of phase $\delta\phi$, in radians, by

$$\delta\phi = (2\pi v_0)\delta\tau. \tag{11}$$

In the following, y is defined as the *fractional frequency fluctuation* or fractional frequency deviation. It is the dimensionless value of δv normalized to the average (nominal) signal frequency v_0,

$$y = \delta v/v_0, \tag{12}$$

where $y(t)$ is the instantaneous fractional frequency deviation from the nominal frequency v_0.

A. NOISE SIDEBANDS

Noise sidebands can be thought of as arising from a composite of low-frequency signals. Each of these signals modulate the carrier-producing components in both sidebands separated by the modulation frequency, as illustrated in Fig. 2. The signal is represented by a pair of symmetrical sidebands (pure AM) and a pair of antisymmetrical sidebands (pure FM).

The basis of measurement is that when noise modulation indices are small, correlation noise can be neglected. *Two signals are uncorrelated if their phase and amplitudes have different time distributions so that they do not cancel in a phase detector.* The separation of the AM and FM components are illustrated as a modulation phenomenon in Fig. 3. Amplitude fluctuations can be measured with a simple detector such as a crystal. Phase or frequency fluctuations can be detected with a discriminator. Frequency modulation (FM) noise or rms frequency deviation can also be measured with an amplitude (AM) detection system after the FM variations are converted to AM variations, as shown in Fig. 3a. The FM–AM conversion is obtained by applying two signals in phase quadrature (90°) at the inputs to a balanced mixer (detector). This is illustrated in Fig. 3 by the 90° phase advances of the carrier.

B. SPECTRAL DENSITY

Stability in the frequency domain is commonly specified in terms of spectral densities. There are several different, but closely related, spectral densities that are relevant to the specification and measurement of stability of the frequency, phase, period, amplitude, and power of signals. Concise, tutorial

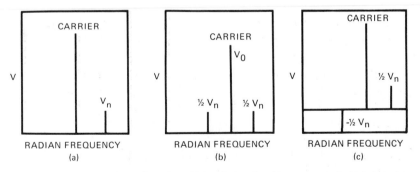

FIG. 2 (a) Carrier and single upper sideband signals; (b) symmetrical sidebands (pure AM); (c) an antisymmetrical pair of sidebands (pure FM).

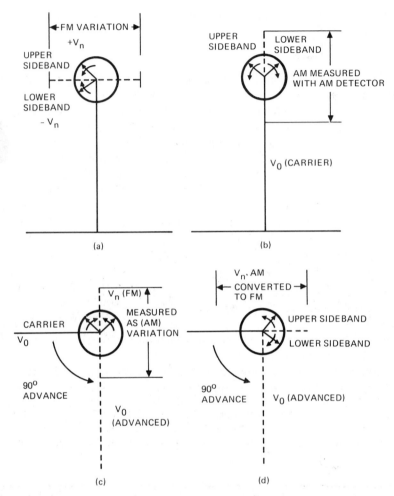

FIG. 3 (a) Relationships of the FM signal to the carrier; (b) relationship of the AM signal to the carrier; (c) carrier advanced 90° to obtain FM–AM conversion; (d) AM–FM conversion.

descriptions of twelve defined spectral densities and the relationships among them were given by Shoaf et al. (1973) and Halford et al. (1973).

Recall that in the perturbed area of the sine wave in Fig. 1 the frequencies are being produced for a given *instant of time*. This *amount of time* the signal spends in producing another frequency is referred to as the *probability density* of the generated frequencies relative to v_0. The frequency domain plot is illustrated in Fig. 4. A graph of these probability densities over a period of time produces a continuous line and is called the *Power spectral density*.

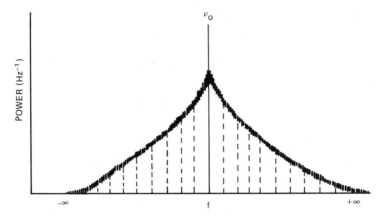

FIG. 4   A power density plot.

The spectral density is the distribution of total variance over frequency. The units of power spectral density are *power per hertz*; therefore, a plot of power spectral density obtained from amplitude (voltage) measurements requires that the voltage measurements be squared.

The spectral density of power versus frequency, shown in Fig. 4, is a *two-sided* spectral density because the range of Fourier frequencies f is from minus infinity to plus infinity.

The notation $S_g(f)$ represents the two-sided spectral density of fluctuations of any specified time-dependent quantity $g(t)$. Because the frequency band is defined by the two limit frequencies of minus infinity and plus infinity, the total mean-square fluctuation of that quantity is defined by

$$G_{\text{sideband}} = \int_{-\infty}^{+\infty} S_g(f) \, df. \tag{13}$$

Two-sided spectral densities are useful mainly in pure mathematical analysis involving Fourier transformations.

Similarly, for the one-sided spectral density,

$$G_{\text{sideband}} = \int_{0}^{+\infty} S_g(f) \, df. \tag{14}$$

The two-sided and one-sided spectral densities are related as follows:

$$\int_{-\infty}^{+\infty} S_{g_2} \, df = 2 \int_{0}^{+\infty} S_{g_2} \, df = \int_{0}^{+\infty} S_{g_1} \, df, \tag{15}$$

where g_1 indicates one-sided and g_2 two-sided spectral densities. It is noted that the one-sided density is twice as large as the corresponding two-sided spectral density. The terminology for single-sideband versus double-sideband signals is totally distinct from the one-sided spectral density versus two-sided spectral terminology. They are totally different concepts. The definitions and concepts of spectral density are set forth in NBS Technical Note 632 (Shoaf *et al.*, 1973).

C. Spectral Densities of Phase Fluctuations in the Frequency Domain

The spectral density $S_y(f)$ of the instantaneous fractional frequency fluctuations $y(t)$ is defined as a measure of frequency stability, as set forth by Barnes *et al.* (1970). $S_y(f)$ is the one-sided spectral density of *frequency fluctuations* on a "per hertz" basis, i.e., the dimensionality is Hz^{-1}. The range of Fourier frequency f is from *zero to infinity*. $S_{\delta v}(f)$, in hertz squared per hertz, is the spectral density of *frequency fluctuations* δv. It is calculated as

$$S_{\delta v}(f) = \frac{(\delta v_{rms})^2}{\text{bandwidth used in the measurement of } \delta v_{rms}}. \tag{16}$$

The range of the Fourier frequency f is from *zero to infinity*.

The spectral density of *phase fluctuations* is a normalized frequency domain measure of phase fluctuation sidebands. $S_{\delta\phi}(f)$, in radians squared per hertz, is the one-sided spectral density of the phase fluctuations on a "per hertz" basis:

$$S_{\delta\phi}(f) = \frac{\delta\phi_{rms}}{\text{bandwidth used in the measurement of } \delta\phi_{rms}}. \tag{17}$$

The power spectral densities of phase and frequency fluctuation are related by

$$S_{\delta\phi}(f) = (v_0^2/f)S_y(f). \tag{18}$$

The range of the Fourier frequency f is from *zero to infinity*.

$S_{\delta\Omega}(f)$, in radians squared Hertz squared per hertz is the spectral density of angular frequency fluctuations $\delta\Omega$:

$$S_{\delta\Omega}(f) = (2\pi)^2 S_{\delta v}(f). \tag{19}$$

The defined spectral densities have the following relationships:

$$S_{\delta v}(f) = v_0^2 S_y(f) = (1/2\pi)^2 S_{\delta\Omega}(f) = f^2 S_{\delta\phi}(f); \tag{20}$$

$$S_{\delta\phi}(f) = (1/\omega)^2 S_{\delta\Omega}(f) = (v_0/f)^2 S_y(f) = [S_{\delta v}(f)/f^2]. \tag{21}$$

Note that Eq. (20) is hertz squared per hertz, whereas Eq. (21) is in radians squared per hertz.

The term $S_{\sqrt{rfP}}(v)$, in watts per hertz, is the spectral density of the (square root of) radio frequency power P. The power of a signal is dispersed over the frequency spectrum owing to noise, instability, and modulation. This concept is similar to the concept of spectral density of voltage fluctuations $S_{\delta v}(f)$. Typically, $S_{\delta v}(f)$ is more convenient for characterizing a baseband signal where voltage, rather than power, is relevant. $S_{\sqrt{rfP}}(v)$ is typically more convenient for characterizing the dispersion of the signal power in the vicinity of the nominal carrier frequency v_0. To relate the two spectral densities, it is necessary to specify the impedance associated with the signal.

A definition of frequency stability that relates the actual sideband power of phase fluctuations with respect to the carrier power level, discussed by Glaze (1970), is called $\mathscr{L}(f)$. For a signal with PM and with no AM, $\mathscr{L}(f)$ is the normalized version of $S_{\sqrt{rfP}}(v)$, with its frequency parameter f referenced to the signal's average frequency v_0 as the origin such that f equals $v - v_0$. If the signal also has AM, $\mathscr{L}(f)$ is the normalized version of those portions of $S_{\sqrt{rfP}}(v)$ that are phase-modulation sidebands.

Because f is the Fourier frequency difference $(v - v_0)$, the range of f is from minus v_0 to plus infinity. Since $\mathscr{L}(f)$ is a normalized density (phase noise sideband power),

$$\int_{-v_0}^{+\infty} \mathscr{L}(f)\, df = 1. \tag{22}$$

$\mathscr{L}(f)$ is defined as the ratio of the power in one sideband, referred to the input carrier frequency on a per hertz of bandwidth spectral density basis, to the total signal power, at Fourier frequency difference f from the carrier, *per one device*. It is a normalized frequency domain measure of phase fluctuation sidebands, expressed in decibels relative to the carrier per hertz:

$$\mathscr{L}(f) = \frac{\text{power density (one phase modulation sideband)}}{\text{carrier power}}. \tag{23}$$

For the types of signals under consideration, by definition the two phase-noise sidebands (lower sideband and upper sideband, at $-f$ and f from v_0, respectively) of a signal are approximately coherent with each other, and they are of approximately equal intensity.

It was previously show that the measurement of phase fluctuations (phase noise) required driving a double-balanced mixer with two signals in phase quadrature so the FM–AM conversion resulted in voltage fluctuations at the mixer output that were analogous to the phase fluctuations. The operation of the mixer when it is driven at quadrature is such that the amplitudes of the two phase sidebands are added linearly in the output of the mixer, resulting in four times as much power in the output as would be present if only one of the phase sidebands were allowed to contribute to the output of the mixer. Hence, for $|f| < v_0$, and considering only the phase modulation portion of the spectral density of the (square root of) power, we obtain

$$S_{\delta V}(|f|)/(V_{rms})^2 \cong 4[S_{\sqrt{rfP}}(v_0 + f)]/(P_{tot}) \tag{24}$$

and, using the definition of $\mathscr{L}(f)$,

$$\mathscr{L}(f) \equiv [S_{\sqrt{rfP}}(v_0 + f)]/(P_{tot}) \cong \tfrac{1}{2}S_{\delta\phi}(|f|). \tag{25}$$

Therefore, for the condition that the phase fluctuations occurring at rates (f) and faster are small compared to one radian, a good approximation in radians squared per hertz for one unit is

$$\mathscr{L}(f) = \tfrac{1}{2}S_{\delta\phi}(f). \tag{26}$$

If the small angle condition is not met, Bessel-function algebra must be used to relate $\mathscr{L}(f)$ to $S_{\delta\phi}(f)$.

The NBS-defined spectral density is usually expressed in decibels relative to the carrier per hertz and is calculated for one unit as

$$\mathscr{L}(f) = 10 \log[\tfrac{1}{2}S_{\delta\phi}(f)]. \tag{27}$$

It is very important to note that the theory, definitions, and equations previously set forth relate to a single device.

D. MODULATION THEORY AND SPECTRAL DENSITY RELATIONSHIPS

Applying a sinusoidal frequency modulation f_m to a sinusoidal carrier frequency v_0 produces a wave that is sinusoidally advanced and retarded in phase as a function of times. The instantaneous voltage is expressed as,

$$V(t) = V_0 \sin(2\pi v_0 t + \Delta\phi \sin 2\pi f_m t), \tag{28}$$

where $\Delta\phi$ is the *peak phase deviation* caused by the modulation signal.

The first term inside the parentheses represents the linearly progressing phase of the carrier. The second term is the phase variation (advancing and retarded) from the linearly progressing wave. The effects of modulation can be expressed as *residual f_m noise* or as *single-sideband phase noise*. For modulation by a single sinusoidal signal, the peak-frequency deviation of the carrier (v_0) is

$$\Delta v_0 = \Delta\phi \cdot f_m, \tag{29}$$

$$\Delta\phi = \Delta v_0/f_m, \tag{30}$$

where f_m is the modulation frequency. This ratio of peak frequency deviation to modulation frequency is called *modulation index m* so that $\Delta\phi = m$ and

$$m = \Delta v_0/f_m. \tag{31}$$

The frequency spectrum of the modulated carrier contains frequency components (sidebands) other than the carrier. For small values of modulation index ($m \ll 1$), as is the case with random phase noise, only the carrier and first upper and lower sidebands are significantly high in energy. The ratio of the amplitude of either single sideband to the amplitude of the carrier is

$$V_{sb}/V_0 = m/2. \tag{32}$$

This ratio is expressed in decibels below the carrier and is referred to as dBc for the given bandwidth B:

$$V_{sb}/V_0 = 20 \log(m/2) = 20 \log(\Delta v_0/2f_m)$$
$$= 10 \log(m/2)^2 = 10 \log(\Delta v_0/2f_m)^2. \tag{33}$$

If the frequency deviation is given in terms of its rms value, then

$$\Delta v_{rms} = \Delta v_0/\sqrt{2}. \tag{34}$$

Equation (33) now becomes

$$V_{sb}/V_0 = \mathcal{L}(f) = 20 \log(\Delta v_{rms}/\sqrt{2f_m})$$
$$= 10 \log(\Delta v_{rms}/\sqrt{2f_m})^2. \tag{35}$$

The ratio of single sideband to carrier power in decibels (carrier) per hertz is

$$\mathcal{L}(f) = 20 \log(\Delta v_{rms}/f_m) - 3 \tag{36}$$

and, in decibels relative to one squared radian per hertz,

$$S_{\delta\phi}(f) = 20 \log(\Delta v_{rms}/f_m). \tag{37}$$

The interrelationships of modulation index, peak frequency deviation, rms frequency, and spectral density of phase fluctuations can be found from the following:

$$\tfrac{1}{2}m = \Delta v_0/2f_m = \Delta v_{rms}/\sqrt{2f_m}, \tag{38}$$
$$= 10 \exp(\mathcal{L}(f)/10) = \tfrac{1}{2}S_{\delta\phi}(f); \tag{39}$$

or

$$\tfrac{1}{2}m = \Delta v_{rms}/\sqrt{2f_m} = \sqrt{10 \exp(\mathcal{L}(f)/10)} = \sqrt{\tfrac{1}{2}S_{\delta\phi}(f)}, \tag{40}$$

and

$$m = \Delta v_0/f_m = 2\Delta v_{rms}/\sqrt{2f_m}$$
$$= 2\sqrt{10 \exp(\mathcal{L}(f)/10)} = 2\sqrt{\tfrac{1}{2}S_{\delta\phi}(f)}. \tag{41}$$

The basic relationships are plotted in Fig. 5.

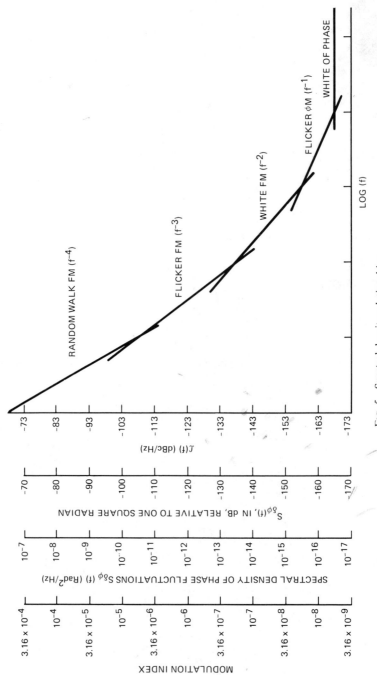

FIG. 5 Spectral density relationships.

E. NOISE PROCESSES

The spectral density plot of a typical oscillator's output is usually a combination of different noise processes. It is very useful and meaningful to categorize these processes because the first job in evaluating a spectral density plot is to determine which type of noise exists for the particular range of Fourier frequencies.

The two basic categories are the *discrete-frequency noise* and the *power-law noise* process. Discrete-frequency noise is a type of noise in which there is a dominant observable probability, i.e., deterministic in that they can usually be related to the mean frequency, power-line frequency, vibration frequencies, or ac magnetic fields, or to Fourier components of the nominal frequency. Discrete-frequency noise is illustrated in the frequency domain plot of Fig. 6. These frequencies can have their own spectral density plots, which can be defined as noise on noise.

Power-law noise processes are types of noise that produce a certain slope on the one-sided spectral density plot. They are characterized by their dependence on frequency. The spectral density plot of a typical oscillator output is usually a combination of the various power-law processes.

In general, we can classify the power-law noise processes into five categories. These five processes are illustrated in Fig. 5, which can be referred to with respect to the following description of each process.

(1) *Random walk FM* (random walk of frequency). The plot goes down as $1/f^4$. This noise is usually very close to the carrier and is difficult to measure. It is usually related to the oscillator's physical environment (mechanical shock, vibration, temperature, or other environmental effects).

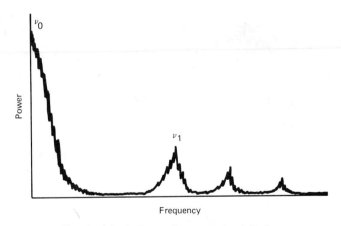

FIG. 6   A basic discrete-frequency signal display.

fig. 7.

(2)   *Flicker FM* (flicker of frequency). The plot goes down as $1/f^3$. This noise is typically related to the physical resonance mechanism of the active oscillator or the design or choice of parts used for the electronic or power supply, or even environmental properties. The time domain frequency stability over extended periods is constant. In high-quality oscillators, this noise may be marked by white FM ($1/f^2$) or flicker phase modulation ϕM ($1/f$). It may be masked by drift in low-quality oscillators.

(3)   *White FM* (white frequency, random walk of phase). The plot goes down as $1/f^2$. A common type of noise found in passive-resonator frequency standards. Cesium and rubidium frequency standards have white FM noise characteristic because the oscillator (usually quartz) is locked to the resonance feature of these devices. This noise gets better as a function of time until it (usually) becomes flicker FM ($1/f^3$) noise.

(4)   *Flicker ϕM* (flicker modulation of phase). The plot goes down as $1/f$. This noise may relate to the physical resonance mechanism in an oscillator. It is common in the highest-quality oscillators. This noise can be introduced by noisy electronics—amplifiers necessary to bring the signal amplitude up to a usable level—and frequency multipliers. This noise can be reduced by careful design and by hand-selecting all components.

(5)   *White ϕM* (white phase). White phase noise plot is flat f^0. Broadband phase noise is generally produced in the same way as flicker ϕM ($1/f$). Late stages of amplification are usually responsible. This noise can be kept low by careful selection of components and by narrow-band filtering at the output.

The power-law processes are illustrated in Fig. 5.

F.   Integrated Phase Noise

The integrated phase noise is a measure of the phase-noise contribution (rms radians, rms degrees) over a designated range of Fourier frequencies. The integration is a process of summation that must be performed on the measured spectral density within the actual IF bandwidth (B) used in the measurement of $\delta\phi_{rms}$. Therefore, the spectral density $S_{\delta\phi}(f)$ must be unnormalized to the particular bandwidth used in the measurement. Define $S_u(f)$, in radians squared, as the unnormalized spectral density:

$$S_u(f) = 2[10 \exp(\mathscr{L}(f) + 10 \log B)/10].\qquad(42)$$

Then, the integrated phase noise over the band of Fourier frequencies (f_1 to f_n) *where measurements are performed using a constant IF bandwidth*, in radians squared, is

$$S_B(f_1 \text{ to } f_n) = \int_{f_1}^{f_n} S_u(f)\, df,\qquad(43)$$

or, in rms radians,

$$S_B(f_1 \text{ to } f_n) = \sqrt{\sum S_u(f_1) + S_u(f_2) = \cdots = S_u(f_n)}, \tag{44}$$

and the integrated phase noise in rms degrees is calculated as

$$S_B(360/2\pi). \tag{45}$$

The integrated phase noise in decibels relative to the carrier is calculated as

$$S_B = 10 \log(\tfrac{1}{2} S_B^2). \tag{46}$$

The previous calculations correspond to the illustration in Fig. 7, which includes two bandwidths (B1 and B2) over two ranges of Fourier frequencies.

In the measurement program, different IF bandwidths are used as set forth by Lance *et al.* (1977). The total integrated phase noise over the different ranges of Fourier frequencies, which are measured at constant bandwidths as illustrated, is calculated in rms radians as follows:

$$S_{Btot} = \sqrt{(S_{B1})^2 + (S_{B2})^2 + \cdots + (S_{Bn})^2}, \tag{47}$$

where it is recalled that the summation is performed in terms of radians squared.

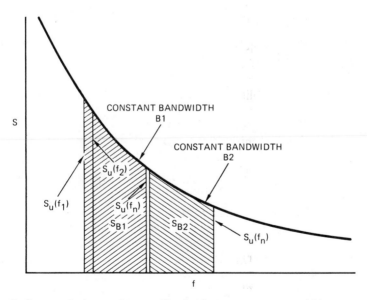

FIG. 7   Integrated phase noise over Fourier frequency ranges at which measurements were performed using constant bandwidth.

G. AM Noise in the Frequency Domain

The spectral density of AM fluctuations of a signal follows the same general derivation previously given for the spectral density of phase fluctuations. Amplitude fluctuations $\delta\varepsilon$ of the signal under test produces voltage fluctuations δA at the output of the mixer. Interpretating the mean-square fluctuations $\delta\varepsilon$ and δ in spectal density fashion, we obtain $S_{\delta\varepsilon}(f)$, the spectral density of amplitude fluctuations $\delta\varepsilon$ of a signal in volts squared per hertz:

$$S_{\delta\varepsilon}(f) = (\tfrac{1}{2}V_0)^2[S_{\delta A}(f)/(A_{\text{rms}})^2]. \tag{48}$$

The term $m(f)$ is the normalized version of the amplitude modulation (AM) portion of $S_{\sqrt{\text{rf}P}}(v)$, with its frequency parameter f referenced to the signal's average frequency v_0, taken as the origin such that the difference frequency f equals $v - v_0$. The range of Fourier frequency difference f is from minus v_0 to plus infinity.

The term $m(f)$ is defined as the ratio of the spectral density of *one amplitude-modulated sideband* to the *total signal power*, at Fourier frequency difference f from the signal's average frequency v_0, for a single specified signal or device. The dimensionality is per hertz. $\mathscr{L}(f)$ and $m(f)$ are similar functions; the former is a measure of phase-modulated (PM) sidebands, the later is a corresponding measure of amplitude-modulated (AM) sidebands. We introduce the symbol $m(f)$ to have useful terminology for the important concept of normalized AM sideband power.

For the types of signals under consideration, by definition the two amplitude-fluctuation sidebands (lower sideband and upper sideband, at $-f$ f from v_0, respectively) of a signal are coherent with each other. Also, they are of equal intensity. The operation of the mixer when it is driven at colinear phase is such that the amplitudes of the two AM sidebands are added linearly in the output of the mixer, resulting in four times as much power in the output as would be present if only one of the AM sidebands were allowed to contribute to the output of the mixer. Hence, for $|f| < v_0$,

$$S_A(|f|)/(A_{\text{rms}})^2 = 4[S_{\sqrt{\text{rf}P}}(v_0 + f)]/P_{\text{tot}}, \tag{49}$$

and, using the definition

$$m(f) \equiv [S_{\sqrt{\text{rf}P}}(v_0 + f)]/P_{\text{tot}}, \tag{50}$$

we find, in decibels (carrier) per hertz,

$$m(f) = (1/2V_0^2)S_{\delta\varepsilon}(|f|). \tag{51}$$

III. Phase-Noise Measurements Using the Two-Oscillator Technique

A functional block diagram of the two-oscillator system for measuring phase noise is shown in Fig. 8. NBS has performed phase noise measurements since 1967 using this basic system. The signal level and sideband levels can be measured in terms of voltage or power. The low-pass filter prevents local oscillator leakage power from overloading the spectrum analyzer when baseband measurements are performed at the Fourier (offset) frequencies of interest. Leakage signals will interfere with autoranging and with the dynamic range of the spectrum analyzer.

The low-noise, high-gain preamplifier provides additional system sensitivity by amplyfying the noise signals to be measured. Also, because spectrum analyzers usually have high values of noise figure, this amplifier is very desirable. As an example, if the high-gain preamplifier had a noise figure of 3 dB and the spectrum analyzer had a noise figure of 18 dB, the system sensitivity at this point has been improved by 15 dB. The overall system sensitivity would not necessarily be improved 15 dB in all cases, because the limiting sensitivity could have been imposed by a noisy mixer.

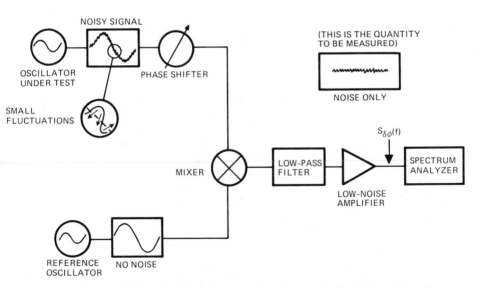

FIG. 8 The two-oscillator technique for measuring phase noise. Small fluctuations from nominal voltage are equivalent to phase variations. The phase shifter adjusts the two signals to quadrature in the mixer, which cancels carriers and converts phase noise to fluctuating dc voltage.

Assume that the reference oscillator is perfect (no phase noise), and that it can be adjusted in frequency. Also, assume that both oscillators are extremely stable, so that phase quadrature can be maintained without the use of an external phase-locked loop or reference. The double-balanced mixer acts as a phase detector so that when two input signals are identical in frequency and are in phase quadrature, the output is a small fluctuating voltage. This represents the phase-modulated (PM) sideband component of the signal because, due to the quadrature of the signals at the mixer input, the mixer converts the amplitude-modulated (AM) sideband components to FM, and at the same time it converts the PM sideband components to AM. These AM components can be detected with an amplitude detector, as shown in Fig. 3.

If the two oscillator signals applied to the double-balanced mixer of Fig. 8 are slightly out of zero beat, a slow sinusoidal voltage with a peak-to-peak voltage V_{ptp} can be measured at the mixer output. If these same signals are returned to zero beat and adjusted for phase quadrature, the output of the mixer is a small fluctuating voltage (δv) centered at zero volts. If the fluctuating voltage is small compared to $\frac{1}{2} V_{ptp}$, the phase quadrature condition is being closely maintained and the "small angle" condition is being met. Phase fluctuations in radians between the test and reference signals (phases) are

$$\delta\phi = \delta(\phi_t - \phi_r). \tag{52}$$

These phase fluctuations produce voltage fluctuations at the output of the mixer,

$$\delta v = \frac{1}{2} V_{ptp} \delta\phi, \tag{53}$$

where phase angles are in radian measure and $\sin \delta\phi = \delta\phi$ for small $\delta\phi$ ($\delta\phi \ll 1$ rad). Solving for $\delta\phi$, squaring both sides, and taking a time average gives

$$\langle(\delta\phi)^2\rangle = 4\langle(\delta v)^2\rangle/(V_{ptp})^2, \tag{54}$$

where the angle brackets represent the time average.

For the sinusoidal beat signal,

$$(V_{ptp})^2 = 8(V_{rms})^2. \tag{55}$$

The mean-square fluctuations of phase $\delta\phi$ and voltage δv interpreted in a spectral density fashion gives the following in radians squared per hertz:

$$S_{\delta\phi}(f) = S_{\delta v}(f)/2(V_{rms})^2. \tag{56}$$

Here, $S_{\delta v}(f)$, in volts squared per hertz, is the spectral density of the voltage fluctuations at the mixer output. Because the spectrum analyzer measures rms voltage, the noise voltage is in units of volts per square root hertz, which means volts per square root bandwidth. Therefore,

$$S_{\delta v}(f) = [\delta v_{rms}/\sqrt{B}] = (\delta v_{rms})^2/B, \tag{57}$$

where B is the noise power bandwidth used in the measurement.

Because it was assumed that the reference oscillator did not contribute any noise, the voltage fluctuations v_{rms} represent the oscillator under test, and the spectral density of the phase fluctuations in terms of the voltage measurements performed with the spectrum analyzer, in radians squared per hertz, is

$$S_{\delta\phi}(f) = \tfrac{1}{2}[(\delta v_{rms})^2/B(V_{rms})^2]. \tag{58}$$

Equation (46) is sometimes expressed as

$$S_{\delta\phi}(f) = S_{\delta v}(f)/K^2, \tag{59}$$

where K is the calibration factor in volts per radian. For sinusoidal beat signals, the peak voltage of the signal equals the slope of the zero crossing in volts per radian. Therefore, $(V_p)^2 = 2(V_{rms})^2$, which is the same as the denominator in Eq. (56).

The term $S_{\delta\phi}(f)$ can be expressed in decibels relative to one square radian per hertz by calculating $10 \log S_{\delta\phi}(f)$ of the previous equation:

$$S_{\delta\phi}(f) = 20 \log(\delta v_{rms}) - 20 \log(V_{rms}) - 10 \log(B) - 3. \tag{60}$$

A correction of 2.5 is required for the tracking spectrum analyzer used in these measurement systems. $\mathscr{L}(f)$ differs by 3 dB and is expressed in decibels (carrier) per hertz as

$$\mathscr{L}(f) = 20 \log(\delta v_{rms}) - 20 \log(V_{rms}) - 10 \log(B) - 6. \tag{61}$$

A. Two Noisy Oscillators

The measurement system of Fig. 6 yields the output noise from both oscillators. If the reference oscillator is superior in performance as assumed in the previous discussions, then one obtains a direct measure of the noise characteristics of the oscillator under test.

If the reference and test oscillators are the same type, a useful approximation is to assume that the measured noise power is twice that associated with one noisy oscillator. This approximation is in error by no more than 3 dB for the noisier oscillator, even if one oscillator is the major source of noise. The equation for the spectral density of measured phase fluctuations in radians squared per hertz is

$$S_{\delta\phi}(f)\Big|_{\#1} + S_{\delta\phi}(f)\Big|_{\#2} = \Big|S_{\delta v}(f)\Big|_{\text{(two devices)}} \div 2(V_{rms})^2 \doteq \Big|2S_{\delta\phi}(f)\Big|_{\text{(one device)}}. \tag{62}$$

The measured value is therefore divided by two to obtain the value for the single oscillator. A determination of the noise of each oscillator can be made if one has three oscillators that can be measured in all pair combinations. The phase noise of each source 1, 2, and 3 is calculated as follows:

$$\mathscr{L}_1(f) = 10 \log[\tfrac{1}{2}(10^{\mathscr{L}_{12}(f)/10} + 10^{\mathscr{L}_{13}(f)/10} - 10^{\mathscr{L}_{23}(f)/10})], \quad (63)$$

$$\mathscr{L}^2(f) = 10 \log[\tfrac{1}{2}(10^{\mathscr{L}_{12}(f)/10} + 10^{\mathscr{L}_{23}(f)/10} - 10^{\mathscr{L}_{13}(f)/10})], \quad (64)$$

$$\mathscr{L}^3(f) = 10 \log[\tfrac{1}{2}(10^{\mathscr{L}_{13}(f)/10} + 10^{\mathscr{L}_{23}(f)/10} - 10^{\mathscr{L}_{12}(f)/10})]. \quad (65)$$

B. AUTOMATED PHASE-NOISE MEASUREMENTS USING THE TWO-OSCILLATOR TECHNIQUE

The automated phase-noise measurement system is shown in Fig. 9. It is controlled by a programmable calculator. Each step of the calibration and measurement sequence is included in the program. The software program controls frequency slection, bandwidth settings, settling time, amplitude ranging, measurements, calculations, graphics, and data plotting. Normally, the system is used to obtain a direct plot of $\mathscr{L}(f)$. The integrated phase noise can be calculated for any selected range of Fourier frequencies.

A quasi-continuous plot of phase noise performance $\mathscr{L}(f)$ is obtained by performing measurements at Fourier frequencies separated by the IF bandwidth of the spectrum analyzer used during the measurement. Plots of other defined parameters can be obtained and plotted as desired.

The IF bandwidth settings for the Fourier (offset) frequency-range selections are shown in the following tabulation:

IF bandwidth (Hz)	Fourier frequency (kHz)	IF bandwidth (kHz)	Fourier frequency (kHz)
3	0.001–0.4	1	40–100
10	0.4–1	3	100–400
30	1–4	10	400–1300
100	4–10		
200	10–40		

The particular range of Fourier frequencies is limited by the particular spectrum analyzer used in the system. A fast Fourier analyzer (FFT) is also incorporated in the system to measure phase noise from submillihertz to 25 kHz.

High-quality sources can be measured without multiplication to enhance the phase noise prior to downconverting and measuring at baseband frequencies. The measurements are not completely automated because the calibration sequence requires several manual operations.

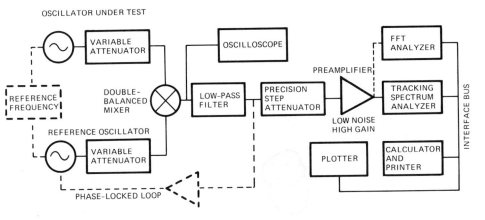

FIG. 9   An automated phase-noise measurement system.

C. CALIBRATION AND MEASUREMENTS USING THE TWO-OSCILLATOR SYSTEM

$\mathscr{L}(f)$ is a normalized frequency domain measure of phase-fluctuation side-band power. The noise power is measured relative to the carrier power level. Correction must be applied because of the type of measurement and the characteristics of the measurement equipment. The general procedure for the calibration and measurement sequence includes the following: measuring the noise power bandwidth for each IF bandwidth setting on the Tracking Spectrum Analyzer (Section III.C.1); establishing a carrier reference power level referenced to the output of the mixer (Section III.C.2); obtaining phase quadrature of the two signals applied to the mixer (Section III.C.3); measuring the noise power at the selected Fourier frequencies (Section III.C.4); performing the calculations and plotting the data (Section III.C.5); and measuring the system noise floor characteristics, usually referred to as the system sensitivity.

1. Noise-Power Bandwidth

Approximations of analyzer-noise bandwidths are not adequate for phase noise measurements and calculations. The IF noise-power bandwidth of the tracking spectrum analyzer must be *known* and used in the calculations of phase noise parameters. Figure 10 shows the results of measurements performed using automated techniques. For example, with a 1-MHz signal input to the tracking spectrum analyzer, the desired incremental frequency changes covering the IF bandwidth are set by calculator control.

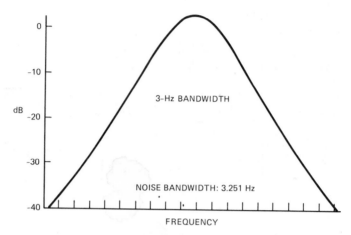

FIG. 10   Plot of automated noise-power bandwidth.

The spectrum analyzer power output is recorded for each frequency setting over the range, as illustrated in Fig. 10. The 40-dB level and the 100 increments in frequency are not the minimum permissible values. The recorded can be plotted for each IF bandwidth, as illustrated, and the noise-power bandwidth is calculated in hertz as

$$\text{noise power bandwidth} = \frac{(P_1 + P_2 + P_3 + \cdots + P^{100})\,\Delta f}{\text{peak power reading}}, \quad (66)$$

where Δf is the frequency increment in hertz and the peak power is the maximum measured point obtained during the measurements. All power values are in watts.

2.   Setting the Carrier Power Reference Level

Recall from Section III.6 that for sinusoidal signals the peak voltage of the signal equals the slope of the zero crossing, in volts per radian. A frequency offset is established, and the peak power of the difference frequency is measured as the carrier-power reference level; this establishes the calibration factor of the mixer in volts per radian.

Because the precision IF attenuator is used in the calibration process, one must be aware that the impedance looking back into the mixer should be 50 Ω. Also, the mixer output signal should be sinusoidal. Fischer (1978) discussed the mixer as the "critical element" in the measurement system. It is advisable to drive the mixer so that the sinusoidal signal is obtained at the mixer output. In most of the TRW systems, the mixer drive levels are 10 dBm for the reference signal and about zero dBm for the unit under test.

System sensitivity can be increased by driving the mixer with high-level signals that lower the mixer output impedance to a few ohms. This presents a problem in establishing the calibration factor of the mixer, because it might be necessary to calibrate the mixer for different Fourier frequency ranges.

The equation sensitivity = slope = beat-note amplitude does not hold if the output of the mixer is not a sine wave. The Hewlett-Packard 3047 automated phase noise measurement system allows accurate calibration of the phase-detector sensitivity even with high-level inputs by using the derivative of the Fourier representation of the signal (the fundamental and its harmonics). The slope at $\phi \cong 0$ radians is given by

$$A \sin \phi - B \sin 3\phi + C \sin 5\phi = A \cos \phi - 3B \cos 3\phi + 5C \cos 5\phi$$

$$= A - 3B + 5C + \cdots. \qquad (67)$$

Referring to Fig. 9, the carrier-power reference level is obtained as follows.

(1)   The precision IF step attenuator is set to a high value to prevent overloading the spectrum analyzer (assume 50 dB as our example).

(2)   The reference and test signals at the mixer inputs are set to approximately 10 dBm and 0 dBm, as previously discussed.

(3)   If the frequency of one of the oscillators can be adjusted, adjust its frequency for an IF output frequency in the range of 10 to 20 kHz. If neither oscillator is adjustable, replace the oscillator under test with one that can be adjusted as required and that can be set to the identical power level of the oscillator under test.

(4)   The resulting IF power level is measured by the spectrum analyzer, and the measured value is corrected for the attentuator setting, which was assumed to be 50 dB. The correction is necessary because this attenuator will be set to its zero decibel indication during the measurements of noise power. Assuming a spectrum analyzer reading of −40 dBm, the carrier-power reference level is calculated as

$$\text{carrier power reference level} = 50 \text{ dB} - 40 \text{ dBm} = 10 \text{ dBm}. \qquad (68)$$

3.   *Phase Quadrature of the Mixer Input Signals*

After the carrier-power reference has been established, the oscillator under test and the reference oscillator are tuned to the same frequency, and the original reference levels that were used during calibration are reestablished. The quadrature adjustment depends on the type of system used. Three possibilities, illustrated in Fig. 9, are described here.

(1)   If the oscillators are very stable, have high-resolution tuning, and are not phase-locked, the frequency of one oscillator is adjusted for *zero dc*

voltage output of the mixer as indicated by the sensitive oscilloscope. Note: Experience has shown that the quadrature setting is not critical if the sources have low AM noise characteristics. As an example, experiments performed using two HP 3335 synthesizers showed that degradation of the phase-noise measurement became noticeable with a phase-quadrature offset of 16 degrees.

(2) If the common reference frequency is used, as illustrated in Fig. 9, then it is necessary to include a phase shifter in the line between one of the oscillators and the mixer (preferably between the attenuator and mixer). The phase shifter is adjusted to obtain and maintain zero volts dc at the mixer output. A correction for a nonzero dc value can be applied as exemplified by the HP 3047 automated phase-noise measurement system.

(3) If one oscillator is phase-locked using a phase-locked loop, as shown dotted in on Fig. 9, the frequency of the unit under test is adjusted for zero dc output of the mixer as indicated on the oscilloscope.

A phase-locked loop is a feedback system whose function is to force a voltage-controlled oscillator (VCO) to be coherent with a certain frequency, i.e., it is highly correlated in both frequency and phase. The phase detector is a mixer circuit that mixes the input signal with the VCO signal. The mixer output is $v_i \pm v_0$, when the loop is locked, the VCO duplicates the input frequency so that the difference frequency is zero, and the output is a dc voltage proportional to the phase difference. The low-pass filter removes the sum frequency component but passes the dc component to control the VCO. The time constant of the loop can be adjusted as needed by varying amplifier gain and RC filtering within the loop.

A *loose phase-locked loop* is characterized by the following.

(1) The correction voltage varies as phase (in the short term) and phase variations are therefore observed directly.

(2) The bandwidth of the servo response is small compared with the Fourier frequency to be measured.

(3) The response time is very slow.

A *tight phase-locked loop* is characterized by the following.

(1) The correction voltage of the servo loop varies as frequency.

(2) The bandwidth of the servo response is relatively large.

(3) The response time is much smaller than the smallest time interval τ at which measurements are performed.

Figure 11 shows the phase-noise characteristics of the H.P. 8640B synthesizer measured at 512 MHz. The phase-locked-loop attenuation characteristics extend to 10 kHz. The internal-oscillator-source characteristics are plotted at Fourier frequencies beyond the loop-bandwidth cutoff at 10 kHz.

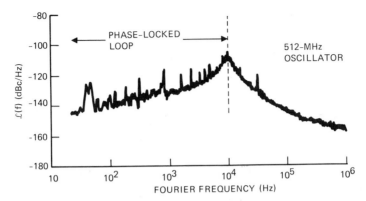

FIG. 11   Phase–locked-loop characteristics of the H.P. 8640B signal generator, showing the normalized phase-noise sideband power spectral density.

4. Measurements, Calculations, and Data Plots

The measurement sequence is automated except for the case where manual adjustments are required to maintain phase quadrature of the signals. After phase quadrature of the signals into the mixer is established, the IF atten- uator is returned to the zero-decibel reference setting. This attenuator is set to a high value [assumed to be 50 dB in Eq. (65)] to prevent saturation of the spectrum analyzer during the calibration process.

The automated measurements are executed, and the direct measurement and data plot of $\mathscr{L}(f)$ is obtained in decibels (carrier) per hertz using the equation

$$\mathscr{L}(f) = - [\text{carrier power level} - (\text{noise power level} - 6 + 2.5$$
$$- 10 \log B - 3)]. \tag{69}$$

The noise power (dBm) is measured relative to the carrier-power level (dBm), and the remaining terms of the equation represent corrections that must be applied because of the type of measurement and the characteristics of the measurement equipment, as follows.

(1)   The measurement of noise sidebands with the signals in phase quadrature requires the −6-dB correction that is noted in Eq. (69).

(2)   The nonlinearity of the spectrum analyzer's logarithmic IF amplifier results in compression of the noise peaks which, when average-detected, require the 2.5-dB correction.

(3)   The bandwidth correction is required because the spectrum analyzer measurements of random or white noise are a function of the particular bandwidth used in the measurement.

(4)   The −3-dB correction is required because this is a direct measure of $\mathscr{L}(f)$ of *two oscillators*, assuming that the oscillators are of a similar type and that the noise contribution is the same for each oscillator. If one oscillator is sufficiently superior to the other, this correction is not required.

Other defined spectral densities can be calculated and plotted as desired. The plotted or stored value of the spectral density of phase fluctuations in decibels relative to one square radian (dBc rad²/Hz) is calculated as

$$S_{\delta\phi}(f) = \mathscr{L}(f) + 3. \tag{70}$$

The spectral density of phase fluctuations, in radians squared per hertz, is calculated as

$$S_{\delta\phi}(f) = 10 \exp(S_{\delta\phi}(f)/10), \tag{71}$$

The spectral density of frequency fluctuations, in hertz squared per hertz, is

$$S_{\delta v}(f) = f^2 S_{\delta\phi}(f). \tag{72}$$

where $S_{\delta\phi}(F)$ is in decibels with respect to 1 radian.

5.   *System Noise Floor Verification*

A plot of the system noise floor (sensitivity) is obtained by repeating the automated measurement procedures with the system modified as shown in Fig. 12. Accurate measurements can be obtained using the configuration shown in Fig. 12a. The reference source supplies 10 dBm to one side of the mixer and 0 dBm to the other mixer input through equal path lengths; phase quadrature is maintained with the phase shifter.

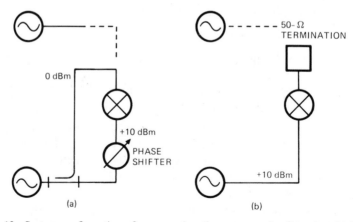

FIG. 12   System configurations for measuring the system noise floor (sensitivity): (a) configuration used for accurate measurements; (b) alternate configuration sometimes used.

The configuration shown in Fig. 12b is sometimes used and does not greatly degrade the noise floor because the reference signal of 10 dBm is larger than the signal frequency. See Sections IV.B and IV.C.4 for additional discussions related to system sensitivity and recommended system evaluation.

Proper selection of drive and output termination of the double-balanced mixer can result in improvement by 15 to 25 dB in the performance of phase-noise measurements, as discussed by Walls et al. (1976). The beat frequency between the two oscillators can be a sine wave, as previously mentioned, with proper low drive levels. This requires a proper terminating impedance for the mixer. With high drive levels, the mixer output waveform will be clipped. The slope of the clipped waveform at the zero crossings, illustrated by Walls et al. (1976), is twice the slope of the sine wave and therefore improves the noise floor sensitivity by 6 dB, i.e., the output signal, proportional to the phase fluctuations, increases with drive level. This condition of clipping requires characterization over the Fourier frequency range, as previously mentioned for the Hewlett-Packard 3047 phase noise measurement system. An amplifier can be used to increase the mixer drive levels for devices that have insufficient output power to drive the double-balanced mixers.

Lower noise floors can be achieved using high-level mixers when available drive levels are sufficient. A step-up transformer can be used to increase the mixer drive voltage because the signal and noise power increase in the same ratio, and the spectral density of phase of the device under test is unchanged, but the noise floor of the measurement system is reduced.

Walls et al. (1976) used a correlation technique that consisted primarily of two phase-noise measurement systems. At TRW the technique is used as shown in Fig. 13. The cross spectrum is obtained with the fast Fourier transform (FFT) analyzer that performs the product of the Fourier transform of one signal and the complex conjugate of the Fourier transform of

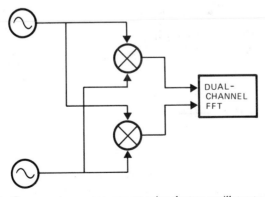

Fig. 13   Cross-spectrum measurement using the two-oscillator technique.

the second signal. This cross spectrum, which is a phase-sensitive character-istic, gives a phase and amplitude sensitivity measure directly. A signal-to-noise enhancement greater than 20 dB can be achieved.

If the double-balanced phase noise measurement system does not provide a noise floor sufficient for measuring a high-quality source, frequency multiplier chains can be used if their inherent noise is 10–20 dB below the measurement system noise. In frequency multiplication the noise increases according to

$$10 \log(\text{final frequency/original frequency}). \qquad (73)$$

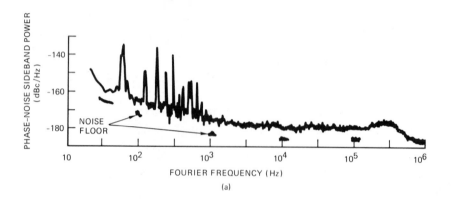

(a)

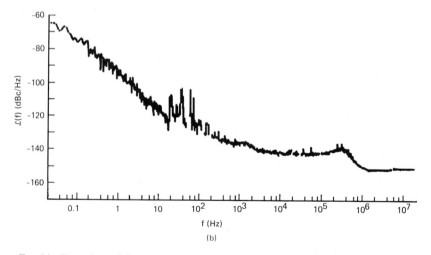

(b)

FIG. 14   Data plots of the automated phase-noise measurement system: (a) a high quality 5-MHz quartz oscillator; (b) combined noise of two H.P. 8662A synthesizers (subtract 3 dB for a single unit).

The following equation is used to correct for noise-floor contribution P_{nf}, in dBc/Hz, if desired or necessary:

$$\mathscr{L}(f)(\text{corrected}) = -\mathscr{L}(f) + 10 \log\left[\frac{P_{\mathscr{L}(f)} - P_{nf}}{P_{\mathscr{L}(f)}}\right]. \tag{74}$$

The correction for noise-floor contribution can also be obtained by using the measurement of $S_{\delta v}(f)$ of Eq. (57). Measurement of $S_{\delta v}(f)$ of the oscillator plus floor is obtained, then $S_{\delta v}(f)$ is obtained for the noise floor only. Then,

$$\left. S_{\delta v}(f) \right|_{cor} = \left. S_{\delta v}(f) \right|_{(osc + nf)} - \left. S_{\delta v}(f) \right|_{nf}. \tag{75}$$

Figure 14a shows a phase noise plot of a very high-quality (5-MHz) quartz oscillator, measured by the two-oscillator technique. The sharp peaks below 1000 Hz represent the 60-Hz line frequency of the power supply and its harmonics and are not part of the oscillator phase noise. Figure 14b shows measurements to 0.02 Hz of the carrier at a frequency of 20 MHz.

IV. Single-Oscillator Phase-Noise Measurement Systems and Techniques

The phase-noise measurements of a single-oscillator are based on the measurement of *frequency fluctuations* using discriminator techniques. The practical discriminator acts as a filter with finite bandwidth that suppresses the carrier and the sidebands on both sides of the carrier. The ideal carrier-suppression filter would provide infinite attentuation of the carrier and zero attenutation of all other frequencies. The effective Q of the practical discriminator determines how much the signals are attenuated.

Frequency discrimination at very high frequencies (VHF) has been obtained using slope detectors and ratio detectors, by use of lumped circuit elements of inductance and capacitance. At ultrahigh frequencies (UHF) between the VHF and microwave regions, measurements can be performed by beating, or heterodyning, the UHF signal with a local oscillator to obtain a VHF signal that is analyzed with a discriminator in the VHF frequency range. Those techniques provide a means for rejecting residual amplitude-modulated (AM) noise on the signal under test. The VHF discriminators usually employ a limiter or ratio detector.

Ashley *et al.* (1968) and Ondria (1968) have discussed the microwave cavity discriminator that rejects AM noise, suppresses the carrier so that the input level can be increased, and provides a high discriminated output to improve the signal-to-noise floor ratio. The delay line used as an FM discriminator has been discussed by Tykulsky (1966), Halford (1975), and

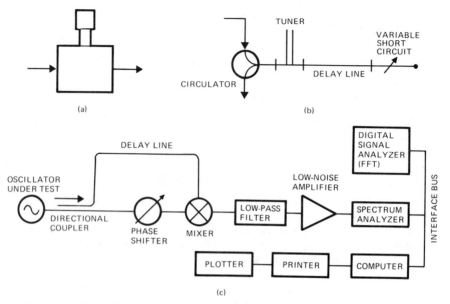

FIG. 15 Single-oscillator phase-noise measurement techniques: (a) cavity discriminator; (b) reflective-type delay-line discriminator; (c) one-way delay line.

Ashley *et al.* (1968). Ashley *et al.* (1968) proposed the reflective-type delay-line discriminator shown in Fig. 15b. The cavity can also be used to replace delay line. The one-way delay line shown in Fig. 15c is implemented in the TRW measurement systems. The theory and applications set forth in this section are based on a system of this particular type.

A. THE DELAY LINE AS AN FM DISCRIMINATOR

1. *The Single-Oscillator Measurement System*

The single-oscillator signal is split into two channels in the system shown in Fig. 15. One channel is called the nondelay or reference channel. It is also referred to as the local-oscillator (LO) channel because the signal in this channel drives the mixer at the prescribed impedance level (the usual LO drive). The signal in the second channel arrives at the mixer through a delay line. The two signals are adjusted for phase quadrature with the phase shifter, and the output of the mixer is a fluctuating voltage, analogous to the frequency fluctuations of the source, centered on approximately zero dc volts.

The delay line yields a phase shift by the time the signal arrives at the balanced mixer. The phase shift depends on the instantaneous frequency of the signal. The presence of frequency modulation (FM) on the signal gives rise to differential phase modulation (PM) at the output of the differential delay and its associated (nondelay) reference line. This relationship is linear if the delay τ_d is nondispersive. This is the property that allows the delay line to be used as an FM discriminator. In general, the conversion factors are a function of the delay (τ_d) and the Fourier frequency f but not of the carrier frequency.

The differential phase shift of the nominal frequency v_0 caused by the delay line is

$$\Delta\phi = 2\pi v_0 \tau_d, \tag{76}$$

where τ_d is the time delay.

The phase fluctuations at the mixer are related to the frequency fluctuations (at the rate f) by

$$\delta\phi = 2\pi\tau_d\, \delta v(f). \tag{77}$$

The spectral density relationships are

$$S_{\delta\phi}(f)\Big|_{\text{mixer}} = (2\pi\tau_d)^2\, S_{\delta v}(f)\Big|_{\text{osc}} \tag{78}$$

and

$$S_{\delta v}(f) = f^2\, S_{\delta\phi}(f). \tag{79}$$

Then,

$$S_{\delta\phi}(f)\Big|_{\text{dlm}} = (2\pi f\tau_d)^2\, S_{\delta\phi}(f)\Big|_{\text{osc}}, \tag{80}$$

where the subscript dlm indicates delay-line method. From Eq. (56), the spectral density of phase for the two-oscillator technique, in radians squared per hertz, is

$$S_{\delta\phi}(f) = 4\frac{S_{\delta v}(f)}{(V_{\text{ptp}})^2} = \frac{S_{\delta v}(f)}{2(V_{\text{rms}})^2} = \left[\frac{(\delta v_{\text{rms}})^2}{2(V_{\text{rms}})B}\right] \tag{81}$$

because

$$(V_{\text{ptp}})^2 = 8(v_{\text{rms}})^2 = 4(V_{\text{p}})^2 = 4[2(v_{\text{rms}})^2]$$

and

$$\mathscr{L}(f) = 2(S_{\delta v}(f)/(V_{\text{ptp}})^2) = (\delta v_{\text{rms}})^2/4(V_{\text{rms}})^2 B \tag{82}$$

per hertz.

The sensitivity (noise floor) of the two-oscillator measurement system includes the thermal and shot noise of the mixer and the noise of the baseband preamplifier (referred to its input). This noise floor is measured with the oscillator under test inoperative. The measurement system sensitivity of the two-oscillator system, on a per hertz density basis (dBc/Hz) is

$$\mathcal{L}(f)_{nf} = 10 \log[2(\delta v_n)^2/(V_{ptp})^2], \tag{83}$$

where δv_n is the rms noise voltage measured in a one-hertz bandwidth.

The two-oscillator system therefore yields the output noise from both oscillators. If the reference oscillator is superior in performance, as assumed in the previous discussions, then one obtains a direct measure of the noise characteristics of the oscillator under test. If the reference and test oscillators are the same type, a useful approximation is to assume that the measured noise power is twice that associated with one noisy oscillator. This approximation is in error by no more than 3 dB for the noisier oscillator. Substituting in Eq. (80) and using the relationships in Eq. (56), we have, per hertz,

$$\mathcal{L}(f)\bigg|_{dlm} = 2[(\delta v_{rms})^2/(V_{ptp})^2](2\pi f \tau_d)^2 \tag{84}$$

Examination of this equation reveals the following.

(1) The term in the brackets represents the two-oscillator response. *Note that this term represents the noise floor of the two-oscillator method.* Therefore, adoption of the delay-line method results in a higher noise by the factor $(2\pi f \tau_d)^2$ when compared with the two-oscillator measurement method. The sensitivity (noise floor) for delay lines with different values of time delay are illustrated in Fig. 17.

(2) Equation (84) also indicates that the measured value of $\mathcal{L}(f)$ is periodic in $\omega = 2\pi f$. This is shown in Fig. 21. The first null in the responses is at the Fourier frequency $f = 1/\tau_d$. The periodicity indicates that the calibration range of the discriminator is limited and that valid measurements occur only in the indicated range, as verified by the discriminator slope shown in Fig. 16. (See Fig. 23.)

(3) The maximum value of $(2\pi f \tau_d)^2$ can be greater than unity (it is 4 at $f = 1/2\tau_d$). This 6-dB advantage is utilized in the noise-floor measurement. However, it is beyond the valid calibration range of the delay-line system. The 6-dB advantage is offset by the line attenuation at microwave frequencies, as discussed by Halford (1975).

The delay-line discriminator system has been analyzed in terms of a power-limited system (a particular idealized system in which the choice of power oscillator voltage, the attenuator of the delay line, and the conversion loss of the mixer are limited by the capability of the mixer) by Tykulsky (1966), Halford (1975), and Ashley *et al.* (1977). For this particular case, Eq. (83) indicates that an increase in the length of the delay line (to increase τ_d for decorrelation of Fourier frequencies closer to the carrier) results in an increase in attenuation of the line, which causes a corresponding decrease in V_{ptp}. The optimum length occurs where τ_d is such that the decrease in V_{ptp} is approximately compensated by the increase in $(2\pi f\tau_d)$, i.e., where

$$\frac{d}{d\tau_d}\frac{2\pi f\tau_d}{V_{ptp}} = 0. \tag{85}$$

This condition occurs where the attenuation of the delay line is 1 Np (8.686 dB). However, when the system is not power limited, the attenuation of the delay line is not limited, because the input power to the delay line can be adjusted to maintain V_{ptp} at the desired value. The optimum delay-line length is determined at a particular selectable frequency. However, since the attenuation varies slowly (approximately proportional to the square root of frequency), this characteristic allows near-optimum operation over a considerable frequency range without appreciable degradation in the measurements.

A practical view of the time delay (τ_d) and Fourier-frequency functional relationship can be obtained by reviewing the basic concepts of the dual-channel time-delay measurement system discussed by Lance (1964). If the differential delay between the two channels is zero, there is no phase difference at the detector output when a swept-frequency cw signal is applied to the system. Figure 16 shows the detected output interference display when a swept-frequency cw signal (zero to 4 MHz) is applied to a system that has

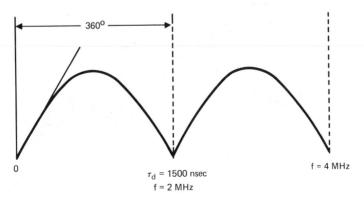

FIG. 16 Swept-frequency interference display at the output of a dual-channel system with a differential delay of 500 nsec.

a differential delay of 500 nsec between the two channels. The signal amplitudes are assumed to be almost equal, thus producing the familiar voltage-
standing-wave pattern or interference display. Because this is a two-channel
system, there is a null every 360°, as shown.

2. *System Sensitivity* (*Noise Floor*) *When Using the
Differential Delay-Line Technique*

Halford (1975) has shown that the sensitivity (noise floor) of the single-
oscillator differential delay-line technique is reduced relative to the two-
oscillator techniques. The sensitivity is modified by the factor

$$S_d = 2(1 - \cos 2\pi f \tau_d). \tag{86}$$

For $\omega \tau_d = 2\pi d \tau_d \ll 1$ a good approximation is

$$S_d^2 = 2(1 - \cos 2\pi f \tau_d) = (\omega \tau_d)^2 [1 - \tfrac{1}{12}(\omega \tau_d)^2] = (2\pi f \tau_d)^2 = \theta^2, \tag{87}$$

where θ is the phase delay of the differential delay line evaluated at the
frequency f. Figure 17 shows the relative sensitivity (noise floor) of the two-
oscillator technique and the single-oscillator technique with different
delay-line lengths. The f^{-2} slope is noted at Fourier frequencies beyond

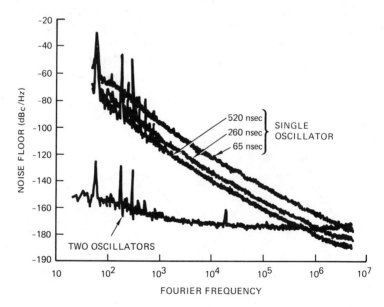

FIG. 17 Relative sensitivity (noise floor) of single-oscillator and two-oscillator phase-
noise measurement systems.

about 1 kHz. For Fourier frequencies closer to the carrier, the slope is f^{-3}, i.e., the sum of the f^{-2} slope of Eq. (87) and the f^{-1} flicker noise.

Phase-locked sources have phase-noise characteristics that cannot be measured at close-in Fourier frequencies using this basic system. The relative sensitivity of the system can be improved by using a dual (two-channel) delay-line system and performing cross-spectrum analysis, which will be presented in this chapter.

Labaar (1982) developed the delay-line rf bridge configuration shown in Fig. 18. At microwave frequencies where a high-gain amplifier is available, suppression of the carrier by the rf bridge allows amplification of the noise going into the mixer. A relative sensitivity improvement of 35 dB has been obtained without difficulty. The limitations of the technique depend on the available rf power and the carrier suppression by the bridge. Naturally, if the rf input to the bridge is high one must use the technique with adequate precautions to prevent mixer damage that can occur by an accidental bridge unbalance. Labaar (1982) indicated the added advantage of using the rf bridge carrier-suppression technique when attempting to measure phase noise close to the carrier when AM noise is present. Figure 19 shows the

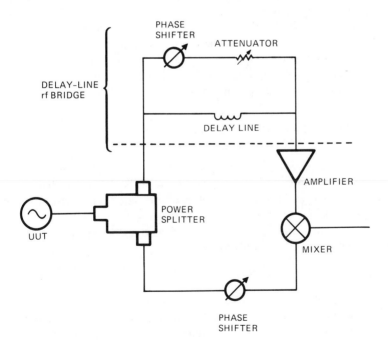

FIG. 18 Carrier suppression using an rf bridge to increase relative sensitivity. (Courtesy Instrument Society of America.)

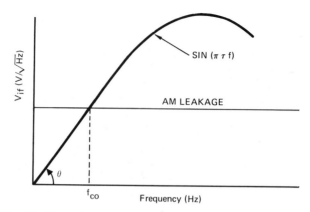

FIG. 19 Phase detector output (AM–PM crossover); τ, delay time.

mixer output for phase (PM) and amplitude (AM) noise in the single-oscillator delay-line FM discriminator system. It is noted that the phase noise and AM noise intersect and that the AM will therefore limit the measurement accuracy near the carrier. Even though AM noise is much lower than phase noise in most sources, and even though the AM is normally suppressed about 20 dB, there is still AM at the mixer output. This output is AM leakage and is caused by the finite isolation between the mixer ports. The two-oscillator technique does not experience this problem to this extent because the phase noise and AM noise maintain their relative relationships at the mixer output independent of the offset frequency from the carrier.

B. CALIBRATION AND MEASUREMENTS USING THE DELAY LINE AS AN FM DISCRIMINATOR

The block diagram of a practical single-oscillator phase noise measurement system is shown in Fig. 20. The signals in the delay-line channel of the system experience the one-way delay of the line. With adequate source power, the system is not limited to the optimum 1 Np (8.686 dB) previously discussed for a power-limited system. Measurements are performed using the following operational procedures.

(1) Measure the tracking spectrum analyzer IF bandwidths as set forth in Section III.C.1.

(2) Establish the system power levels (Section IV.B.1).

(3) Establish the discriminator calibration factor (Section IV.B.2).

(4) Measure and plot the oscillator characteristics in the automatic system used (Section IV.B.3).

(5) Measure the system noise floor (sensitivity) (Section IV.B.4).

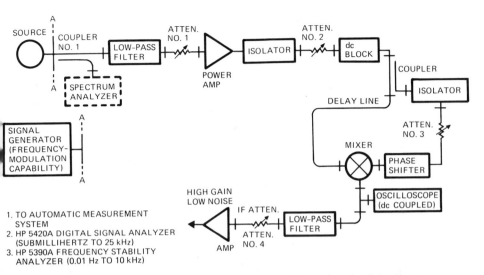

FIG. 20 Single-oscillator phase noise measurement system using the delay line as an FM discriminator. (From Lance *et al.*, 1977a.)

1. System Power Levels

The system power levels are set using attenuators, as shown in Fig. 20. Because the characteristic impedance of attenuator No. 4 is 50 Ω, mismatch errors will occur if the mixer output impedance is not 50 Ω. As previously discussed, the mixer drive levels are set so that the mixer output signal, as observed during calibration, is sinusoidal. This has been accomplished in TRW systems with a reference (LO) signal level of 10 dBm and a mixer input level of about 0 dBm from the delay line.

A power amplifier can be used to increase the source signal to the measurement system. This amplifier must not contribute appreciable additional noise to the signal.

2. Discriminator Calibration

The discriminator characteristics are measured as a function of frequency and voltage. The hertz-per-volt sensitivity of the discriminator is defined as the *calibration factor* (CF). The calibration process involves measuring the effects of intentional modulation of the source (carrier) frequency. A known modulation index must be obtained to calculate the calibration factors of the discriminator. The modulation index is obtained by using amplitude modulation to establish the carrier-to-sideband ratio when there is considerable instability of the source or when the source cannot be frequency modulated.

It is convenient to consider the system equations and calibration techniques in terms of frequency modulation of stable sources. If the source to be measured cannot be frequency modulated, it must be replaced, during the calibration process, with a modulatable source. The calibration process will be described using a modulatable source and a 20-kHz modulation frequency. However, other modulation frequencies can be used. The calibration factor of this type discriminator has been found to be constant over the usable Fourier frequency range, within the resolution of the measuring technique. The calibration factor of the discriminator is established after the system power levels have been set with the unit under test as the source.

The discriminator calibration procedures are as follows.

(1)　Set attenuator No. 4 (Fig. 20) to 50 dB.

(2)　Replace the oscillator under test with a signal generator or oscillator that can be frequency modulated. *The power output and operating frequency of the generator must be set to the same precise frequency and amplitude values that the oscillator under test will present to the system during the measurement process.*

(3)　Select a modulation frequency of 20 kHz and increase the modulation until the carrier is reduced to the first Bessel null, as indicated on the spectrum analyzer connected to coupler No. 1. This establishes a modulation index ($m = 2.405$).

(4)　Adjust the phase shifter for zero volts dc at the output of the mixer, as indicated on the oscilloscope connected as shown in Fig. 20. *This establishes the quadrature condition for the two inputs to the mixer.* This quadrature condition is continuously monitored and is adjusted if necessary.

(5)　Tune the tracking spectrum analyzer to the modulation frequency of 20 kHz. The power reading at this frequency is recorded in the program and is corrected for the 50-dB setting of attenuator No. 4, which will be set to zero decibel indication during the automated measurements.

$$P(\text{dBm}) = (-\text{dBm power reading}) + 50 \text{ dB} \tag{88}$$

This power level is converted to the equivalent rms voltage that the spectrum analyzer would have read if the total signal had been applied:

$$V_{\text{rms}} = \sqrt{10^{P/10}/1000 + R}. \tag{89}$$

(6)　The discriminator calibration factor can now be calculated because this power in dBm can be converted to the corresponding rms voltage using the following equation:

$$V_{\text{rms}} = \sqrt{(10^{P/10}/1000) \times R}, \tag{90}$$

where $R = 50 \, \Omega$ in this system.

(7)   The discriminator calibration factor is calculated in hertz per volt as

$$CF = mf_m/\sqrt{2}\,V_{rms} = 2.405 f_m/\sqrt{2}\,V_{rms}. \tag{91}$$

The modulation index m for the first Bessel null as used in this technique is 2.405. The modulation frequency is f_m.

3.   Measurement and Data Plotting

After the discriminator is calibrated, the modulated signal source is replaced with the frequency source to be measured. Quadrature of the signals into the mixer is reestablished, attenuator No. 4 (Fig. 20) is set to 0 dB, and the measurement process can begin.

The measurements, calculations, and data plotting are completely automated. The calculator program selects the Fourier frequency, performs autoranging, and sets the bandwidth, and measurements of Fourier frequency power are performed by the tracking spectrum analyzer. Each Fourier frequency noise-power reading P_n (dBm) is converted to the corresponding rms voltage by

$$v_{1rms} = \sqrt{10^{(P_n + 2.5)/10}/1000 \times R}. \tag{92}$$

The rms frequency fluctuations are calculated as

$$\delta v_{rms} = v_{1rms} \times CF. \tag{93}$$

The spectral density of frequency fluctuations in hertz squared per hertz is calculated as

$$S_{\delta v}(f) = (\delta v_{rms})^2/B, \tag{94}$$

where B is the measured IF noise-power bandwidth of the spectrum analyzer. The spectral density of phase fluctuations in radians squared per hertz is calculated as

$$S_{\delta\phi}(f) = S_{\delta v}(f)/f^2. \tag{95}$$

The NBS-designated spectral density in decibels (carrier) per hertz is calculated as

$$\mathscr{L}(f)_{dB} = 10 \log \tfrac{1}{2} S_{\delta\phi}(f). \tag{96}$$

Spectral density is plotted in real time in our program. However, the data can be stored and the desired spectral density can be plotted in other forms. Integrated phase noise can be obtained as desired.

4. Noise Floor Measurements

The relative sensitivity (noise floor) of the single-oscillator measurement system is measured as shown in Fig. 12a for the two-oscillator technique. The delay line must be removed and equal channel lengths constructed, as in Fig. (12a). The same power levels used in the original calibration and measurements are reestablished, and the noise floor is measured at specific Fourier frequencies, using the same calibration–measurement technique, or by repeating the automated measurement sequence.

A correction for the noise floor requires a measurement of the rms voltage of the oscillator ($v_{1\,rms}$) and a measurement of the noise floor rms voltage ($v_{2\,rms}$). These voltages are used in the following equation to obtain the corrected value:

$$v_{rms} = \sqrt{(v_{1\,rms})^2 - (v_{2\,rms})^2}. \tag{97}$$

The value v_{rms} is then used in the calculation of frequency fluctuations. If adequate memory is available, each value of $v_{1\,rms}$ can be stored and used after the other set of measurements are performed at the same Fourier frequencies.

The following technique was developed by Labaar (1982). Carrier suppression is obtained using the rf bridge illustrated in Fig. 18. One can easily improve sensitivity more than 40 dB. At 2.0 and 3.0 GHz 70-dB carrier suppression was realized. In general, the improvement in sensitivity will depend on the availability of an amplifier or adequate input power.

Figure 21 shows the different noise floors in a delay-line bridge discriminator. It is good measurement discipline to always determine these noise floors; also, the measurements, displayed in Fig. 21, give a quick understanding of the physical process involved. The first trace is obtained by terminating the input of the baseband spectrum analyzer. The measured output noise power is then a direct measure of the spectrum analyzer's noise figure (NF). The input noise is thermal noise and is usually indicated by "KTB," which is short for "the thermal noise power at absolute temperature of T degrees K(elvin) per one hertz bandwidth (B). This KTB number is, at 18°C, about -174 dBm/Hz.

Figure 21a shows that trace number 1 for frequencies above about 1 kHz is level with a value of about -150 dBm $= -(174-24)$ dBm, which means that the spectrum analyzer has an NF of 24 dB. At 20 Hz the NF has gone up to about 48 dB. To improve the NF, a low-noise (NF, 2dB), low-frequency (10 Hz–10 MHz) amplifier is inserted as a preamplifier. Terminating its input now results in trace number 2. At the high frequency end, the measured power goes up by about 12–13 dB, and the amplifiers gain is 34 dB. This means that the NF is improved by $34 - 12$–$13 = 21$–22 dB, which is an

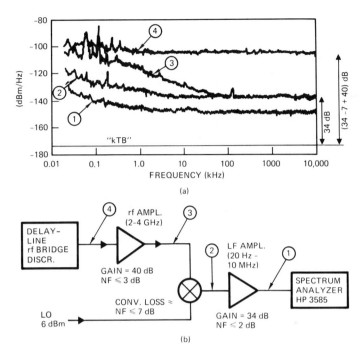

FIG. 21 (a) Noise contribution analysis; (b) phase noise test setup using a delay-line rf bridge discriminator (rf = 2.8 GHz; ○, termination points. NF, noise figure; LF, low frequency. (From Seal and Lance, 1981.)

NF ≃ 2–3 dB as expected, i.e., the first stage noise predominates. The low-frequency end at 20 Hz gives an NF of 26 dB, which overall is quite an improvement.

In trace number 3 the mixer is included with it's rf (signal) port terminated. It is clear from this trace that certainly up to 100 kHz, the noise generated by the mixer diodes being "pumped" by the LO signal dominates. This case represents the "classic" delay-line discriminator. The last trace (number 4) includes the low-noise, high-gain rf amplifier that can be used because the carrier is suppressed in the delay-line rf bridge discriminator, in contrast to the classic delay-line discriminator case. This trace shows that from 1 kHz on up the measured output power is flat, representing a 2–3-dB NF.

At about 20–40 Hz, trace numbers 3 and 4 begin nearing their crossover floor. In this particular case, which is discussed in full by Labaar (1982), the measurement systems noise floor (resolution) has been improved by 40 dB.

Figure 22 shows plots of phase noise as measured at two frequencies using delay lines of different lengths. The delay line used measure at 600 MHz

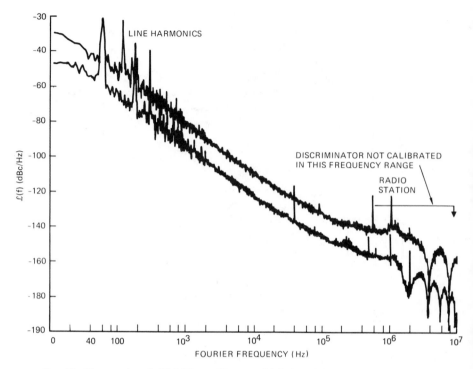

FIG. 22 Phase noise of 600-MHz oscillator multiplied to 2.4 GHz (From Lance *et al.*, 1977a.)

was about 500 nsec long, as noted by the first null, i.e., the reciprocal of the Fourier frequency of 2 MHz is the approximate differential time delay. Note that a shorter delay line (approximately 250 nsec differential) is used to measure the higher frequency because the delay-line discriminator calibration is valid only to a Fourier frequency at approximately 35% of the Fourier frequency at which the first null occurs, if a linear transfer function is assumed.

The actual transfer function of a delay-line discriminator (classic and rf bridge types) is sinusoidal, as shown in Fig. 23a. The baseband spectrum analyzer measures power in a finite bandwidth, and as a consequence it is possible to measure through a transfer-function null if the noise power does not change substantially over a spectrum–analyzer bandwidth. The following power relations then hold:

$$P_{\text{meas}}(\omega) = 1/\Delta\omega \int_{\omega - \Delta\omega/2}^{\omega + \Delta\omega/2} P(\omega') \, d\omega' \simeq \frac{P(\omega)}{\Delta\omega} \int_{\omega - \Delta\omega/2}^{\omega + \Delta\omega/2} d\omega' = P(\omega). \quad (98)$$

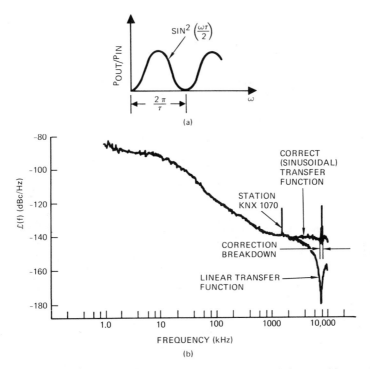

FIG. 23 Transfer functions for a delay-line rf bridge discriminator: (a) actual; (b) approximate (linear) and "correct" (sinusoidal). Phase noise: H.P. 8672A at 2.4 GHz.

Figure 23b shows the results using a linear approximation and the "correct" transfer function for a delay-line rf bridge discriminator. The correct transfer function breaks down close to the null because the signal level drops below the system's noise floor, as explained by Labaar (1982).

Using the sinusoidal transfer function in the calculator software gives correct results barring frequency intervals of 5 to 10 spectrum analyzer's bandwidths (10 × 30 = 300 kHz) centered at the transfer function nulls. These particular data were selected to illustrate the characteristics of the system. Recall that one can easily make the noise floor 40 dB lower using the rf bridge shown in Fig. 18.

C. Dual Delay-Line Discriminator

1. Phase Noise Measurements

The dual delay-line discriminator is shown in Fig. 24. This system was suggested by Halford (1975) as a technique for lowering the noise floor of the delay-line phase noise measurement system. The system consists of

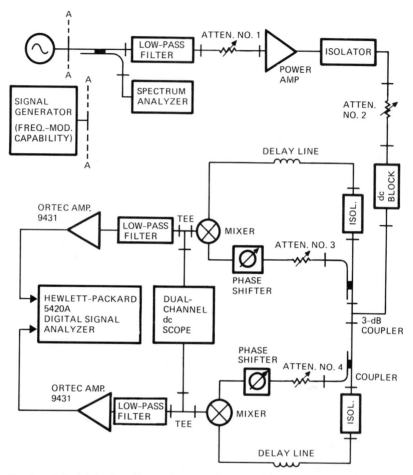

FIG. 24 A dual delay-line phase noise measurement system. (Courtesy Instrument Society of America.)

two differential delay-line systems. The single-oscillator signal is applied to both systems and cross-spectrum analysis is performed on the signal output from the two delay-line systems. Signal processing is performed with the Hewlett-Packard 5420A digital signal analyzer. The *cross spectrum* is obtained by taking the product of the Fourier transform of one signal and the complex conjugate of the Fourier transform of a second signal. It is a phase-sensitive characteristic resulting in a complex product that serves as a measurement of the relative phase of two signals. Cross spectrum gives a phase- and amplitude-sensitive measurement directly. By performing the product $Sy(f) \cdot Sx(f)^*$, a certain signal-to-noise enhancement is achieved.

The low-noise amplifiers preceding the digital signal analyzer are used when performing measurements at Fourier frequencies from 1 Hz to 25 kHz. The amplifiers are not used when performing measurements below the Fourier frequency of 1 Hz.

2. Calibrating the Dual Delay-Line System

Each delay line in the system is calibrated separately following the same basic procedure set forth in Section IV.B. The Hewlett-Packard 5420 measures the one-sided spectral density of frequency fluctuations in hertz squared per hertz. The spectral density of phase fluctuation in radians squared per hertz can be calculated as

$$S_{\delta\phi}(f) = S_{\delta v}(f)/f^2, \tag{99}$$

and

$$\mathscr{L}(f) = S_{\delta v}(f)/2f^2, \tag{100}$$

per hertz. The Hewlett-Packard 5420 measurement of $S_{\delta v}(f)$ in Hz^2/Hz must, therefore, be corrected by $1/2f^2$. However, the f^2 correction must be entered in terms of radian frequency ($\omega = 2\pi f$). This conversion is accomplished by

$$\mathscr{L}(f) = S_{\delta v}(f)(1/2f^2)(4\pi^2/4\pi^2) = [2\pi^2 S_{\delta v}(f)]/(\omega)^2 \tag{101}$$

per hertz since Eq. (100) can be stated in the following terms:

$$[2\pi^2 S_{\delta v}(f)]/4\pi^2 f^2.$$

Signal-to-noise enhancement greater than 20 dB has been obtained using the dual-channel delay-line system.

D. MILLIMETER-WAVE PHASE-NOISE MEASUREMENTS

1. Spectral Density of Phase Fluctuations

The delay line used as an FM discriminator is based, in principle, on a *nondispersive* delay line. However, a waveguide can be used as the delay line because the Fourier frequency range of interest is a small percentage of the operating bandwidth (seldom over 100 MHz), and the dispersion can be considered negligible.

The calibration and measurement are performed as set forth in Section IV.B. The modulation index m is usually established using the carrier-to-sideband ratio that uses amplitude modulation because millimeter sources are either unstable or cannot be modulated. The two approaches to measurements at millimeter frequencies are shown in Figs. 25 and 26. Figure 25 shows the *direct measurement* using a waveguide delay line. This system offers improved sensitivity if adequate input power is available. The rf bridge and delay-line portion of the system differs from Fig. 18 because pre- and

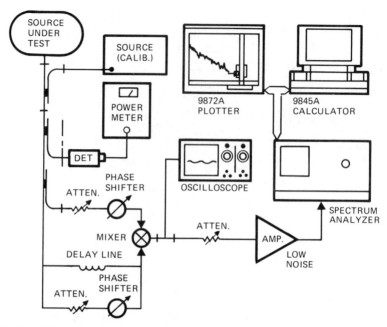

FIG. 25   Millimeter-wave phase noise measurements using a waveguide delay line. (From Seal and Lance, 1981.)

post-bridge amplifiers with appropriate gain are not available, so the sensitivity can equal the amount of carrier suppression.

Figure 26 shows the use of a harmonic mixer to downconvert to the convenient lower frequency where post-bridge amplifiers are available. The relatively low sensitivity to frequency drift that is characteristic of delay-line discriminators becomes an advantage here. A separate calibration generator is required, as shown in Fig. 25, and a power meter is used to assure proper power levels during the calibration process.

2.   SPECTRAL DENSITY OF AMPLITUDE FLUCTUATIONS

AM noise measurements require equal electrical length in the two channels that supply the signals to the mixer. The delay line must be replaced with the necessary length of transmission line to establish the equal-length condition when the systems shown in Figs. 25 and 26 are used. The AM noise measurement system is calibrated and the noise measurements are performed directly in units of power for a direct measurement of $m(f)$ in dBc/Hz. $m(f)$ is the spectral density of one modulation sideband divided by the total signal power at a Fourier frequency difference f from the signal's average frequency v_0. The system calibration establishes the detection characteristics in terms of total power output at the IF port of the mixer (detector).

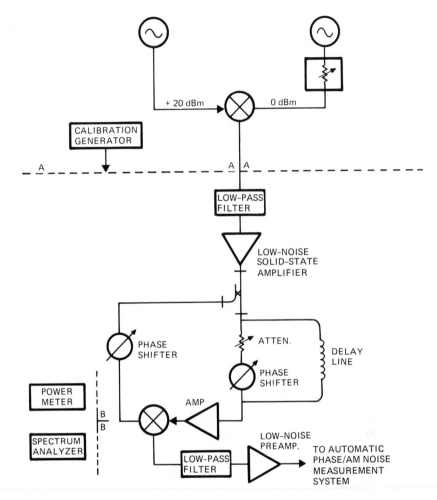

FIG. 26 A millimeter-wave hybrid phase noise measurement system that produces IF frequency and uses a delay-line discriminator at IF frequency. (From Seal and Lance, 1981.)

The AM noise measurements are performed according to the following.

(1) A known AM modulation (carrier–sideband ratio) must be established to calibrate this detector in terms of total power output at the IF port. The modulation must be low enough so that the sidebands are at least 20 dB below the carrier. This is to keep the total added power due to the modulation small enough to cause an insignificant change in the detector characteristics.

(2) The rf power levels are adjusted for levels of approximately 10 dBm at the reference port and 0 dBm at the test port of the mixer.

(3)   Approximately 40 dB is set in the precision IF attenuator. The system is adjusted for an *out-of-phase quadrature condition.*

(4)   The modulation frequency and power level are measured by the automatic baseband spectrum analyzer. The total carrier-power reference level is measured power, plus the carrier–sideband modulation ratio, plus the IF attenuator setting.

(5)   The AM modulation is removed, the IF attenuator set to 0 dB, and the system re-checked to verify the out-of-phase quadrature (maximum dc output from the mixer IF port). Noise (V_n) is measured at the selected Fourier frequencies. A direct calculation of $m(f)$ in dBc/Hz is

$$m(f) = [(\text{modulation power (dBm)} + \text{carrier–sideband ratio (dB)}$$

$$+ \text{ IF attenuation (dB)} - \text{noise power (dBm)} + 2.5 \text{ dB}$$

$$- 10 \log(\text{BW})]. \tag{102}$$

Figure 27 illustrates the measurements of AM and phase noise of two GUNN oscillators that were offset in frequency by 1 GHz. The measurements were performed using the coaxial delay-line system.

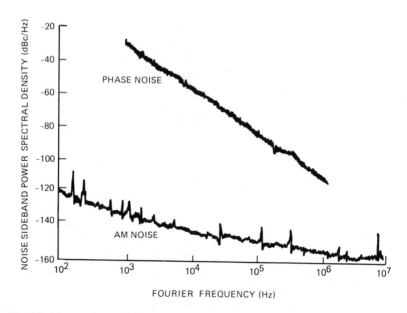

FIG. 27   Phase noise and AM noise of 40- and 41-GHz Gunn oscillators. (From Seal and Lance, 1981.)

V. Conclusion

The fundamentals and techniques for measurement of phase noise have been set forth for two basic systems. The two-oscillator technique provides the capability for measuring high-performance cw sources. The system sensitivity is superior to the single-oscillator technique for measuring phase noise very close to the carrier.

High-stability sources such as those used in frequency standards applications can be measured without using phase-locked loops. However, most microwave sources exhibit frequency instability that requires phase-locked loops to maintain the necessary quadrature conditions. The characteristics of the phase-locked loops must be evaluated to obtain the source phase noise characteristics. Also, in principle, one must have three sources at the same frequency to characterize a given source. If three sources are not available, one must assume that either one source is superior in performance or that they have equal phase noise contributions.

The single-oscillator technique employing the delay line as an FM discriminator has adequate sensitivity for measuring most microwave sources. The economic advantages of using this system include the fact that only one source is required, phase-locked loops are not required, system configuration is relatively inexpensive, and the system is inherently insensitive to oscillator frequency drift.

The single-oscillator technique using the delay-line discriminator can be adapted to measure the phase noise of pulsed sources. Pulsed sources have been measured at 94 GHz by F. Labaar at TRW, Redondo Beach, California.

ACKNOWLEDGMENTS

Our initial preparations for developing a phase noise measurement capability were the result of discussions with Dr. Jorg Raue of TRW. Our first phase noise development effort was assisting in the evaluation of phase noise measurement systems designed and developed by Bill Hook of TRW (Hook, 1973). The efforts of Don Leavitt of TRW were vital in initiating the measurement program.

We are very grateful to Dr. Donald Halford of the National Bureau of Standards in Boulder, Colorado, for his interest, consultations, and valuable suggestions during the development of the phase noise measurement systems at TRW.

We appreciate the measurement cross-checks performed by C. Reynolds, J. Oliverio, and H. Cole of the Hewlett-Packard Company, Dr. J. Robert Ashley of the University of Colorado, Colorado Springs, and G. Rast of the U.S. Army Missile Command, Redstone Arsenal, Huntsville, Alabama.

REFERENCES

Allen, D. W. (1966). *Proc. IEEE* **54**, 221–230.
Ashley, J. R., Searles, C. B., and Palka, F. M. (1968). *IEEE Trans. Microwave Theory Tech.* **MTT-16**(9), 753–760.

Ashley, J. R., Barley, T. A., and Rast, G. J. (1977). *IEEE Trans. Microwave Theory Tech.* **MTT-25**(4), 294–318.

Baghdady, E. J., Lincoln, R. N., and Nelin, B. D. (1965). *Proc. IEEE* **53**, 704–722.

Barnes, J. A. (1969). "Tables of Bias Functions, B_1 and B_2, for Variances Based on Finite Samples of Processes with Power Law Spectral Densities," National Bureau of Standards Technical Note 375.

Barnes, J. A., Chie, A. R., and Cutler, L. S. (1970). "Characterization of Frequency Stability," National Bureau of Standards Technical Note 394; also published in *IEEE Trans. Instrum. Meas.* **IM-20**(2), 105–120 (1971).

Brandenberger, H., Hadorn, F., Halford, D., and Shoaf, J. H. (1971). *Proc. 25th Ann. Symp. Freq. Control*, Fort Monmouth, New Jersey, pp. 226–230.

Culter, L. S., and Searle, C. L. (1966). *Proc. IEEE* **54**, 136–154.

Fisher, M. C. (1978). "Frequency Domain Measurement Systems," Paper presented at the 10th Annual Precise Time and Time Interval Applications and Planning Meeting, Goddard Space Flight Center, Greenbelt, Maryland.

Halford, D. (1968). *Proc. IEEE* **56**(2), 251–258.

Halford, D. (1975). "The Delay Line Discriminator," National Bureau of Standards Technical Note 10, pp. 19–38.

Halford, D., Wainwright, A. E., and Barnes, J. A. (1968). *Proc. 22nd Ann. Symp. Freq. Control*, Fort Monmouth, New Jersey, 340–341.

Halford, D., Shoaf, J. H., and Risley, A. S. (1973). *Proc. 27th Ann. Symp. Freq. Control*, Cherry Hill, New Jersey, pp. 421–430.

Hook, W. R. (1973). "Phase Noise Measurement Techniques for Sources Having Extremely Low Phase Noise," TRW Defense and Space Systems Group Internal Document, Redondo Beach, California.

Hewlett-Packard Company (1965). "Frequency and Time Standards," Application Note 52. Hewlett-Packard Co., Palo Alto, California.

Hewlett-Packard Company (1970). "Computing Counter Applications Library," Application Notes 7, 22, 27, and 29. Hewlett-Packard Co., Palo Alto, California.

Hewlett-Packard Company (1976). "Understanding and Measuring Phase Noise in the Frequency Domain," Application Note 207. Hewlett-Packard Co., Loveland, Colorado.

Labaar, F. (1982). "New Discriminator Boosts Phase Noise Testing." *Microwaves* **21**(3), 65–69.

Lance, A. L. (1964). "Introduction to Microwave Theory and Measurements." McGraw-Hill, New York.

Lance, A. L., Seal, W. D., Mendoza, F. G., and Hudson, N. W. (1977a). *Microwave J.* **20**(6), 87–103.

Lance, A. L., Seal, W. D., Mendoza, F. G., and Hudson, N. W. (1977b). *Proc. 31st Annu. Symp. Freq. Control*, Atlantic City, New Jersey, pp. 463–483.

Lance, A. L., Seal, W. D., Halford, D., Hudson, N., and Mendoza, F. (1978). "Phase Noise Measurements Using Cross-Spectrum Analysis." Paper presented at the IEEE Conference on Electromagnetic Measurements, Ottawa, Canada.

Lance, A. L., Seal, W. D., and Labaar, F. (1982). *ISA Trans.* **21**(4), 37–44.

Meyer, D. G. (1970). *IEEE Trans.* **IM-19**(4), 215–227.

Ondria, J. (1968). *IEEE Trans. Microwave Theory Tech.* **MTT-16**(9), 767–781.

Ondria, J. G. (1980). *IEEE-MTT-S Int. Microwave Symp. Dig.*, pp. 24–25.

Payne, J. B., III (1976). *Microwave System News* **6**(2), 118–128.

Scherer, D. (1979). "Design Principles and Measurement Low Phase Noise RF and Microwave Sources," Hewlett-Packard Co., Palo Alto, California.

Seal, W. D., and Lance, A. L. (1981). *Microwave System News* **11**(7), 54–61.

Shoaf, J. H., Halford, D., and Risley, A. S. (1973). "Frequency Stability Specifications and Measurement," National Bureau of Standards Technical Note 632.

Tykulsky, A. (1966). *Proc. IEEE* **54**(2), 306.

Van Duzer, V. (1965). *Proc. IEEE-NASA Symp. Definition Meas. Short-Term Freq. Stability*, NASA SP-80, 269–272.

Walls, F. L., and Stein, S. R. (1977). *Proc. 31st Ann. Symp. Freq. Control*, Atlantic City, New Jersey, pp. 335–343.

Walls, F. L., Stein, S. R., Gray, J. E., Glaze, D. J., and Allen, D. W. (1976). *Proc. 30th Ann. Symp. Freq. Control*, Atlantic City, New Jersey, p. 269.

CHAPTER 8

Basic Design Considerations for Free-Electron Lasers Driven by Electron Beams from rf Accelerators†

A. Gover*, H. Freund‡, V. L. Granatstein§, J. H. McAdoo¶, and Cha-Mei Tang

Naval Research Laboratory
Washington, D. C.

† Supported in part by the Office of Naval Research and in part by AFOSR Grant 82-0239.
* Present address: Science Applications, Inc., Mclean, Virginia 22102 and Tel Aviv University, Faculty of Engineering, Tel Aviv, Israel.
‡ Present address: Science Applications, Inc., Mclean, Virginia 22102.
§ Present address: Electrical Engineering Department, University of Maryland, College Park, Maryland 20742.
¶ Present address: Electrical Engineering Department, University of Maryland, College Park, Maryland 20742.

ISBN 0-12-147711-8

I. Introduction

A. BACKGROUND

The first demonstration of a free-electron laser oscillator (FEL) driven by an rf accelerator was made at Stanford University in 1977 (Deacon *et al.*, 1977; Eckstein *et al.*, 1982). In this experiment coherent infrared radiation was generated by using a beam of electrons as the active medium as in a microwave tube; this suggested the potential for developing coherent sources at short wavelengths (namely, infrared, visible, and ultraviolet) with other desirable properties of microwave tubes such as continuous tunability and high efficiency. A very substantial research and development effort has been launched in a number of countries to realize this potential.

Much of the ongoing research is concentrated on the issue of optimizing efficiency, and a number of theoretical (Kroll *et al.*, 1981; Tang and Sprangle, 1982) and experimental (Brau and Cooper, 1980) studies of the effect of wiggler-magnet tapering on nonlinear FEL behavior have been reported. In contrast, the present chapter treats the linear regime of FEL operation and stresses techniques and design procedures that may be used to accelerate the buildup time of the laser oscillation. In the infrared FEL experiment at Stanford University, the radiation took several tens of microseconds to grow from ambient noise levels to saturation (Eckstein, 1982). Such a long rise time could be accommodated because the experiment was performed on a superconducting linear accelerator with quasi-cw operation. However, most other rf accelerators, especially those that are relatively compact and inexpensive, have macropulse durations less than about 10 μsec (Kapitza and Melekhin, 1978, Varian Radiation Division, 1980†). Thus, designing FELs to achieve a more rapid buildup of radiation is of considerable practical value and may be crucial in making FELs relevant to applications that re-require compact systems. Furthermore, slow buildup may be exacerbated by the wiggler-taper techniques used to enhance efficiency, and innovations to compensate for this effect are of general interest.

The rise time to saturation (oscillation buildup time) will obviously be decreased by increasing the linear gain of the FEL. Much of this chapter is

---

† A descriptive brochure is available from Varian Radiation Division, 611 Hansen Way, Palo Alto, California 94303.

devoted to a consideration of the various phenomena that influence gain. The various equations and inequalities that will be derived display the dependence of gain on myriad interrelated factors (namely, electron micropulse duration, micropulse current, wiggler field intensity, wiggler period, wiggler length, wavelenght of the FEL radiation, and radius of the radiation beam). A step-by-step procedure for achieving an optimum design is described.

One technique for enhancing gain that has not been extensively considered elsewhere is using a waveguide configuration for the FEL instead of the usual open optical cavity defined by two coaxial concave mirrors. The optical-cavity configuration is contrasted with the waveguide configuration in Fig. 1. Larger gain might be achieved in the waveguide configuration for a number of reasons. First, the radius of the radiation beam can be decreased to more closely approximate the radius of the electron beam, thereby ensuring a stronger interaction. Second, the region of the cavity between the end of the wiggler magnet and the mirror, where no gain occurs, can be decreased or even eliminated in the waveguide configuration. On the other hand, waveguide wall losses will diminish the gain, and mode control may be more difficult in a waveguide. This chapter will clarify when a waveguide configuration can be used to advantage.

In addition to improving the gain of FELs, other techniques are available for decreasing the rise time to saturation. One that is considered in this

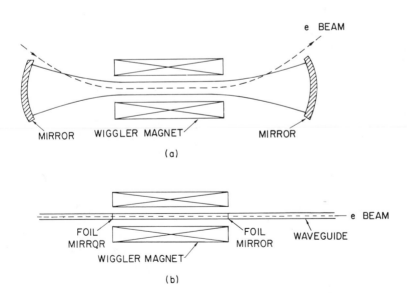

FIG. 1 (a) An open optical cavity configuration compared with (b) the waveguide cavity configuration for a free-electron laser (FEL).

chapter is injecting a signal at the start of the electron macropulse that is above the ambient noise or spontaneous emission level. Tunable lasers are available for this purpose in many parts of the spectrum so that the tunability of the FEL need not be sacrificed. Signal injection can also improve coherence and mode control. A signal injection scheme is described in detail in Section VI.

B. PARAMETRIC DEPENDENCE OF FEL GAIN

Our analysis of the FEL oscillator will be based on the existent small-signal theory of FEL amplifier gain. In a Fabry–Perot resonator the single-pass gain of the traveling wave in the oscillator is taken to be equal to the gain of the same length. A fraction of the radiation is coupled out of the front mirror, and the rest is retro-reflected to the back mirror. In a FEL there is no gain in the return trip to the back mirror, but when the wave is reflected forward again, the same fraction of its energy is added back to it as on the previous pass. This process of single-pass gain continues until the wave power grows so large that the energy added on a single pass becomes comparable to the maximum energy that can be extracted from the electron beam in a single pass. The gain now becomes a function of the wave power and is said to be in the nonlinear or saturation regime. This picture is applicable to the oscillation buildup process from incoherent spontaneous emission or from a pulse of coherent seed radiation that is injected initially into an FEL resonator.

In many cases the duration of the rf accelerator macropulse is so short that the oscillation buildup process does not arrive to saturation at all. In this case the FEL with injected radiation can be regarded as a regenerative amplifier rather than an oscillator. Certainly the use of the small-signal gain expression to describe this device is fully justified. When the pulse is long enough to reach saturation, then the use of the small-signal single-path gain expression is correct only for the earlier stages of the oscillation buildup. But it can be used approximately for estimating the oscillation threshold and the oscillation buildup time when we have an expression for the maximum power that can be extracted from the electron beam in a single path.

The single-path small-signal (linear) gain expression and gain regimes of the FEL amplifier were analyzed in detail by a number of authors (Gover and Sprangle, 1981; Kroll and McMullin, 1978; Sprangle *et al.*, 1979). In this work we will follow the parametrization of Gover and Sprangle (1981), wherein it was shown that the FEL gain and gain regimes can be completely defined in terms of only four parameters: the normalized gain parameter $\bar{Q}$, the normalized space-charge parameter $\bar{\theta}_p$, the normalized axial beam-velocity spread parameter $\bar{\theta}_{th}$, and the normalized detuning parameter $\bar{\theta}$.

These are defined in MKS units by

$$\bar{Q} \equiv 2(r_e J_0 A_b/ceA_{em})(e\bar{B}_w/mc)^2 \lambda L^3(\gamma_{0z}^2/\gamma_0^3), \tag{1}$$

$$\bar{\theta}_p \equiv (\omega_p/\gamma_0^{1/2}\gamma_{0z}c)L, \tag{2}$$

$$\bar{\theta}_{th} \equiv 2\pi(v_{zth}'/c)(L/\lambda), \tag{3}$$

$$\bar{\theta} \equiv (\omega/v_{0z} - k - k_w)L, \tag{4}$$

where

$$r_e \equiv e^2/4\pi\varepsilon_0 mc^2 = 2.818 \times 10^{-15} \text{ m} \tag{5}$$

is the classical electron radius,

$$\omega_p \equiv (e^2 n_0/m\varepsilon_0)^{1/2} = [(e/m\varepsilon_0)(J_0/v_{0z})]^{1/2} \tag{6}$$

is the beam plasma frequency,

$$\gamma_{0z} \equiv (1 - \beta_{0z})^{-1/2} = \gamma_0/(1 + \bar{a}_w^2)^{1/2}, \tag{7}$$

$$\bar{a}_w \equiv \bar{p}_{0\perp}/mc = e\bar{A}_w/mc = e\bar{B}_w/k_w mc. \tag{8}$$

We also define the parameters used in the above equations: J_0, the beam current density; c, the speed of light; $e = |e|$, the electron charge; m, the electron rest mass; A_b, the effective electron-beam cross-sectional area; A_{em}, the effective electromagnetic-wave cross-section area ($A_{em} \geq A_b$); $\bar{B}_w$, the rms value of the periodic magnetic wiggler field; λ, the lasing wavelength; L, the interaction length; $\gamma_0 = (1 - \beta_0^2)^{-1/2}$, the Lorenz relativistic beam-energy parameter; $\beta_0 = v_0/c$, the average beam velocity; $\beta_{0z} = v_{0z}/c$, the axial component of the average beam velocity; v_{zth}, the axial-velocity spread; ω, the laser frequency; $k = \omega/c = 2\pi/\lambda$, the laser wave number; $k_w = 2\pi/\lambda_w$, the wiggler wave number; and λ_w, the wiggler wavelength.

In these expressions the extreme relativistic limit

$$\gamma_0, \gamma_{0z} \gg 1 \tag{9}$$

was assumed wherever taking the limit makes little difference to the parameter value. Also, operation near synchronizm $\bar{\theta} \simeq 0$ is assumed. Hence the approximate radiation wavelength condition follows from Eq. (4):

$$\lambda/\lambda_w \simeq \beta_{0z}^{-1} - 1 \simeq (2\gamma_{0z}^2)^{-1}. \tag{10}$$

The gain of the FEL amplifier can be calculated in terms of the four parameters, $\bar{Q}, \bar{\theta}_p, \bar{\theta}_{th}, \bar{\theta}$, by means of numerical computation (Gover and Sprangle, 1981; Jarby and Gover, 1984; Livni and Gover, 1979). In the cold beam limit $\bar{\theta}_{th} = 0$ an explicit analytic expression for the single-pass gain can be found by the following:

$$P(L)/P(0) = G(\bar{Q}, \bar{\theta}_p, \bar{\theta}), \tag{11}$$

where $P(z)$ is the power of the amplified electromagnetic wave. The explicit analytic expression of G is cumbersome and is given in Appendix C only in terms of a computer program that calculates its value for given parameters $\bar{Q}$, $\bar{\theta}_p$, and $\bar{\theta}$.

In certain limits that define gain regimes, the expression for the gain can be considerably simplified. We will summarize the approximations of the gain expression following the analysis of Gover and Sprangle (1981).

In the cold, tenuous-beam low-gain regime

$$\bar{Q}, \bar{\theta}_p, \bar{\theta}_{th} \ll \pi, \tag{12}$$

the gain is given by

$$\frac{\Delta P}{P} = \bar{Q} F(\bar{\theta}) \equiv \bar{Q} \frac{d}{d\bar{\theta}} \left[\frac{\sin^2(\bar{\theta}/2)}{(\bar{\theta}/2)^2} \right], \tag{13}$$

where $\Delta P \equiv P(L) - P(0)$. The S-shaped gain curve $F(\bar{\theta})$ shown in Fig. 2 attains its maximum-value point for a detuning parameter value $\bar{\theta} = -2.6$, for which

$$(\Delta P/P)_{max} = 0.27\,\bar{Q}. \tag{14}$$

In the cold-beam space–charge-dominated (collective) low-gain regime

$$\pi, \bar{\theta}_{th}, \bar{Q} \ll \bar{\theta}_p, \tag{15}$$

the maximum gain is

$$(\Delta P/P)_{max} = \bar{Q}/2\bar{\theta}_p.$$

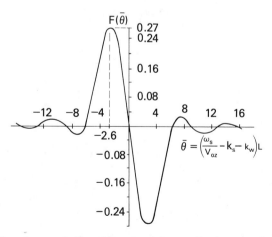

FIG. 2   The gain curve in the cold tenuous beam small-gain regime for an FEL.

In the warm-beam low-gain regime

$$\pi, \bar{\theta}_p, \bar{Q} \ll \theta_{th},$$ (16)

the maximum gain is

$$(\Delta P/P)_{max} = 3\bar{Q}/\theta_{th}^2.$$ (17)

High-gain regimes that require $\bar{Q} \gg \pi, \theta_{th}$ are not likely to be obtained with low-current rf linacs and will not be discussed here. The asymptotic behavior of the maximum-gain expression as a function of the parameter $\bar{Q}$ is depicted in Fig. 3. In most cases of relevance to the present work, the cold, tenuous-beam low-gain expression (13) is applicable. In cases when $\bar{Q}$ is of the order of unity, we would be in an intermediate-gain regime (broken line in Fig. 3). In such cases one needs to use the exact expression (11) with the aid of program COLD (Appendix C).

It is evident from comparison of Eqs. (14) and (17) that operation in the warm-beam regime $\theta_{th} \gg \pi$ is undesirable, because in this regime the gain is considerably attenuated. Also, in most practical cases of rf accelerator FELs the space-charge effects are small ($\bar{\theta}_p \simeq 0$). Hence in this limit the FEL gain function (11) is only a function of $\bar{Q}$ and $\bar{\theta}$, and the maximum-gain value at $\bar{\theta} = \bar{\theta}_m$ is a function of the gain parameter $\bar{Q}$ only:

$$G_{max}(\bar{Q}) = G(\bar{Q}, 0, \bar{\theta}_m).$$ (18)

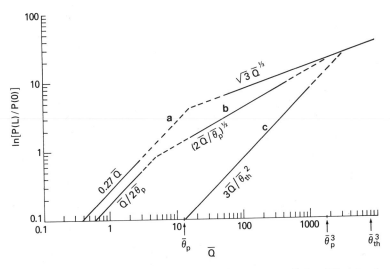

FIG. 3 The dependence of FEL gain on the gain parameter $\bar{Q}$ for different values of beam parameters. Curve a, $\bar{\theta}_p, \bar{\theta}_{th} < 1$; curve b, $\bar{\theta}_p > 1, \bar{\theta}_{th}$; curve c, $\bar{\theta}_{th} > 1, \bar{\theta}_p$. Curve b is drawn for $\bar{\theta}_p = 12$ and curve c for $\bar{\theta}_{th} = 20$.

This maximum-gain function is given asymptotically by $G_{max}(\bar{Q}) \simeq 1 + 0.27\bar{Q}$ in the low-gain limit $\bar{Q} \ll 1$ [Eq. (14)], and by $G_{max}(\bar{Q}) \simeq \exp \bar{Q}^{1/3}/9$ in the high-gain limit $\bar{Q} \gg \pi$ (Gover and Sprangle, 1981). Although operating in the high-gain regime is not very likely with rf accelerator parameters, the intermediate regime $\bar{Q} \gtrsim \pi$ may be obtainable. For this reason we display in Fig. 4a the curve $G_{max}(\bar{Q})$ as calculated from program COLD for $\theta_p \to 0$. This curve is useful for determining the FEL gain in the intermediate regime $\bar{Q} \simeq \pi$.

The scaling laws of the gain parameter $\bar{Q}$ as a function of the FEL parameters are indicative of the maximum-gain scaling in all gain regimes and in particular in the low-gain regimes where the maximum gain is proportional to $\bar{Q}$. In considering the scaling laws of $\bar{Q}$ as a function of the operating parameters [Eq. (1)], it should be remembered that the parameters λ, λ_w, γ_0, and $\bar{a}_w$ are dependent on each other through the synchronism condition (10) and Eq. (7). Because the available beam energy in rf accelerator usually varies over quite a wide range, and also because present-day technology allows obtaining quite high magnetic fields, we prefer to keep γ_0 and $\bar{B}_w$ as

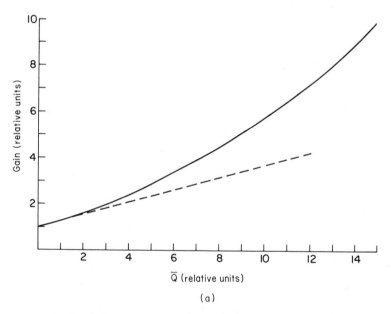

(a)

FIG. 4   (a) The maximum gain in the cold tenuous beam limit $G_{max}(\bar{Q}) \equiv G(\bar{\theta}_m, Q)$ in the intermediate regime $0 < \bar{Q} < 15$: ——, $G(\bar{\theta}_m, \bar{Q})$, – –, $1 + 0.27\,\bar{Q}$. (b) The second derivative with respect to $\bar{\theta}$ of the gain curve at the maximum gain point $\bar{\theta}_m$ in the cold tenuuous beam limit: – –, $G''(\bar{\theta}_m, \bar{Q})$; – – –, $0.08845\,\bar{Q}$.

dependent variables; by substituting Eqs. (7), (8), and (10) in Eq. (1), we express the gain parameter $\bar{Q}$ in terms of λ, λ_{w}, and $\bar{a}_{\mathrm{w}}$ only:

$$\bar{Q} = 8\sqrt{2}\,\pi^2(r_e/c)(I_0/eA_{\mathrm{em}})(\lambda^{3/2}L^3/\lambda_{\mathrm{w}}^{5/2})[\bar{a}_{\mathrm{w}}^2/(1 + \bar{a}_{\mathrm{w}}^2)^{3/2}]$$

$$= 6.55 \times 10^{-3}\,I_0(\lambda^{3/2}L^3/A_{\mathrm{em}}\lambda_{\mathrm{w}}^{5/2})[\bar{a}_{\mathrm{w}}^2/(1 + \bar{a}_{\mathrm{w}}^2)^{3/2}], \qquad (19)$$

where in the second part of Eq. (19) I_0 is expressed in units of amperes.

Evidently, for fixed wavelength λ, $\bar{Q}$ reaches a maximum value when $\bar{a}_{\mathrm{w}} = \sqrt{2}$, for which case

$$\bar{Q} = (2.52 \times 10^{-3})I_0(\lambda^{3/2}L^3/A_{\mathrm{em}}\lambda_{\mathrm{w}}^{5/2}). \qquad (20)$$

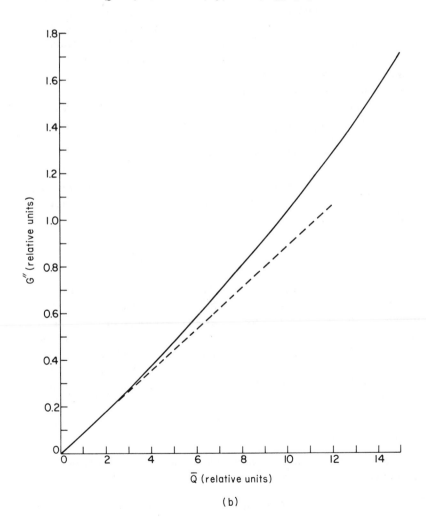

(b)

This simple scaling law should be taken with some caution. For example, the scaling law correctly represents the dependence on λ and L only if the parameter A_{em} in Eqs. (19) and (20) can be considered constant (as is the case in a lossless waveguide or when diffraction effects are negligible). Furthermore, in optimizing the gain expression, choice of the parameters is restricted both by the expression's domain of validity and by technological constraints. These restrictions lead to scaling laws different from (19) and (20), which will be investigated in the following sections.

Before proceeding to the next section, we point out that the linear gain expressions given here were derived assuming that the electromagnetic wave is a single eigenmode of an overmoded waveguide or a free-space TEM plane wave (which can also approximate a finite-cross-section light beam at lengths for which diffraction is negligible). The periodic wiggler field is taken to be purely transverse and transversely uniform. The finite cross section of the electron beam is taken care of by the filling factor A_b/A_{em} of (1), which should always satisfy $A_b/A_{em} \leq 1$. The gain expressions are equally valid for both linear and helical wigglers, provided that higher-order harmonics are neglected and that $\bar{B}_w$ is the rms value of the magnetic wiggler field. For the linear wiggler this rms value is $1/\sqrt{2}$ of the maximum amplitude of the periodic magnetic field B_w. In a helical wiggler the rms and maximum-amplitude fields are identical. In the following sections we will refine some of the idealized assumptions of the basic FEL gain theory and present design formulas that take into account realistic experimental and theoretical constraints.

II.   Electron Orbits and *E*-Beam Propagation

Realizable periodic wiggler fields are required to be free of both curl and divergence in a vacuum. Consequently, the usual assumption of purely transverse and transversely uniform axially periodic magnetic fields cannot be satisfied. Transverse gradients and axial field components of the magnetic wiggler modify the electron trajectories in the wiggler region and consequently affect the FEL operation. We consider separately here linear and helical wigglers (Blewett and Chasman, 1977; Sprangle and Tang, 1981; Tang, 1982).

A.   Linear Wiggler

An appropriate vector potential field configuration of a realizable linear wiggler can be

$$\mathbf{A}_w(\mathbf{r}) = A_w \cosh(k_w y) \cos(k_w z)\hat{e}_x. \tag{21}$$

The magnetic field that corresponds to this configuration is

$$\mathbf{B}_w(\mathbf{r}) = -B_w[\cosh(k_w y) \sin(k_w z) \,\hat{e}_y + \sinh(k_w y) \cos(k_w z) \,\hat{e}_z]. \quad (22)$$

Only near the wiggler axis in the limit

$$k_w y \ll 2\pi \quad (23)$$

is the magnetic field transversely uniform and purely polarized in the y direction. Within this limit the electrons perform a sinusoidal wiggling motion (synchrotron oscillation) only in the x direction with a period λ_w.

Solution of the motion equations with the field [Eq. (21)] to second order in $k_w y$ produces the electron trajectories (Sprangle and Tang, 1981; Tang 1982). In the x direction, integration results in

$$p_x = eA_w(x, y, z) + \gamma m \bar{v}_{x0}, \quad (24)$$

$$y_x = v_{\perp w} \cosh(k_w y) \cos(k_w z) + \bar{v}_{x0}, \quad (25)$$

$$x = (y_{\perp w}/k_w) \cosh(k_w y) \sin(k_w z) + x_0 + (\bar{v}_{x0}/\bar{v}_z)z, \quad (26)$$

where

$$v_{\perp w} \equiv (eA_w/\gamma m). \quad (27)$$

The electron trajectory, averaged over the fast synchrotron oscillation, is then given by

$$\begin{pmatrix} \bar{x} \\ \bar{v}_x \end{pmatrix} = \begin{pmatrix} 1 & z/\bar{v}_z \\ 0 & 1 \end{pmatrix} \begin{pmatrix} \bar{x}_0 \\ \bar{v}_{x0} \end{pmatrix}. \quad (28)$$

In the y direction, $p_y = \gamma m \bar{v}_y$, and the trajectories are given by

$$\begin{pmatrix} \bar{y} \\ \bar{v}_y \end{pmatrix} = \begin{pmatrix} \cos(k_\beta z) & (\bar{v}_z k_\beta)^{-1} \sin(k_\beta z) \\ -\bar{v}_z k_\beta \sin(k_\beta z) & \cos(k_\beta z) \end{pmatrix} \begin{pmatrix} y_0 \\ v_{y0} \end{pmatrix}, \quad (29)$$

where

$$k_\beta \equiv v_{\perp w} k_w / \sqrt{2}\bar{v}_z \quad (30)$$

is the betatron oscillation wave number.

The axial electron velocity averaged over the synchrotron oscillation is

$$\bar{v}_z^2 \equiv v_z^2 - v_x^2 - v_y^2$$
$$= v^2 - \tfrac{1}{2}v_{\perp w}^2 - v_{x0}^2 - (v_{y0}^2 + \bar{v}_z^2 k_\beta^2 y_0^2). \quad (31)$$

The electron trajectories in the x and y directions are shown in Fig. 5. In the x direction the electron executes fast synchrotron oscillation while drifting off axis in a straight line away from the entrance point $x = x_0$ at a

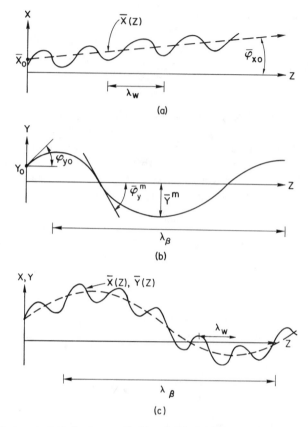

FIG. 5 Electron trajectories in a realizable wiggler field: (a) synchrotron oscillation and transverse drift in the x direction (perpendicular to the magnetic field of a linear wiggler); (b) betatron oscillation in the y direction (parallel to the magnetic field lines of a linear wiggler); (c) simultaneous synchrotron and betatron oscillation in an arbitrary direction in a helical wiggler.

constant drift angle relative to the z axis of $\phi_{x0} = \bar{v}_{x0}/\bar{v}_z$. In the y direction the electron executes long-period ($\lambda_\beta = 2\pi/k_\beta \gg \lambda_w$) betatron oscillations that focus the electron trajectories and confine them near the axis. The maximum amplitudes of the displacement ($\bar{y}$), velocity ($\bar{v}_y$), and angular deviation ($\bar{\phi}_y \equiv \bar{v}_y/\bar{v}_z$) depend only on the initial conditions of the electron (y_0, v_{y0}) and can be found from Eq. (29) as follows:

$$\bar{y}^m = [y_0^2 + (v_{y0}/\bar{v}_z k_\beta)^2]^{1/2}, \tag{32}$$

$$\bar{v}_y^m = [v_{y0}^2 + (\bar{v}_z k_\beta y_0)^2]^{1/2} = \bar{v}_z k_\beta \bar{y}^m, \tag{33}$$

$$\bar{\phi}_y^m = \bar{v}_y^m/\bar{v}_z = [\phi_{y0}^2 + (k_\beta y_0)^2]^{1/2} = k_\beta \bar{y}^m. \tag{34}$$

It is instructive to note that the average axial velocity [Eq. (31)] of any electron orbit stays constant throughout the betatron oscillation period and depends only on the initial conditions of the electrons or on the value of its betatron oscillation amplitude [Eq. (33)]. This feature can be explained by examining Eq. (25), which shows that the amplitude of the synchrotron oscillation $v_{\perp w} \cosh(k_w y)$ increases the more the electron is displaced in the y direction while executing its betatron oscillation. Its betatron oscillation velocity $\bar{v}_y$ then decreases concurrently, so that the average transverse velocity in Eq. (31) stays constant. Because the total velocity v is a constant of the motion, the average axial velocity $\bar{v}_z$ of each individual electron remains constant.

The preceding orbit equations can be used to study the phase-space evolution of a realistic electron beam occupying a finite area in phase space. We assume that the phase-space distribution of the electron beam at the start of the interaction region is similar to the initial distribution prior to its entrance into the wiggler. The model distribution we chose is shown in Fig. 6 in the phase space defined by $(k_\beta^{-1}\phi_y, \bar{y})$. For simplicity, the initial distribution is uniform within a bounding ellipse having major and minor axes of $k_\beta^{-1}\phi_{by0}$ and y_{b0} that coincide with the principal axes of the phase-space coordinates. This distribution corresponds to an electron beam focused to its waist (point of minimum width) in the y direction right at the entrance to the wiggler.

The beam emittance is defined as the phase-space area occupied by the beam that, for this case, is related to y_{b0} and ϕ_{by0} via

$$\varepsilon_y = \pi y_{b0} \phi_{by0}. \tag{35}$$

In an ideal (i.e., conservative) system the phase-space area is conserved. Thus, if losses caused by collisions and radiation can be neglected, ε_y will remain constant as the beam propagates through the wiggler. Furthermore, since the emittance is a constant property of the beam, with ideal electron lenses we can change at will the beam waist size (half-width) y_{b0} and the angular spread (half-maximum opening angle) ϕ_{by0} of a given beam as long as we keep ε_y [Eq. (35)] constant.

Based upon Eq. (29), points in phase space transform along the wiggler length z in the following manner:

$$\bar{y} = y_0 \cos(k_\beta z) + k_\beta^{-1}\phi_{y0} \sin(k_\beta z), \tag{36}$$

$$k_\beta^{-1}\phi_y = -y_0 \sin(k_\beta z) + k_\beta^{-1}\phi_{y0} \cos(k_\beta z). \tag{37}$$

Electrons follow a circular trajectory through phase space as they traverse the wiggler, and the ellipse shown in Fig. 6 goes through one rotation as $z \to z + 2\pi/k_\beta$. Evidently, only those electrons confined to a circle of radius

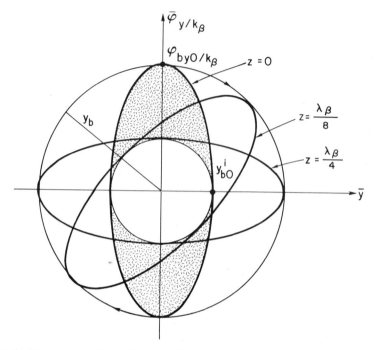

FIG. 6   Phase-space evolution diagram of an electron beam in the y (betatron oscillation) dimension of a linear wiggler. The diagram shows the phase-space areas (ellipses) occupied by the beam at different distances z. At $z = 0$ the beam is focused to its waist point (erect ellipse). The case shown corresponds to over-focusing. The electrons in the shaded part of the initial phase-space distribution have high angular deviation off-axis and produce beam envelope expansion (scalloping) after a quarter betatron oscillation period.

y_{bo} about the origin of phase space remain within the original beam envelope. The other electrons (shaded region) produce a scalloping effect in the beam envelope, and its width oscillates in the y direction between y_{bo} and $k_\beta^{-1}\phi_{byo}$ with a period $\lambda_\beta/2$. In the example illustrated in Fig. 6, evidently the electron beam was not focused at the wiggler entrance to its optimum waist size. Because of overfocusing, the beam width grows to a larger value after one-quarter betatron oscillation period. Having the freedom to adjust y_{bo} and $k_\beta^{-1}\phi_{byo}$ at will, keeping the emittance (15) constant, it is evidently optimal to choose them equal:

$$y_{bo}^{opt} = k_\beta^{-1}\phi_{byo}^{opt} = (\varepsilon_y/\pi k_\beta)^{1/2}. \tag{38}$$

In this case the initial configuration of the electron-beam distribution in phase space is a circle that transforms to itself under the transformation [Eqs. (36), (37)]. The beam width stays constant along the wiggler length and is the minimum width beam that can be propagated in this wiggler with the given emittance value.

In general the maximum beam width y_b in the wiggler will be the constraining parameter in the FEL design. In the general case (nonoptimal focusing but still the injection of the beam at its waist point), the beam width is

$$y_b = \max(y_{b0}, (\varepsilon_y/\pi)/k_\beta y_{b0}). \tag{39}$$

The beam maximum angular spread and transverse velocity spread in the y direction can then be related to the beam width by

$$\phi_{by} = k_\beta y_b, \tag{40}$$

$$v_{by} = \bar{v}_z k_\beta y_b. \tag{41}$$

Clearly, injection of the electron beam at its optimum waist-size width is advantageous for obtaining not only minimum beam width but also minimum angular spread and transverse-velocity spread. This in turn helps to keep the axial-velocity spread minimal, as evidenced from Eqs. (31), (33).

In Figure 7 we illustrate the electron-beam distribution as it evolves along the wiggler length in the phase space of the x coordinate (parallel to the

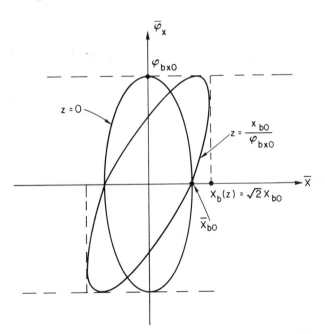

FIG. 7 Phase-space diagram of an electron beam in the x dimension of a linear wiggler. At $z = 0$ the beam is focused to a waist point (erect elipse). At $z = x_{b0}/\phi_{bx0}$ the ellipse stretches and the beam width increases by a factor of $\sqrt{2}$.

wiggling plane). Again we start with an initial elliptical distribution, with the ellipse major and minor axes ϕ_{bx0} and x_{b0} that coincide with the principal axes of the phase-space coordinates. This corresponds to an electron focused to its waist in the x direction at the entrance to the beam.

Based on Eq. (28), points in phase space transform along the wiggler length z in the following manner:

$$\bar{x} = x_0 + z\bar{\phi}_{x0}, \tag{42}$$

$$\bar{\phi}_x = \bar{\phi}_{x0}, \tag{43}$$

where $\bar{\phi}_x \equiv \bar{v}_x/v_{z0}$ is the angular deviation of the averaged electron trajectory in the x direction off the z axis, where averaging is over the synchrotron oscillations (see Fig. 5a). As the beam propagates along the wiggler axis, the ellipse in Fig. 7 stretches in the $\bar{x}$ coordinate of phase space. Consequently, the beam width is monotonously increasing; however, the angular spread (half-maximum opening angle) remains constant $\phi_{bx} = \phi_{bx0}$.

The maximum extent of the ellipse in the $\bar{x}$ direction, x_b, at some position z within the wiggler region is given by

$$x_b(z) = \sqrt{x_{b0}^2 + (z\phi_{bx0})^2}. \tag{44}$$

If the initial half-opening angle of the beam is limited by the beam emittance $\varepsilon_x = \pi x_{b0}\phi_{bx0}$, then at $z = L$ we have

$$x_b(L) = \sqrt{x_{b0}^2 + [(\varepsilon_x/\pi)(L/x_{b0})]^2}. \tag{45}$$

Thus, for a given interaction length L and emittance ε_x, the minimum drift in the x direction occurs when

$$x_{b0}^{opt} = [(\varepsilon_x L/\pi)]^{1/2}. \tag{46}$$

The minimum beam spread that can be obtained at the end of the wiggler is then

$$[x_b(L)]^{min} = \sqrt{2}\, x_{b0}^{opt} = [(2\varepsilon_x L/\pi)]^{1/2}. \tag{47}$$

Slightly better focusing of the beam is possible if it is initially focused so that its waist point is at the center of the wiggler $x_b(L/2) = x_{b0}$. Of course, this requires a separate means of focusing in the x and in the y directions. In this case the beam width as a function of z is given by

$$x_b(z) = \{x_{b0}^2 + [(\varepsilon_x/\pi)(z - L/2)/x_{b0}]^2\}^{1/2}. \tag{48}$$

The optimal waist size for a given wiggler length L and emittance ε_x is then

$$x_{b0}^{opt} = [(\varepsilon_x L/2\pi)]^{1/2}. \tag{49}$$

The minimal beam spread at the wiggler ends is

$$[x_b(0)]^{\min} = [x_b(L)]^{\min} = \sqrt{2}\, x_{b0}^{\text{opt}} = [(\varepsilon_x L/\pi)]^{1/2}. \tag{50}$$

Of course, better focusing is always possible if electron lenses are placed along the wiggler to provide beam confinement in the x direction.

B. HELICAL WIGGLER

An appropriate field configuration of a realizable helical wiggler can be

$$A_w(r) = A_w[-I_0(k_w r)\cos(k_w z) + I_2(k_w r)\cos(k_w z - 2\alpha)]\hat{e}_x$$
$$\quad - A_w[I_0(k_w r)\sin(k_w z) + I_2(k_w r)\sin(k_w z - 2\alpha)]\hat{e}_y, \tag{51}$$

where I_m is a modified Bessel function of the m order. The magnetic field that corresponds to this configuration is

$$\mathbf{B}_w(\mathbf{r}) = B_w[I_0(k_w r)\cos(k_w z) + I_2(k_w r)\cos(k_w z - 2\alpha)]\hat{e}_x$$
$$\quad + B_w[I_0(k_w r)\sin(k_w z) - I_2(k_w r)\sin(k_w z - 2\alpha)]\hat{e}_y$$
$$\quad - 2B_w I_1(k_w r)\sin(k_w z - \alpha)\hat{e}_z, \tag{52}$$

where $\mathbf{B}_w = k_w A_w$.

The solution of the electron equations of motion in the field (32) to second order in $k_w r$ is given by Tang (1982). The electron trajectories are described by the following equations:

$$p_x(\mathbf{r}) = eA_{wx}(x, y, z) + \gamma m\bar{v}_x, \tag{53}$$

$$p_y(\mathbf{r}) = eA_{wy}(x, y, z) + \gamma m\bar{v}_y, \tag{54}$$

$$\begin{bmatrix} \bar{x} \\ \bar{v}_x \end{bmatrix} = \begin{bmatrix} \cos(k_\beta z) & (\bar{v}_z k_\beta)^{-1}\sin(k_\beta z) \\ -\bar{v}_z k_\beta \sin(k_\beta z) & \cos(k_\beta z) \end{bmatrix} \begin{bmatrix} \bar{x}_0 \\ \bar{v}_{x0} \end{bmatrix}, \tag{55}$$

$$\begin{bmatrix} \bar{y} \\ \bar{v}_y \end{bmatrix} = \begin{bmatrix} \cos(k_\beta z) & (\bar{v}_z k_\beta)^{-1}\sin(k_\beta z) \\ -\bar{v}_z k_\beta \sin(k_\beta z) & \cos(k_\beta z) \end{bmatrix} \begin{bmatrix} \bar{y}_0 \\ \bar{v}_{y0} \end{bmatrix}, \tag{56}$$

where k_β and $v_{\perp w}$ are still defined by Eqs. (30) and (27), respectively. The resulting orbits are therefore similar to the orbits in the linear wiggler, except that in the helical wiggler there are synchrotron and betatron oscillations in both the x and y directions, as illustrated in Fig. 5c.

The electron axial velocity, averaged over the fast synchrotron oscillation, is given in the case of the helical wiggler by an expression slightly different from Eq. (31):

$$\bar{v}_z^2 = v^2 - v_{\perp w}^2 - (\bar{v}_{r0}^2 + \bar{v}_z^2 k_\beta^2 \bar{r}_0^2), \tag{57}$$

where $\bar{r}_0^2 \equiv \bar{x}_0^2 + \bar{y}_0^2$ and $\bar{v}_{r0}^2 = \bar{v}_{x0}^2 + \bar{v}_{y0}^2$.

The betatron oscillation trajectories (averaged over the synchrotron oscillations) in both the x and the y directions are similar to the betatron oscillation trajectories of the linear wiggler in the y direction. Hence Eqs. (32–34) for the oscillation amplitudes in the y direction ($\bar{y}^m$, $\bar{v}_y^m$, $\bar{\phi}_y^m$) also apply for the helical wiggler. Furthermore, similar expressions apply to the oscillation amplitudes in the x direction ($\bar{x}^m$, $\bar{v}_x^m$, $\bar{\phi}_x^m$) and also to the resultant transverse-oscillation amplitudes

$$\bar{r}^m = [(\bar{x}^m)^2 + (\bar{y}^m)^2]^{1/2}, \quad \bar{v}_\perp^m = [(\bar{v}_x^m)^2 + (\bar{v}_y^m)^2]^{1/2}$$

and

$$\bar{\phi}_\perp^m = [(\bar{\phi}_x^m)^2 + (\bar{\phi}_y^m)^2]^{1/2}.$$

The transformation of the beam distribution in phase space, as illustrated in Fig. 6, applies for the helical wiggler also in the x-coordinate phase space. If the beam is azimuthally symmetric, which means that it has the same elliptical phase-space distribution in the x and y coordinates (and in any other transverse direction), then we define the emittance of the beam as

$$\varepsilon = \pi r_{b0} \phi_{b\perp 0}, \tag{58}$$

where r_{b0} and $\phi_{b\perp 0}$ are the beam waist radius and the velocity spread half-opening angle, respectively. Then, if the beam is injected into the wiggler entrance at its waist point it will exhibit, in the general case, periodic envelope scalloping with a period $\lambda_\beta/2$. Corresponding to Eqs. (39)–(41), and Eq. (38), the maximum beam radius, averaged angular spread, and transverse-velocity spread are given by

$$r_b = \max[r_{b0}, \varepsilon/\pi k_\beta r_{b0}], \tag{59}$$

$$\bar{\phi}_{b\perp} = k_\beta r_b, \tag{60}$$

$$\bar{v}_{b\perp} = \bar{v}_z k_\beta r_b. \tag{61}$$

Minimum values for these parameters are obtained for an electron beam of a given emittance value when it is initially focused to an optimal waist size:

$$r_{b0}^{\text{opt}} = k_\beta^{-1} \phi_{\perp 0}^{\text{opt}} = [\varepsilon/\pi k_\beta]^{1/2}. \tag{62}$$

In this particular case the beam propagates along the wiggler with a uniform cross section and with minimum axial-velocity spread [as evidenced from Eq. (57)].

C.   SELF-FIELD EFFECTS

In a dense electron beam, the electron trajectories may be affected by the repelling radial electric field that is generated by the beam's uncompensated space charge and by the azimuthal self-magnetic field that is generated by the beam current. To account for the self-field effects on an electron beam that propagates in a wiggler, it is useful to consider the beam envelope equation that applies to an azimuthally symmetric beam propagating in a radially focusing field (Lawson, 1977; Neil, 1979):

$$\frac{d^2 r_{\rm b}}{dz^2} + k_\beta^2 r_{\rm b} - \frac{K}{r_{\rm b}} - \frac{(\varepsilon/\pi)^2}{r_{\rm b}^3} = 0. \tag{63}$$

The parameter k_β specifies the strength of the radial focusing field. In the case of a helical wiggler, it is the betatron oscillation wave number defined Eq. (30). The parameter K is the generalized perveance that measures the strength of the self-field

$$K = \frac{1}{2\pi} \frac{(\mu_0/\varepsilon_0)^{1/2} I_0}{mc^2/e} \frac{1/\gamma_{0z}^2 - f}{\gamma_0 \beta_z^2}, \tag{64}$$

where f is the charge–compensation ratio. For an uncompensated beam ($f = 0$), and also assuming $\beta_z = 1$ and using Eq. (7),

$$K = \frac{1}{2\pi} \frac{(\mu_0/\varepsilon_0)^{1/2} I_0}{mc^2/e} \frac{1 + \bar{a}_{\rm w}^2}{\gamma_0^3}. \tag{65}$$

It is easy to verify that for the case of a beam propagating in free space without focusing ($k_\beta = 0$), and with negligible self-field effects ($K = 0$), the solution of Eq. (63) is given by Eqs. (44), (45) (replace $x_{\rm b}$ by $r_{\rm b}$). Also, in Eq. (63), by setting $K = 0$ and $dr_{\rm b}(z)/dz = 0$, one finds that in the case of propagation in a radial focusing field with negligible self-field effects, the uniform-envelope beam solution $r_{\rm b}(z) = r_{\rm b0}^{\rm opt}$ is given by Eq. (62). The uniform-envelope beam solution in a focusing field can also be readily found for the case where $K \neq 0$:

$$r_{\rm b}^{\min} = \{K/2 + [(K/2)^2 + (k_\beta \varepsilon/\pi)^2]^{1/2}\}^{1/2}/k_\beta. \tag{66}$$

Clearly the condition for neglecting self-field effects in this case is

$$K \ll (2/\pi) k_\beta \varepsilon, \tag{67}$$

when Eq. (66) reduces to Eq. (62).

In more general cases it is hard to find analytical solutions to Eq. (63). In most cases one must resort to numerical solutions. For the case of a beam propagating in free space without focusing ($k_\beta = 0$), we can derive directly from Eq. (63) a sufficient condition for neglecting self-field effects by requiring that the third term in the equation (expansion rate caused by self-fields) will be smaller than the fourth term (expansion rate caused by finite emittance):

$$|K| \ll (\varepsilon/\pi r_{\rm b})^2. \tag{68}$$

The condition for neglecting beam expansion caused by self-field effects in cases when they dominate over the finite emittance and focusing effects can be found by series expansion of $r_b(z)$ around r_{b0} and substitution in Eq. (63), assuming $k_\beta = \varepsilon = 0$. The condition for negligible self-field expansion beyond a length L is then found to be

$$K \ll 2r_{b0}^2/L^2. \tag{69}$$

In the consideration of beam propagation in an FEL wiggler, inequality (68) is a sufficient condition for neglecting self-field effects altogether. In the helical wiggler and in the y (betatron oscillation) dimension of the linear wiggler it may be sufficient to satisfy the more lenient condition (67) when $L/\lambda_\beta > 1$ and $r_b = r_b^{opt}$, because then the focusing forces dominate. In the x (synchrotron oscillation) direction of the linear wiggler, it is necessary to keep $x_{b0} \geq x_{b0}^{opt}$ [Eq. (69)] to avoid beam expansion caused by finite emittance. Simultaneous satisfaction of Eq. (69) is a necessary and sufficient condition for neglecting beam expansion caused by either finite emittance or self-field effects. In most cases the current in rf accelerator beams is too low to dissatisfy conditions (67), (68), and (69). However, in low-emittance, low-energy beams these conditions, and particularly Eqs. (68) and (69), should be taken into account.

Finally, in this section we consider the effect of the space-charge radial field on the energy spread of the electron beam. By integrating the radial space-charge field across the beam, it is straightforward to calculate the potential depression in the center of the beam relative to its envelope:

$$V_{sc} = (-en_0/4\varepsilon_0)r_b^2 = 30I_0/\beta_z, \tag{70}$$

where I_0 is measured in amperes and V_{sc} in volts. In the highly relativistic limit $\beta_z = 1$, we write the expression for the space-charge-related energy spread in the form

$$(\Delta E/E)_{sc} = (6 \times 10^{-5})(I_b/\gamma), \tag{71}$$

where I_b is given in amperes. This energy-spread contribution us usually negligible with rf accelerator parameters.

D. LIMITATIONS ON FEL DESIGN SET BY ELECTRON-BEAM PARAMETERS

The careful consideration of the electron orbits and the electron-beam propagation characteristics in the wiggler fields was necessary in the first place to provide design criteria that insure good transport of the beam along an interaction region with finite transverse dimensions. The limitations on the transverse dimensions of the electron beam are also set by the design parameters of the laser resonator, keeping in mind that the electron-beam width should always be smaller than the EM wave width. These considerations will be elaborated in Section III.

In addition to proper beam propagation, the design of an FEL should also take into consideration the effect of the beam parameters on the FEL interaction characteristics. Clearly, the FEL gain parameter $\bar{Q}$ [Eq. (1)] depends on only two beam parameters: the current $I_0 = J_0 A_b$ and the beam energy parameter γ_0. Because the beam energy in rf accelerators can be adjusted at will within a wide range of values, we prefer in this work to derive scaling laws in which the beam energy appears as an implicit parameter that is determined by other parameters through the synchronization condition (10), (7). Thus the gain parameter $\bar{Q}$ appears in Eqs. (19), (20) to depend on only the current parameter I_0 of the beam.

Only in the low-gain, cold, tenuous-beam regime [Eqs. (13), (14)] and in the high-gain, strong-pump regime (Gover and Sprangle, 1981) does the FEL gain function depend on only the parameter $\bar{Q}$. In the warm-beam gain regime, the FEL gain [Eq. (17)] depends also on the beam-velocity spread parameter $\bar{\theta}_{th}$, which is a function of the beam axial-velicity spread parameter (v'_{zth}). In the collective gain regimes, the gain depends on the beam space-charge parameter $\bar{\theta}_p$, which is a function of the beam current density [Eqs. (2), (6)].

In most cases of interest we will try to stay away from FEL design for operation in the warm beam regime $\bar{\theta}_{th} \gg \pi$. In this regime the maximum gain [Eq. (17)] is strongly reduced in comparison with the cold beam gain, by a factor $\bar{\theta}_{th}^2$. Thus the beam axial-velocity spread parameter v'_{zth} will usually appear in FEL design problems under the restriction $\bar{\theta}_{th} \ll \pi$, which can be written in the form

$$2v'_{zth}/c \ll \lambda/L. \tag{72}$$

The significance of this condition is that the allowed spread in axial electron velocities in the beam should be small enough to keep the detuning parameter $\bar{\theta}$ [Eq. (4)] fixed with an uncertainty smaller than π, i.e., $\Delta\bar{\theta} \ll \pi$, and it can be directly derived by differentiating Eq. (4).

Axial-velocity spread in the beam can result from spread in the total acceleration energy of the beam electrons or from their transverse-velocity spread. To estimate the axial-velocity spread of an electron beam in a linear wiggler, let us expand the velocity of a single electron in the beam [Eq. (31)] into a Taylor series around the velocity of the ideal electron trajectory for which $v = v_0$ (the average beam velocity) and $v_{x0} = v_{y0} = 0$:

$$\bar{v}_z = v_{z0} + (1 + \bar{a}_w^2)\, \Delta v - \bar{v}_z$$
$$= v_{z0} + (1 + \bar{a}_w^2)\, \Delta v - (v_{x0}^2/2v_{z0}) - [(\bar{v}_y^m)^2/2v_{z0}], \tag{73}$$

where $v_{z0} = (v_0^2 - v_{\perp w}^2/2)^{1/2}$, $\bar{v}_y^m$ is given by Eq. (33), and we used the identity

$$\partial \bar{v}_z/\partial v = (\gamma^2/\gamma_z^2)(\beta/\beta_z) \simeq 1 + \bar{a}_w^2. \tag{74}$$

The second, third, and fourth terms on the right-hand side of Eq. (73) are limited by the beam energy spread and the beam angular spread in the x and y directions, respectively. It follows that the three contributions to axial-velocity spread are

$$(v_{zth}')_{\Delta E/E} = (1 + \bar{a}_w^2)v_{th} = c[(1 + \bar{a}_w^2)/\gamma^2](\Delta E/2E), \tag{75}$$

where v_{th} is defined as the half-width of the beam velocity spread and ΔE is the full width of the beam energy spread,

$$(v_{zth}')_{\varepsilon_x} = v_{bx0}^2/2v_{z0} = \tfrac{1}{2}v_{z0}(\varepsilon_x/\pi)^2/x_{b0}^2, \tag{76}$$

$$(v_{zth}')_{\varepsilon_y} = [(\bar{v}_{by}^m)^2/2v_{z0}] = \tfrac{1}{2}v_{z0}k_\beta^2 y_b^2, \tag{77}$$

where y_b is given by Eq. (19).

Substituting the axial-velocity contributions [Eqs. (75)–(77)] into the inequality (72), we get the following three conditions for operating the FEL in the cold beam regime:

$$\Delta E/E \ll \tfrac{1}{2}\lambda_w/L = 1/2N_w, \tag{78}$$

where N_w is the number of wiggler periods along the interaction length,

$$[(\varepsilon_x/\pi)/x_{b0}]^2 \ll \lambda/L, \tag{79}$$

$$y_b^2 \ll (1/k_\beta^2)(\lambda/L) = 1/4\pi^2\,(1 + \bar{a}_w^2/a_w^2)(\lambda_w^3/L), \tag{80}$$

where in Eqs. (78) and (80) we used the synchronization conditions (10) and (7) to eliminate λ.

For a finite emittance beam, there is a limit to which the electron beam width can be reduced in the y direction, $y_b = y_b^{opt}$, given by Eq. (18). For the optimal value, Eq. (80) reduces to

$$\varepsilon_{ny} \ll (1/2\sqrt{2})[(1 + \bar{a}_w^2)/a_w]\lambda_w^2/L, \tag{81}$$

where $\varepsilon_{ny} \equiv \beta_0\gamma_0\varepsilon_y$ is the normalized emittance of the electron beam in the y direction. It is thus desirable to have separate focusing of the beam in the x and y directions. In the y direction, the beam should be injected at its waist point with width $y_b = y_b^{opt}$, and in the x direction it can be injected with a wider waist size x_{b0} that would be limited only by the transverse dimension of the radiation wave. Notice that if there is no separate focusing in both dimensions and the beam is azimuthally symmetric ($x_{b0} = y_{b0}$, $\varepsilon_x = \varepsilon_y$), then condition (79) is automatically satisfied when Eq. (80) is satisfied [use Eq. (39) for the comparison], and it is then sufficient to consider only inequalities (78) and (80).

In the case of a helical wiggler, the derivation of conditions for the cold beam regime is very similar. Series expansion of Eq. (57) instead of Eq. (31) results in

$$\bar{v}_z = v_{z0} + (1 + \bar{a}_w^2)\,\Delta v - [(\bar{v}_\perp^m)^2/2v_{z0}],\tag{82}$$

instead of Eq. (73), where $v_{z0} = (v_0^2 - v_{\perp w}^2)^{1/2}$. The energy spread condition (78) applies in this case without change. Instead of Eqs. (79) and (80) the conditions on the maximum radius and transverse emittance of the beam are

$$r_b^2 \ll (1/k_\beta^2)(\lambda/L) = (1/4\pi^2)[(1 + \bar{a}_w^2)/\bar{a}_w^2](\lambda_w^3/\lambda),\tag{83}$$

where r_b is given by Eq. (59). For $r_b = r_b^{opt}$, Eq. (62) gives

$$\varepsilon_n \ll (1/2\sqrt{2})[(1 + \bar{a}_w^2)/(\bar{a}_w)](\lambda_w^2/L),\tag{84}$$

where $\varepsilon_n = \beta_0\gamma_0\varepsilon$ is the normalized emittance of the electron beam. The slight difference between the numerical coefficients of Eqs. (83), (84), and (80), (81) is because in the helical wiggler $\bar{a}_w = a_w$ and in the linear wiggler $\bar{a}_w = a_w/\sqrt{2}$, which produces different expressions for the betatron wave number k_β [Eqs. (10) and (7)] when expressed in terms of the rms normalized vector potential $\bar{a}_w$ [Eq. (8)].

Another beam parameter that should be evaluated to find the FEL operating regime is the space-charge parameter $\bar{\theta}_p$ [Eq. (2)]. A sufficient condition for operating out of the collective gain regime is

$$\bar{\theta}_p \ll \pi.\tag{85}$$

This condition is usually satisfied for practical design parameters of rf accelerator FELs in the optical regime.

It is interesting to note that the generalized perveance parameter K Eqs. (65)] can be written in terms of $\bar{\theta}_p$, as given here:

$$K = \tfrac{1}{2}(\bar{\theta}_p r_b/L)^2.\tag{86}$$

The condition for neglecting beam expansion caused by space-charge repulsion [Eq. (68)] that is applicable for the linear wiggler in the x direction reduces, then, for $r_b = x_b^{opt}$ [Eq. (66)] into

$$\bar{\theta}_p \ll 2,\tag{87}$$

hence it is not possible to design an FEL experiment that operates in the collective regime with a linear wiggler unless there are means of focusing the beam along the wiggler to prevent its expansion in the x direction.

In the helical wiggler, substitution of Eq. (85) in Eq. (68) for $r_b = r_{b0}^{opt}$, Eq. (62) produces the following condition for neglecting beam expansion caused by the space-charge field:

$$\bar{\theta}_p \ll 4\pi(L/\lambda_\beta).\tag{88}$$

Hence, in a helical wiggler that has a number of betatron oscillation periods along its length, it is possible in principle to keep the beam confined to the wiggler axis and still operate in a collective regime $\bar{\theta}_p > \pi$. This can happen, of course, only if the rf accelerator can provide enough current to violate condition (85). Substituting $\bar{\theta}_p$ Eq. (85) in terms of the basic FEL parameters Eqs. (2), (6), (7), and (10) and assuming $r_b = r_{b0}^{opt}$ [Eq. (62)] we can express Eq. (85) as a condition on the wavelength:

$$\lambda \ll \left[\frac{\pi}{4} \frac{mc^2/e}{\sqrt{\mu_0/\varepsilon_0}\, I_0} \frac{1 + \bar{a}_w^2}{\bar{a}_w} \frac{\varepsilon_n \lambda_w^{5/2}}{L^2} \right]^{2/3} . \tag{89}$$

With practicable rf-accelator FEL parameters, this condition is usually well satisfied in the visible and near-infrared regime.

III. The Optical Cavity

It is clear from the expression for the gain parameter [Eq. (19)] that for a given beam current I_0 it is desirable to decrease the effective electromagnetic-wave area A_{em} to the smallest size consistent with the restriction $A_{em} \geq A_b$, where A_b, the beam cross-sectional area, is limited by the beam transport conditions of Section II. In a laser resonator that consists of two concave mirrors, the eigenmodes are free-space-propagating modes that exhibit larger diffraction the smaller their beam-waist size is. It thus may be impossible to keep the wave cross-sectional area A_{em} equal to its optimal value A_b (filling factor $A_{em}/A_b = 1$) for a long interaction length. In this case, having an electromagnetic-wave cross-sectional area A_{em} larger than A_b may be necessary to keep efficient interaction along the entire interaction length. The alternative that assures an electromagnetic wave with a small uniform cross section along the entire interaction length is to use a waveguide. However, the drawback with such an approach is that waveguide losses increase as the waveguide cross-sectional dimensions decrease. In this section, we consider both approaches to the design of an optical resonator.

The effective cross-sectional area for an electromagnetic wave is defined in the following way:

$$A_{em}(z) \equiv \left[\iint dx\, dy\, |\mathbf{E}(x, y, z)|^2 \right] / |\mathbf{E}(0, 0, z)|^2, \tag{90}$$

and we define the effective radius as $r_{em} \equiv (A_{em}/\pi)^{1/2}$.

A. OPEN RESONATOR

In a Fabry–Perot open resonator, the eigenmodes that propagate between the back and front concave mirrors are Gaussian radiation beams. The fundamental mode is (Lawson, 1977)

$$\mathbf{E}(r, z) = \mathbf{E}(0)[w_0/w(z)] \exp\{-ikz + i\eta(z) - ikr^2/2R(z) - r^2/w^2(z)\}, \tag{91}$$

where

$$w(z) = w_0[1 + (z - z_w)^2/z_R^2]^{1/2}$$

is the beam spot size, w_0 the beam waist size, z_w the location of the beam waist, and $z_R \equiv \pi w_0^2/\lambda$ the Rayleigh length that indicates the length of almost uniform beam spot size:

$$w_0 < w(z) < \sqrt{2}\,w_0 \qquad \text{for} \quad |z - z_w| < z_R.$$

Substitution of Eq. (22) into the definition of the effective wave area shows that

$$A_{em}'(z) = \tfrac{1}{2}\pi w^2(z). \tag{92}$$

For a given interaction length L, the overall beam cross section is minimized when the beam waist occurs at the center of the interaction region and the interaction length is two Rayleigh lengths:

$$L = 2z_R = 2\pi(w_0^2/\lambda). \tag{93}$$

As a consequence, $w(z) \simeq w_0 = (\lambda L/2\pi)^{1/2}$ and

$$A_{em} = \lambda L/4, \qquad r_{em} = (\lambda L/4\pi)^{1/2}. \tag{94}$$

Substitution of this result into the expression for the gain parameter [Eq. (18)] results in a new scaling law:

$$\bar{Q} = (2.62 \times 10^{-2})I_0(\lambda^{1/2}L^2/\lambda_w^{5/2})[\bar{a}_w^2/(1 + \bar{a}_w^2)^{3/2}]. \tag{95}$$

This expression is valid only as long as the optimal effective wave radius r_{em} given by Eq. (25) is larger than the electron-beam radius

$$r_{em} = (\lambda L/4\pi)^{1/2} > r_b. \tag{96}$$

When condition (27) is not satisfied, it means that we can choose

$$A_{em} = A_b = \pi r_b^2, \qquad r_{em} = r_b, \tag{97}$$

which corresponds to a choice of Gaussian waist size $w_0 = \sqrt{2}r_b$ and a unity filling factor $A_{em}/A_b = 1$. This is the case of negligible wave diffraction when the Rayleigh length is larger than the interaction length. The gain parameter scaling is then given by Eq. (19) with Eq. (28) used for the effective electromagnetic-wave area.

B. WAVEGUIDE RESONATOR

When inequality (27) is satisfied, because of the diffraction in the open resonator we have to use an electromagnetic-wave effective area that is larger than the obtainable electron beam cross-sectional area. In such a case, it may be advantageous to decrease the effective electromagnetic cross section by use of a waveguide. Hollow metallic or dielectric ("leaky") waveguides have been investigated in the past as possible waveguide structures in the infrared and far-infrared regimes (Garmire *et al.*, 1976; Marcatili and Schmeltzer, 1964; Nakahara and Kurauchi, 1967), and their use in optical resonators has been successfully demonstrated (Smith *et al.*, 1981; Yamanaka, 1977). Various kinds of "leaky" waveguides have been investigated, including circular waveguides, parallel-plate waveguides, rectangular waveguides, and composite metal–dielectric rectangular waveguides. Our attention will be focused first on circular waveguides, which is the most appropriate geometry when an electron beam with a circular cross section is used. Rectangular and parallel-plate waveguide structures may be useful mostly in FELs with linear wigglers, wherein the smallest electron-beam cross section is usually rectangular. We will consider rectangular waveguide in this application. The parallel-plate (open) waveguide structure is also endowed with some additional technical advantages, specifically, ease of vacuum pumping and convenience in electron-beam insertion and ejection. The interested reader is referred to the review and research articles (Adam and Kreubühl, 1975; Garmire *et al.*, 1976, Nishihara *et al.*, 1974; Smith *et al.*, 1981; Yamanaka, 1977) for further details of rectangular waveguides. It is worth mentioning that the attenuation laws are similar for both circular and rectangular waveguides.

The most appropriate mode for use in an FEL resonator with a circular waveguide is the EH_{11} hybrid mode, which has relatively low attentuation and is characterized by a maximum amplitude on the axis of the waveguide. The field structure of this mode is given by

$$E(r, z) = E_0 \hat{e}_x J_0(ur/a) \exp(-ikz), \tag{98}$$

where J_0 is the zero-order Bessel function of the first kind, a the waveguide radius, and $u \simeq 2.405$ the first zero of the Bessel function. A schematic of the field-line geometry of this mode is shown in Fig. 8a. The power attenuation constant of this mode is

$$2\alpha_{11} \equiv \frac{dP}{dz}\frac{1}{P} = 2\left(\frac{u}{2\pi}\right)^2 \frac{\lambda^2}{a^3} \operatorname{Re} \frac{v^2 + 1}{2(v^2 - 1)^{1/2}}, \tag{99}$$

where v is the complex index of refraction of the waveguide wall material. The expression for the integrated power attenuation of any mode over the length L of the waveguide is

$$L_G = 2\alpha L = A(\lambda^2 L/a^3), \tag{100}$$

where A is a parameter characteristic of the waveguide geometry, material, and the particular mode of interest. For example, for an EH_{11} hybrid mode

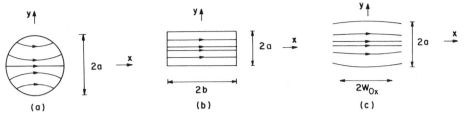

FIG. 8   The cross-section geometry and the mode E-field line configuration in a number of waveguides considered for FEL applications: (a) EH_{11} mode in a circular waveguide; (b) TE_{01} mode in a rectangular waveguide; (c) TE_{01} mode in a parallel-plate waveguide.

with a 10.6 μm wavelength in a circular copper waveguide, $v \simeq 14.2 - i64.5$ (Lenham and Terhune, 1966), and consequently $A = 4.16$.

The effective electromagnetic-wave cross section is found by substitution of Eq. (98) in Eq. (90) and yields

$$A_{em} = 0.27\pi a^2, \qquad r_{em} = 0.52a. \tag{101}$$

Hence the net single-pass gain of the FEL is

$$\left(\frac{\Delta P}{P}\right)_{net} = \frac{\Delta P}{P} - L_G = 7.72 \times 10^{-3} I_0 \frac{\bar{a}_w^2}{(1 + \bar{a}_w^2)^{3/2}} \frac{\lambda^{3/2} L^3}{a^2 \lambda_w^{5/2}} F(\theta) - A \frac{\lambda^2 L}{a^3}, \tag{102}$$

where I_0 is the current in amperes. Evidently, it is desirable to optimize the waveguide radius to provide maximum net gain. At peak gain ($F(\bar{\theta}) = 0.27$) we find the optimal waveguide bore radius to be

$$a_{opt} = 7.2 \times 10^2 (A/I_0)(\lambda^{1/2} \lambda_w^{5/2}/L^2)[(1 + \bar{a}_w^2)^{3/2}/\bar{a}_w^2]. \tag{103}$$

For this optimal bore radius, the maximum net gain is obtained:

$$\left(\frac{\Delta P}{P}\right)_{net}^{max} = 1.34 \times 10^{-9} \frac{I_0^3}{A^2} \left(\frac{\lambda}{\lambda_w}\right)^{1/2} \left(\frac{L}{\lambda_w}\right)^7 \left(\frac{\bar{a}_w^2}{(1 + \bar{a}_w^2)^{3/2}}\right)^3. \tag{104}$$

It should be noted that these expressions for gain are valid subject to certain restrictions. In the first place, the effective radius of the radiation cross section (103) must be more than the electron beam radius:

$$r_{em}^{opt} = 0.52 a_{opt} > r_b. \tag{105}$$

When this condition is not satisfied, one should use a bore radius larger than a_{opt}, to keep the filling factor $A_{em}/A_b < 1$. In this case the optimal choice of the waveguide radius is such that $A_{em} = A_b$ (filling factor $= 1$). Considering Eq. (101), the optimal waveguide radius choice is then

$$a = 1.92 r_b. \tag{106}$$

In this case, and also when a is different from a_{opt} for technical or other reasons, the net single-pass gain should be calculated from Eq. (102). Also, since the cold, tenuous-beam low-gain expression (14) was used to find the optimal values Eq. (103), (104), these optimum values may be modified in cases when Eq. (102) does not apply, e.g., when higher gain ($\bar{Q} \gtrsim \pi$) is available or when short optical pulse effects (see Section VIII) modify the gain expression.

In the case of FELs with linear wigglers, it may be advantageous to use rectangular or parallel-plate waveguides. Normally, the electron beam width in the synchrotron wiggling direction $-2x_b$ would be larger than its thickness in the perpendicular (betatron oscillation) direction $-2y_b$, because of the focusing effect in the latter case. Consequently, we would prefer to use rectangular or parallel-plate waveguide with thickness $-2a$ in the y dimension that is smaller than their width in the x direction. Fortunately, the lowest attenuation modes in such waveguides are TE modes with electric-field lines parallel to the long (x) dimension of the waveguide cross section. Because this happens to be also the synchrotron wiggling direction, such modes are most appropriate for the FEL interaction.

The lowest attenuation mode in parallel-plate and rectangular waveguides is the TE_{01} mode, which is given approximately by

$$\mathbf{E}(x, y, z) = E_0 \hat{e}_x \cos[\pi(x/2b)] \cos[\pi(y/2a)] \exp(-ikz) \qquad (107)$$

in the rectangular waveguide configuration (Adam and Kreubühl, 1975), and by

$$\mathbf{E}(x, y, z) = E_0 \hat{e}_x \exp(-x^2/w_{0x}^2) \cos(\pi y/2a) \exp(-ikz) \qquad (108)$$

in the parallel-plate waveguide configuration (Nishihara et al., 1974). A schematic diagram of the field-line geometry of these modes is shown in Fig. 8b and 8c. The width of the rectangular waveguide is $2b$, and the width of the mode waist in the parallel-plate open waveguide is $2w_{0x}$. This latter width is determined by the spacing and radius of curvature of the parallel plates. Here it was assumed that the radius of curvature of the plates is large and their spacing small so that only minor focusing effect is exhibited by the wave in the x dimension, and the radiation mode width is uniform across the waveguide, $w_x(y) \simeq w_{0x}$.

The power attenuation of the TE_{01} mode in both cases is

$$2\alpha_{01} = (\lambda^2/8a^3) \, \text{Re}[1/(v^2 - 1)^{1/2}], \qquad (109)$$

and the expression for the integrated power attenuation is again given by Eq. (100). Expression (100) is approximately correct for all TE modes in both parallel-plate and rectangular waveguides. The approximation is well justified, especially when the aspect ratio of the waveguide cross section (b/a or w_{0x}/a) is large. Losses are then dominated by the closely spaced waveguide walls, and losses caused by absorption in the other two walls (in rectangular waveguide) or by diffraction losses in the x dimension (in parallel-plate waveguide) are negligible. For the example of a TE_{01} mode with a 10 μm wavelength in a rectangular or parallel-plate waveguide with copper walls ($v = 14.2 - i64.5$), one obtains from Eqs. (109) and (100) $A = 4.05 \times 10^{-4}$. This is a significantly smaller parameter value than the value we calculated for the EH_{11} mode in the circular waveguide.

The effective cross-sectional area of the waveguide electromagnetic mode is found by substituting Eq. (107) or (108) in Eq. (90). We find that

$$A_{em} = 4x_{em}y_{em}, \tag{110}$$

where the effective radiation mode thickness in the y dimension y_{em} is, in both cases,

$$y_{em} = a/2, \tag{111}$$

and its width is

$$x_{em} = b/2 \tag{112}$$

in the case of the rectangular waveguide or

$$x_{em} = w_{0x}\sqrt{\pi}/2^{3/2} \tag{113}$$

in the parallel-plate waveguide configuration.

For both rectangular and parallel-plate waveguides, we can now write the single-path net gain in a form similar to Eq. (102):

$$\left(\frac{\Delta P}{P}\right)_{net} = (1.21 \times 10^{-2})I_0 \frac{\bar{a}_w^2}{(1 + \bar{a}_w^2)^{3/2}} \frac{\lambda^{3/2}L^3}{ax_{em}\lambda_w^{5/2}} F(\bar{\theta}) - A\frac{\lambda^2 L}{a^3}. \tag{114}$$

At peak gain $[F(\bar{\theta}) = 0.27]$ we find that the optimal waveguide thickness a_{opt} for which the net gain [Eq. (114)] is maximal is

$$a_{opt} = 30.3 \left(\frac{A}{I_0}\right)^{1/2} \frac{x_{em}^{1/2}\lambda_w^{5/4}\lambda^{1/4}}{L} \frac{(1 + \bar{a}_w^2)^{3/4}}{\bar{a}_w}, \tag{115}$$

and the corresponding maximal net gain is

$$\left(\frac{\Delta P}{P}\right)_{net}^{max} = (7.2 \times 10^{-5}) \frac{I_0^{3/2}}{A^{1/2}} \frac{\lambda^{5/4}L^4}{x_{em}^{3/2}\lambda_w^{15/4}} \frac{\bar{a}_w^3}{(1 + \bar{a}_w^2)^{9/4}}. \tag{116}$$

As in the case of the circular waveguide, we should note that Eqs. (115) and (116) are valid only as long as the effective optimal thickness of the radiation mode is not smaller than the electron beam thickness:

$$y_{em}^{opt} = \tfrac{1}{2}a_{opt} > y_b. \tag{117}$$

When this condition is not satisfied, one should use a waveguide thickness $2a$ larger than $2a_{opt}$ to keep the whole electron beam within the radiation-mode cross section. In this case, the optimal choice for the waveguide thickness is that for which $y_{em} = y_b$. Considering Eq. (111), the optimal waveguide half-thickness is then

$$a = 2y_b. \qquad (118)$$

In this case, and also when a is different from a_{opt} for technical or other reasons, the net gain should be calculated directly from Eq. (114).

Because we assumed that the waveguide losses are independent of the radiation-mode width in the x direction x_{em} in both rectangular and parallel-plate waveguides, evidently the optimal choice of this parameter is

$$x_{em}^{opt} = x_b. \qquad (119)$$

Considering Eqs. (112) and (113), we get for the rectangular waveguide optimal half-width and the parallel-plate radiation-mode half-waist width, respectively,

$$b_{opt} = 2x_b, \qquad (120)$$

$$w_{0x}^{opt} = 2^{3/2} x_b / \sqrt{\pi}. \qquad (121)$$

One more important point should be made with regard to waveguide resonator FELs. The waveguide dispersion characteristic can modify the synchronization (10), and one should verify that this does not reduce the FEL radiation frequency. In a waveguide, the axial wave number of an electromagnetic mode is $k_z = k \cos \Theta_{em}$, where $k \equiv \omega/c$ and Θ_{em} is the zigzag angle of the mode. The detuning parameter definition (4) should be modified into

$$\bar{\theta} = [(\omega/v_{0z}) - k_z - k_w]L, \qquad (122)$$

and the synchronization condition (10) that is obtained from substituting $\bar{\theta} = 0$ is modified into

$$\lambda/\lambda_w \simeq \beta_{0z}^{-1} - \cos \Theta_{em}. \qquad (123)$$

By expanding $\cos \Theta_{em} \simeq 1 - \frac{1}{2}\Theta^2$, this condition can be written in the highly relativistic limit ($\gamma_{0z} \gg 1$) in the form

$$\lambda/\lambda_w \simeq (1 + \gamma_{0z}^2 \Theta_{em}^2)/2\gamma_{0z}^2. \qquad (124)$$

The zigzag angle Θ_{em} is defined in general by $\sin \Theta_{em} \equiv k_\perp/k$, where $k_\perp$ is the transverse wave number of the mode. For the circular waveguide EH_{11} mode $k_\perp = u/a$, where $u = 2.405$ is the first root of the zero-order Bessel function. Consequently,

$$\Theta_{em} = (1.2/\pi)(\lambda/a). \qquad (125)$$

In the rectangular and parallel-plate waveguides $k_\perp \simeq k_x = \pi/(2a)$ and k_y is negligible relative to k_x. Hence in both cases

$$\Theta_{em} = \tfrac{1}{4}(\lambda/a). \tag{126}$$

The condition for neglecting the waveguide dispersion effect is

$$\Theta_{em}^2 \ll 1/\gamma_{0z}^2 = 2\lambda/\lambda_w, \tag{127}$$

when Eq. (124) reduces back into Eq. (10). This is not a trivial condition, but it is usually satisfied in the short infrared-wavelength regime and with highly overmoded waveguides. In the present work we will assume condition (127) and always use the radiation condition (10).

C. Waveguide or Open Resonators for FELs?

Comparison of the maximum-gain expressions of the open resonator [Eqs. (95), (13)] with the circular waveguide resonator [Eq. (104)] *in the regimes where they both apply* shows that the waveguide resonator will produce larger gain than the open resonator if

$$N_w \equiv \frac{L}{\lambda_w} > 22.1 \left[\frac{A}{I_0} \frac{(1 + \bar{a}_w^2)^{3/2}}{\bar{a}_w^2} \right]^{2/5}. \tag{128}$$

Note that this criterion for the usefulness of waveguides is independent of wavelength in the regimes where it applies. Therefore, for an FEL design with a large number of wiggles there is an advantage in using a waveguide resonator.

In the case of rectangular or parallel-plate waveguides, the comparison of Eq. (116) with Eqs. (95), (13) would result in a criterion for the usefulness of a waveguide resonator that is different and more involved than (128). In this case, when the transverse dimensions of the electron beam are different from each other, it would be more correct to compare the rectangular waveguide resonator with an open resonator configuration having a different optimized radiation-beam waist in the x and y dimensions. Practically, this may be difficult to realize, and therefore it is advisable to compare the net gain of particular waveguide and open resonator designs without *a priori* considerations. It would be expected that in all cases when the open resonator design is diffraction limited [Eq. (96)] the use of a rectangular or parallel-plate waveguide will be advantageous, because the absorption losses given by Eq. (109) are usually miniscule and would lead to an optimal choice [Eqs. (118) and (119)] for which the filling factor is $A_b/A_{em} = 1$. In practice, however, these conclusions should be reexamined to verify that they still hold when additional loss factors (i.e., surface roughness loss, diffraction loss, and waveguide bending losses) are included.

A waveguide laser has the advantage that conventional flat mirrors can be used at its ends instead of curved mirrors. Curved mirrors are used in open resonators to account for the wave curvature from the Gaussian beam diffraction. In Fig. 1b we suggest using two flat "electron transparent" mirrors at the ends of the waveguide. Such mirrors can be employed using suspended metallic foils or metal-coated pellicles, or (in the far-infrared and submillimeter regimes) using fine metallic meshes. The advantage of such a scheme is that electron-beam deflectors at the input and output ends of the interaction region can be eliminated. Normally these deflectors add length to the resonator, thereby increasing the bounce time. Because the added length is outisde the wiggler, it does not contribute to the gain. The time the light spends in the added sections is called dead time. The larger the dead time, the slower the power buildup rate in the resonator. With electron beams of limited duration, the dead time can prevent the gain from reaching saturation. Electron-transparent mirrors can reduce the dead time, but their practicality has yet to be demonstrated experimentally. Problems are anticipated in possible damage or distortion of the mirror and in the degradation of the electron beam.

Waveguide cavities can also be operated with external curved mirrors placed away from the waveguide ends. Abrams (1972) has calculated that a Gaussian wave can be coupled into a circular waveguide input and excite the EH_{11} mode with efficiency as high as 98%. He also showed that the EH_{11} mode losses in gaps in the waveguide smaller than $0.07 \, a^2/\lambda$ will be smaller than 2% (Abrams, 1972; Smith *et al.*, 1981). Operation of a rectangular-waveguide cavity with external mirrors can require separate focusing in both transverse dimensions by nonspheric or complex mirror systems and is evidently less attractive.

Waveguide lasers of various kinds have been successfully operated in the whole infrared–far-infrared-wavelength regime. FEL lasers were operated with waveguide structures in the millimeter-wavelength regime (Parker *et al.*, 1982), but the only infrared FEL oscillator demonstrated so far was operated at Stanford with an open resonator (Deacon *et al.*, 1977). Attempts to operate this laser as a waveguide laser were unsuccessful (Elias, 1982). The reason for this is yet unknown, and further investigation of this scheme is required.

IV. Spontaneous Emission

To determine whether the free-electron laser oscillator will have time to reach saturation, we require knowledge of both the initial signal level and the growth rate. The coherent amplification process will be discussed in Section V. In this section we will treat the spontaneous (noise) radiation emitted by individual electrons to determine whether there is sufficient power in the noise spectrum to obtain saturation over the duration of the interaction time (the electron-beam macropulse duration). The spontaneous emission for a low-density beam is given by Freund (1981):

$$\eta(\omega, \Omega_{\mathbf{k}}) \equiv \frac{1}{V} \frac{d^2 P}{d\omega \, d\Omega_{\mathbf{k}}} = \frac{(2\pi)^6}{4\pi\varepsilon_0} \frac{\omega^2}{Vc^3} \frac{1}{T} \left[|\mathbf{J}_{\mathbf{k},\omega}|^2 - \frac{c^2}{\omega^2} |\mathbf{k} \cdot \mathbf{J}_{\mathbf{k},\omega}|^2 \right]_{k=\omega/c}, \quad (129)$$

where the emissivity $\eta(\omega, \Omega_{\mathbf{k}})$ is the power radiated per unit frequency ω per unit solid angle $\Omega_{\mathbf{k}}$ subtended by the wave vector $\mathbf{k}$ per unit volume V, $T = L/v_z$ the time required for an electron to traverse the wiggler, $\mathbf{J}_{\mathbf{k},\omega}$ the Fourier transform of the microscopic source current

$$\mathbf{J}(\mathbf{r}, t) = -e \sum_{i=1}^{N_b} \mathbf{v}_i(t) \delta[\mathbf{r} - \mathbf{r}_i(t)], \quad (130)$$

and the sum in Eq. (130) is overall N_b electrons in the beam. Equations (129) and (130) describe the general noise spectrum for some arbitrary ensemble of electrons. The detailed characteristics of particular configurations are included by means of the electron orbits in the source current [Eq. (130)].

Detailed analysis of the spontaneous emission of an electron beam in a magnetic wiggler was carried out earlier by a number of authors [e.g., Colson (1977) and Moltz (1951)]. Here we use the results of the recent analysis of Freund (1981) for a helical and a linear wiggler with axial magnetic fields. We reduce these results to the case of zero-axial magnetic field and cold beam.

A. LINEAR WIGGLER

The finite gradient effects of the wiggler discussed in Section II play a relatively minor role in the spontaneous emission. Hence, we substitute $k_w y = 0$ in Eq. (21) to Eq. (26) and write for the electron trajectories

$$v_x = v_{\perp w} \cos(k_w v_z t) + v_{x0},$$
$$v_y = v_{y0}, \quad v_z = v_{z0}, \quad (131)$$

where $v_{\perp w}$ is defined in Eq. (27), and

$$x = (v_{\perp w}/k_w v_z) \sin(k_w v_z t) + v_{x0} t + x_0,$$
$$y = v_{y0} t + y_0, \quad z = v_z t + z_0. \quad (132)$$

Substitution of Eqs. (131), (132) in Eqs. (129), (130) yields, after some manipulation,

$$\eta(\omega, \Theta, \Psi) = \frac{e^2 \eta_b \omega^2 L v_{\perp w}^2}{16\pi^3 \varepsilon_0 c^4 \beta_z} \int dP \, G_b(P) \sum_{l=-\infty}^{\infty} \frac{\sin^2 \frac{1}{2} \bar{\theta}_l}{(\frac{1}{2}\bar{\theta}_l)^2} l^2 \frac{J_l^2(b_x)}{b_x^2}$$

$$\times \left[\sin^2 \Psi + \cos^2 \Psi \left(\cos \Theta - \frac{\omega}{l k_w c} \sin^2 \Theta \right)^2 \right], \quad (133)$$

where $n_b \equiv N_b/V$ is the beam density and $J_l(x)$ the regular Bessel function of the first kind:

$$\bar{\theta}_l \equiv ((\omega/v_z) - k_z - lk_w)L, \tag{134}$$

$$\mathbf{k} \equiv k(\sin \Theta \cos \Psi, \sin \Theta \sin \Psi, \cos \Theta), \tag{135}$$

$$b_x \equiv (k_x v_{\perp w}/k_w v_z) = (a_w \omega/\gamma k_w c \beta_z) \sin \Theta \cos \Psi. \tag{136}$$

This result was obtained using a random phase approximation and replacing the discrete sum over individual electrons in the microscopic source current by a continuous integral over a beam distribution $F_b(P_x, P_y, P)$. The beam was assumed to have zero transverse-momentum distribution, so that $F_b(P_x, P_y, P) = \delta(P_x)\,\delta(P_y)G_b(P)$, where $P_x = \gamma m v_{x0}$ and $P_y = \gamma m v_{y0}$ are the transverse canonical momenta and P is the electron total momentum.

We now specify Eq. (133) for the case of a cold beam: $\bar{\theta}_{th} \ll \pi$, which is the regime of interest for FELs. In this case we can also substitute $G_b(P)$ with a delta function and obtain

$$\eta(\omega, \Theta, \Psi) = \frac{e^2 n_b a_w^2 \omega^2 L}{16\pi^3 \varepsilon_0 c^2 \gamma^2 \beta_z} \sum_{l=-\infty}^{\infty} \frac{\sin^2(\tfrac{1}{2}\bar{\theta}_l)}{(\tfrac{1}{2}\bar{\theta}_l)^2} l^2 \frac{J_l^2(b_x)}{b_x^2}$$

$$\times \left[\sin^2 \Psi + \cos^2 \Psi \left(\cos \Theta - \frac{\omega}{lk_w c} \sin^2 \Theta \right)^2 \right] \tag{137}$$

Near the axis, $\sin \Theta \to 0$ and therefore $b_x \to 0$. We note that all the harmonics in the summation in Eq. (137) are vanishing except $l = 1$. We also notice that $\bar{\theta}_1 = \bar{\theta}$, where $\bar{\theta}$ is the FEL detuning parameter [Eq. (4)]. Thus

$$\eta(\omega, \Theta, \Psi) = (e^2 n_b a_w^2 \omega^2 L/16\pi^3 \varepsilon_0 c^2 \gamma^2 \beta_z)[\sin^2(\tfrac{1}{2}\bar{\theta})/(\tfrac{1}{2}\bar{\theta})^2][J_1^2(b_x)/b_x^2]$$

$$\times [\sin^2 \psi + \cos^2 \Psi (\cos \Theta - (\omega/k_w c) \sin^2 \Theta)^2]. \tag{138}$$

On axis $\Theta = 0$, $b_x = 0$, and $J_1(b_x)/b_x \to \tfrac{1}{2}$, we have

$$\eta(\omega, \hat{e}_k = \hat{e}_z) = (e^2 n_b a_w^2 \omega^2 L/64\pi^3 \varepsilon_0 c^2 \gamma^2 \beta_z)[\sin^2(\tfrac{1}{2}\bar{\theta})/(\tfrac{1}{2}\bar{\theta})^2]. \tag{139}$$

The radiation curve $\sin^2(\tfrac{1}{2}\bar{\theta})/(\tfrac{1}{2}\bar{\theta})^2$ is illustrated in Fig. 9.

The parameter needed for later use is the spectral radiant intensity of the spontaneous emission $I(\omega, \hat{e}_k = \hat{e}_z)$, which is the power emitted on axis per unit solid angle per unit frequency by the entire beam. By multiplying Eq. (139) by the beam volume $V = A_b L$, we obtain on axis

$$I(\omega, \hat{e}_k = \hat{e}_z) = \frac{1}{4\pi} \left(\frac{\mu_0}{\varepsilon_0} \right)^{1/2} \frac{e I_0 L^2}{\lambda_w \lambda} \frac{\bar{a}_w^2}{1 + \bar{a}_w^2} \frac{\sin^2(\tfrac{1}{2}\bar{\theta})}{(\tfrac{1}{2}\bar{\theta})^2}, \tag{140}$$

where we used the synchronism condition (10) in the limit $\beta \to 1$, and expressed a_w in terms of its rms value $a_w = \tfrac{1}{2}\bar{a}_w$. Note that even though the

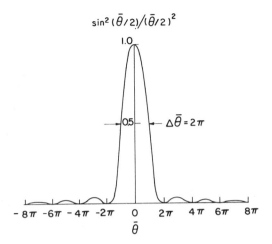

FIG. 9   The normalized spontaneous-emission radiation curve of the FEL as a function of the detuning parameter $\bar{\theta}$.

radiant intensity was derived with the assumption of a continuous beam in the wiggler, it applies as well for a pulsed beam shorter than L. In this case, Eq. (140) corresponds to the peak radiant intensity that will be measured for a duration equal to the duration of the electron beam pulse.

The emission curve $\sin^2(\tfrac{1}{2}\bar{\theta})/(\tfrac{1}{2}\bar{\theta})^2$ that is shown in Fig. 9 has half-amplitude full width of $\Delta\bar{\theta} = 2\pi$. By differentiating Eq. (134) with $k_z = \omega/c$ (on-axis emission) we obtain

$$\Delta\omega_{sp}/\omega = 1/N_w, \tag{141}$$

where $\Delta\omega_{sp}$ is the half-amplitude full width of the spontaneous emission line.

We now derive an expression for the angular spread of the spontaneous emission around the axis for fixed frequency ω_0 the center frequency of the on-axis radiation). The angular spread of the radiation pattern of the fundamental frequency given by Eqs. (136), (138) is determined mostly by the width of the sinc function $\sin^2(\tfrac{1}{2}\bar{\theta})/(\tfrac{1}{2}\bar{\theta})^2$, which is $\Delta\bar{\theta} = 2\pi$ for the full width. Differentiating $\bar{\theta} = \bar{\theta}_1$ [Eq. (134)] *for fixed* ω, using the expansion

$$k_z = k\cos\Theta \simeq k(1 - \tfrac{1}{2}\Theta^2),$$

we find that the half-opening angle of the angular spread is

$$\Theta_{sp} = (\lambda/L)^{1/2}. \tag{142}$$

Note that this angular spread corresponds to the radiation pattern at a fixed frequency $\omega = \omega_0$ (see Fig. 10).

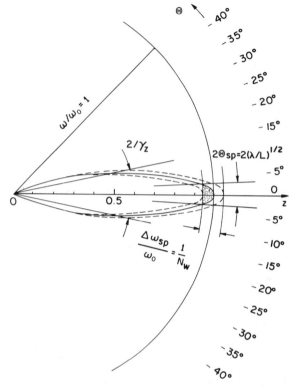

FIG. 10   Spontaneous emission frequency versus radiation direction pattern. The curves are drawn in polar coordinates:———, peak emission frequencies versus azimuthal angle Θ.

To verify that the sinc function controls the angular spread, we check the angular spread introduced by the $J_1^2(b_x)/b_x^2$ function in Eq. (138). The half-amplitude half-width of this function is approximately $\Delta b_x \simeq 1$. From Eq. (136), this corresponds to an angular spread $\Delta(\Theta \cos \Psi) = (1 + \bar{a}_w^2)/(2\gamma a_w) = [(1 + \bar{a}_w^2)^{1/2}/2\bar{a}_w](\lambda/\lambda_w)^{1/2}$. Evidently this angular spread is larger than Eq. (142) because $L \gg \lambda_w$.

In Fig. 10 we display in polar coordinates the spontaneous-emission radiation frequencies ω/ω_0 as a function of the radiation angle Θ, as determined by the sinc-function radiation curve (Fig. 9), where ω_0 is the emission peak frequency in the forward direction. The solid curve is the peak emission frequency as a function of Θ and corresponds to detuning parameter value $\bar{\theta} = 0$:

$$\frac{\omega(\Theta)}{\omega_0} = \frac{1 - \beta_z}{1 - \beta_z \cos \Theta} \simeq \frac{1/\gamma_z^2}{1/\gamma_z^2 + \Theta^2}. \tag{143}$$

In Fig. 10, the area between the broken-line curves corresponds to frequencies ω and directions Θ for which the detuning parameter is within the range $-\pi < \bar{\theta} < \pi$ (i.e., in the range of the emission peak). The spacing between these curves $\Delta\omega_{sp}/\omega$ is given by Eq. (141). The dotted section (Fig. 10) corresponds to frequencies and directions within the angular spread [Eq. (142)] and half the frequency spread [Eq. (141)] below the peak frequency ω_0, for which $-\pi < \bar{\theta} < 0$, and amplification by the stimulated emission curve peak (Fig. 2) can take place. The spontaneous emission frequency falls off to $\omega = \frac{1}{2}\omega_0$ at an angle $\Theta = 1/\gamma_z$. Because the emissivity [Eq. (138)] is proportional to ω^2, this is also the angle at which the total spontaneous emission power falls off by a factor 4. This angle is always larger than the fixed-frequency angular spread [Eq. (142)].

We have now equipped ourselves with the expressions for the spontaneous-emission spectral radiant intensity on axis [Eq. (140)], frequency linewidth [Eq. (141)], and angular spread [Eq. (142)]. All of these parameters are necessary to estimate the power that can be collected by the finite acceptance aperture and gain linewidth of the FEL; they will be used in Section V to estimate the power buildup in the FEL during the oscillation buildup period.

B. HELICAL WIGGLER

Analysis of the spontaneous emission in a helical wiggler follows the same argument as the linear wiggler and produces similar results. The electron trajectories in a helical wiggler, neglecting transverse gradient effects, are obtained by substituting $k_w r = 0$ in Eqs. (51)–(54):

$$v_x = -v_{\perp w}\cos(k_w v_z t) + v_{x0},$$
$$v_y = -v_{\perp w}\sin(k_w v_z t) + v_{y0}, \tag{144}$$
$$v_z = v_{z0},$$

$$x = (v_{\perp w}/k_w v_z)\sin(k_w v_z t) + v_{x0}t + x_0,$$
$$y = -(v_{\perp w}/k_w v_z)\cos(k_w v_z t) + v_{y0}t + y_0, \tag{145}$$
$$z = v_{z0}t + z_0.$$

Substituting Eqs. (144), (145) in Eqs. (129), (130) yields, instead of Eq. (133), the following:

$$\eta(\omega, \Omega_k) = \frac{e^2 n_b \omega^2 L v_{\perp w}^2}{16\pi^3 \varepsilon_0 c^4 \beta_z}\int dP\, G_b(P) \sum_{l=-\infty}^{\infty} \frac{\sin^2(\frac{1}{2}\bar{\theta}_l)}{(\frac{1}{2}\bar{\theta}_l)^2}$$
$$\times \left\{ l^2\frac{J_l^2(b)}{b^2}\left[\cos\Theta - \frac{k_\perp}{lk_w}\sin\Theta\right]^2 + J_l'^2(b)\right\}, \tag{146}$$

where J'_l is the first derivative of the Bessel function, $k_\perp = (\omega/c)\sin\Theta$, and

$$b = (k_\perp v_{\perp w}/k_w v_z) = (a_w \omega/\gamma k_w c\beta_z)\sin\Theta. \tag{147}$$

Again we take the cold-beam limit $\bar\theta_{th} \ll 1$ and note that in the present case $a_w = \bar a_w$:

$$\eta(\omega, \hat e_k = \hat e_z) = \frac{e^2 n_b \bar a_w^2 \omega^2 L}{16\pi^3 \varepsilon_0 c^2 \gamma^2 \beta_z} \sum_{l=-\infty}^{\infty} \frac{\sin^2(\frac{1}{2}\bar\theta_l)}{(\frac{1}{2}\bar\theta_l)^2}$$

$$\times \left\{ l^2 \frac{J_l^2(b)}{b^2}\left[\cos\Theta - \frac{\omega}{lk_w c}\sin^2\Theta\right]^2 + J'^2_l(b)\right\}. \tag{148}$$

The emissivity in the first-order harmonic frequency is

$$\eta(\omega, \Theta, \Psi) = \frac{e^2 n_b \bar a_w^2 \omega^2 L}{16\pi^3 \varepsilon_0 c^2 \gamma^2 \beta_z}\sin^2(\frac{1}{2}\bar\theta_l)$$

$$\times \left\{\frac{J_1^2(b)}{b^2}\left[\cos\Theta - \frac{\omega}{k_w c}\sin^2\Theta\right]^2 + J'^2_1(b)\right\}. \tag{149}$$

On axis we substitute $b = 0$, $J^1(b)/b \to \frac{1}{2}$, $J'_1(b) \to \frac{1}{2}$. After multiplying by the beam volume $V = A_b L$, we find that the radiant intensity on axis reduces exactly to Eq. (140), the same expression as for the linear wiggler.

Examination of Eq. (149) shows right away that, in the case of a helical wiggler, also the emission line width $\Delta\omega_{sp}$ and the angular spread of the radiation pattern are dominated by the sinc function $\sin^2(\frac{1}{2}\bar\theta)/(\frac{1}{2}\bar\theta)^2$, and therefore they are both given by the same equations (141), (142) that we derived earlier.

V. Power Buildup

To calculate the power buildup in the FEL resonator, it is necessary to know the power amplification of a single-resonator mode within the amplification linewidth frequency and the resonator losses. It is also necesary to know the characteristics of the initial source radiation, its bandwidth, and its angular spread. In general only a fraction of the initial radiation power will be amplified, because not all of it will always lie within the gain bandwidth of the FEL and the acceptance aperture of its resonator. We start the analysis of the power buildup by first deriving the multipath spectral-gain curve of the FEL.

Given the single-pass amplification in the free-electron laser, as well as the reflection coefficients and cavity loss rates for the lowest loss mode at a single frequency ω, the power buildup in the oscillator prior to saturation can be calculated by feeding back as the input the power reflected after each traversal through the cavity. The net power amplification per traversal can be written as $G_{net}(\bar{\theta}) = RG(\bar{\theta})$, where $G(\bar{\theta})$ is the single-pass gain (without loss) for the FEL interaction, and R is the round trip transmission factor that includes the reflectivity of the mirrors, transmission factors of any additional optical components (if applicable), and either diffraction or waveguide losses. In the low-gain, tenuous-beam limit ($\bar{Q} < \pi$), the single-path gain function is given by

$$G(\bar{\theta}) = 1 + \bar{Q}F(\bar{\theta}),\tag{150}$$

but the following formulation is general and applies to any gain regime. When $G(\bar{\theta})$ is defined by Eq. (11), it can be evaluated by program COLD (see Appendix C).

If an initial inserted-signal power is given by $P_{in}(\omega)$ with the appropriate polarization, then the spectral output power buildup after N passes is given by $P_N(\omega) = TG_N(\bar{\theta})P_{in}(\omega)$, where

$$G_N(\bar{\theta}) = G_{net}^N(\bar{\theta})\tag{151}$$

and T describes the transmission coefficient of the front mirror. Of course, this implicitly assumes that the signal injection time is much shorter than the round-trip transit time in the cavity, so that on the second and subsequent passes there is no contribution from outside sources. In contrast, when the signal is allowed to build up from spontaneous emission or an incoherent continuous-injection radiation source, the power after N traversals is found by adding up, after each round-trip bounce, the injected input power $P_{in}(\omega)$ and multiplying by $G_{net}(\bar{\theta})$, which gives

$$P_N(\omega) = TP_{in}(\omega) \sum_{n=1}^{N} [G_{net}(\bar{\theta})]^n,\tag{152}$$

The summation in Eq. (152) can be explicitly performed, and we find $P_N(\omega) = T\tilde{G}_N(\bar{\theta})P_{in}(\omega)$, where

$$\tilde{G}_N(\bar{\theta}) = \{[G_{net}(\bar{\theta})]^N - 1\}/[G_{net}(\bar{\theta}) - 1] \simeq G_N(\bar{\theta})/G_{net}(\bar{\theta}) - 1].\tag{153}$$

In the case where coherent radiation is continuously injected into the cavity, instead of Eq. (152) we need to sum up the field *amplitudes* when adding the injected radiation contribution in each bounce:

$$P_N(\omega) = TP_{in}(\omega) \left| \sum_{n=1}^{N} [g_{net}(\bar{\theta})e^{i\psi}]^n \right|^2,\tag{154}$$

where $g_{net}(\bar{\theta}) = \sqrt{G(\bar{\theta})}\sqrt{R}$ is the modulus of the round-trip net amplitude gain and ψ is the phase accumulated in one round trip owing to the phase contribution of the net gain factor, the round-trip optical path, and the optical components. The summation of the geometrical series results in a multipath gain function $G_N(\bar{\theta}) = [(g_{net}e^{i\psi})^N - 1]^2/|g_{net}e^{i\psi} - 1|^2$ instead of Eq. (153). This gain function attains a maximum value when $\psi = 2m\pi$ (m is an integer), i.e., when a cavity longitudinal mode is excited. In this case,

$$\tilde{G}_N(\theta) = \left(\frac{[G_{net}(\bar{\theta})]^{N/2} - 1}{[G_{net}(\bar{\theta})]^{1/2} - 1}\right)^2 \simeq \frac{G_N(\bar{\theta})}{\{[G_{net}(\bar{\theta})]^{1/2} - 1\}^2}. \qquad (155)$$

We now consider the question of the bandwidths implicit in each of the amplification processes discussed previously. If $\Delta\bar{\theta}$ describes the full width at half maximum of the multipath gain curve $G_N(\bar{\theta})$ about the peak gain point $\bar{\theta}_m$, then we write

$$G_{net}(\bar{\theta}_m \pm \tfrac{1}{2}\Delta\bar{\theta}) \simeq G_{net}(\bar{\theta}_m) + \tfrac{1}{2}G''_{net}(\bar{\theta}_m)(\tfrac{1}{2}\Delta\bar{\theta})^2, \qquad (156)$$

where $[G_{net}(\bar{\theta}_m \pm \tfrac{1}{2}\Delta\bar{\theta})]^N = \tfrac{1}{2}[G_{net}(\bar{\theta}_m)]^N$ by definition. As a result, we see that

$$\Delta\bar{\theta} = 2\sqrt{-(2 \ln 2/N)[G(\theta_m)/G''(\bar{\theta}_m)]}, \qquad (157)$$

where we have made use of the approximation $1 - 1/2^{1/N} \simeq \ln 2/N$ and assumed that R is not a sharply peaked function of the frequency.

The derivation of Eqs. (151)–(157) applies to any gain regime. In the case of low-gain tenuous beam [Eqs. (150), (13)] we observe that $\bar{\theta}_m = -2.6$, $G(\bar{\theta}_m) = 1 + 0.27\bar{Q}$, $G''(\theta_m) = (-8.845 \times 10^{-2})\bar{Q}$, and

$$\Delta\bar{\theta} = 4.11\{[-(0.27\bar{Q})^{-1}]/N\}^{1/2}. \qquad (158)$$

The normalized full-width frequency spread is related to this value by $\Delta\omega_N/\omega \simeq \Delta\bar{\theta}/(2\pi N_w)$, as can be verified by differentiating Eq. (4) and substituting Eq. (10). Hence in the regime where Eq. (158) applies we have

$$\Delta\omega_N/\omega = (0.654/N_w)\{[1 + (0.27\bar{Q})^{-1}]/N\}^{1/2}, \qquad (159)$$

which is clearly different from the single-pass linewidth $(\Delta\omega/\omega)_1 \simeq 1/(2N_w)$. Evidently, the linewidth narrows down with increasing N. Therefore, for large N the width of the gain curves (153), (155) is determined mostly by the numerator. Hence Eq. (159) for the linewidth applies for the short-pulse injection as well as for the continuous injection cases.

In a general gain regime in the cold beam limit, one can use program COLD (Appendix C) to evaluate $G(\bar{\theta}_m)$ and $G''(\bar{\theta}_m)$ and plug it into Eq. (8). For the tenuous beam limit $(\bar{\theta}_p = 0)$ intermediate gain regime $(0 < \bar{Q} < 15)$, we have calculated these parameters and displayed them in Figs. 4a, b. The drawn curves start to deviate significantly from the asymptotic low-gain values only for $\bar{Q} > 3$.

To find out the FEL power output after N traversals, we need to integrate $P_N(\omega)$ over frequencies using Eqs. (151) or (153). We also need to specify the effective power input P_{in}, which is the amount of input power that couples into the FEL resonator mode. We will consider three kinds of input power sources: coherent radiation injected into the cavity from an external source (another laser); spontaneous emission generated in the FEL; and incoherent radiation injected into the cavity (e.g., from a flash lamp).

To estimate the effective power input P_{in} that will be coupled into the FEL cavity mode and be amplified in the power-buildup process, we characterize the FEL cavity with an acceptance function that is defined by the linewidth of the FEL gain $\Delta\omega_N$ and by the volume in phase space occupied by the FEL electromagnetic mode. If we consider the excitation of a spatially coherent single transverse mode, then this phase-space volume is defined in terms of the effective wave area A_{em} and the effective solid angle that the mode occupies $\Delta\Omega_{em}$. A_{em} is given by Eqs. (92), (101), and (110) for the open-resonator circular and rectangular waveguide resonators, respectively. $\Delta\Omega_{em}$ is defined for the case of an open-cavity Gaussian mode in terms of the mode half-diffraction angle $\Theta_{em} = \lambda/(\pi w_0)$, as follows:

$$\Delta\Omega_{em} = \tfrac{1}{2}\pi\Theta_{em}^2 = \lambda^2/2\pi w_0^2, \tag{160}$$

where, with the factor $1/2$ in the definition (160), $\Delta\Omega_{em}$ is the solid angle that contains the mode power. For the circular waveguide cavity, $\Delta\Omega_{em}$ is defined in terms of the fundamental mode zigzag angle $\Theta_{em} = k_\perp/k = u\lambda/(2\pi a)$:

$$\Delta\Omega_{em} = (0.72\,\lambda^2/\pi a^2), \tag{161}$$

where $k_\perp$ is the transverse wave number of the waveguide mode, $k = 2\pi/\lambda$. In Eq. (161) we assumed the particular case of the EH_{11}-mode in a circular waveguide, for which $u = 2.405$ is the first root of the zero-order Bessel function [Eq. (98)].

In the case of the rectangular waveguide TE_{01}-mode [Eq. (107)], the mode zigzag angles are $\Theta_{em_x} = k_x/k = \lambda/(4b)$ and $\Theta_{em_y} = \lambda/(4a)$. Consequently, the solid angle occupied by the mode *power* is

$$\Delta\Omega_{em} = \tfrac{1}{2}(2\Theta_{em_x})(2\Theta_{em_y}) = \lambda^2/8ab. \tag{162}$$

Similarly, in the parallel-plate waveguide [Eq. (108)] $\Theta_{em_x} = k_{x0}/k = \lambda(\pi w_{0x})$ and $\Theta_{em_y} = \lambda/(4a)$, where $k_{x0} = 2/w_{0x}$ is the wave-number width of Eq. (108) when Fourier transformed in the x dimension. Consequently, for this mode,

$$\Delta\Omega_{em} = \lambda^2/2\pi a w_{0x}. \tag{163}$$

In all cases the effective input power P_{in} that couples into the resonator mode is the fraction of the total source-radiation power with frequency spread equal or smaller than $\Delta\omega_N$, with the effective cross-sectional area A_{em} and diffraction solid angle $\Delta\Omega_{em}$ of the node.

A. INJECTED COHERENT RADIATION

If the source radiation for the oscillation-buildup process is spatially coherent radiation injected into the cavity from another laser source, then the input radiation occupies the minimum phase-space volume possible. Assuming a Gaussian mode, this phase-space volume is

$$A_{em} \times \Delta\Omega_{em} = \tfrac{1}{2}\pi w_0^2 (\lambda^2/2\pi w_0^2) = \tfrac{1}{4}\lambda^2. \tag{164}$$

If this radiation is coupled into an open-resonator Gaussian mode, then with ideal optical components (lenses, mirrors) it is possible to transform the injected radiation wave into a Gaussian mode with the same waist size and waist location as the cavity mode. This can be done theoretically with 100% efficiency. If the injected radiation is coupled into a circular waveguide, e.g., in the case of an EH_{11} mode of a circular waveguide, the coupling can be made with 98% efficiency (Abrams, 1972). A coherent radiation source like a Gaussian beam can also be coupled with quite high efficiency into the TE_{01} mode of the rectangular and parallel-plate waveguides, but that may require separate focusing means in the x and the y dimensions.

In most cases, the frequency linewidth of the FEL $\Delta\omega_N$ is considerably larger than the linewidth $\Delta\omega_{inj}$ of conventional lasers that can be used for injecting seed radiation into the FEL cavity. In these cases, the entire injected power P_{inj} is an effective input power for the power-buildup process. In the general case, we define the FEL output linewidth by

$$\Delta\omega_{out} = \min[\Delta\omega_{inj}, \Delta\omega_N]. \tag{165}$$

The effective input power for the power-buildup process is

$$P_{in} = P_{inj}(\Delta\omega_{out}/\Delta\omega_{inj}). \tag{166}$$

The total FEL power output after N traversals is then

$$P_{out} = \int P_{out}(\omega)\, d\omega = T\begin{Bmatrix} G_N(\bar\theta_m) \\ \tilde{G}_N(\bar\theta_m) \end{Bmatrix} P_{in}, \tag{167}$$

where $G_N(\bar\theta)$ and $\tilde{G}_N(\bar\theta)$ are defined by Eqs. (151) and (153) for the cases of pulse injection and continuous injection, respectively.

B. SPONTANEOUS-EMISSION RADIATION SOURCE

When the input radiation source for the oscillation-buildup process is the FEL spontaneous emission, we observe that the radiation is always emitted within the cavity-mode cross-sectional area, because we always keep $A_b \leq A_{em}$. The spontaneous-emission frequency spread $\Delta\omega_{sp}$ [Eq. (141)] is always larger than the FEL linewidth $\Delta\omega_N$ [Eq. (157)]. If also the angular spread of the spontaneous emission $\pi\Theta_{sp}^2$ [Eq. (142)] is wider than the cavity-mode solid angle $\Delta\Omega_{em}$ [Eqs. (160)–(163)], then the effective input power is only the fraction of spontaneous-emission radiation within the solid angle $\Delta\Omega_{em}$ and frequency band $\Delta\omega_N$:

$$P_{in} = I(\omega_m, \hat{e}_k = \hat{e}_z)\,\Delta\Omega_{em}\,\Delta\omega_N, \tag{168}$$

where $I(\omega)$, the spontaneous-emission radiant intensity, is given in Eq. (140).

It can easily be verified that the condition $\pi\Theta_{sp}^2 \gg \Delta\Omega_{em}$ is satisfied for the case of an open resonator [$\Delta\Omega_{em}$ given by Eq. (160)] only if the interaction length L is shorter than the Rayleigh length. In the case of a waveguide resonator, the condition $\Theta_{sp} \gg \Theta_{em}$ [Θ_{em} given by Eqs. (125), (126)] is satisfied only when L is smaller than a waveguide mode zigzag-bounce length $2a/\tan\Theta_{em}$. Although these conditions are practical in the open-resonator case, they are of course entirely impractical in the case of a waveguide. In other words, in the case of a waveguide cavity where typically a number of zigzag bounces will take place along the interaction length L, the fixed-frequency spontaneous-emission angular spread is narrower than the angular spread defined by a single waveguide radiation mode. This peculiar result indicates an inaccuracy in our formulation. In this case the spatial Fourier transform analysis taken in Section IV should be modified to account for the finite transverse dimensions of the waveguide. Spontaneous emission in the waveguide at a fixed frequency will be emitted only with quantized transverse wave numbers that correspond to the waveguide modes. We point out the need of further theoretical analysis in this case, but we will continue to use Eq. (168) as a crude estimate of the spontaneous-emission effective input power in both open- and waveguide-resonator configurations.

The total FEL power output after N traversals is calculated by plugging Eq. (168) into Eq. (167). In Eq. (167) the incoherent continuous-source-radiation gain curve $\tilde{G}_N(\bar{\theta})$ [Eq. (153)] should be used because the spontan-

C. INJECTED INCOHERENT RADIATION

We consider the possibility of injecting seed radiation into the cavity from an incoherent source such as a flash tube or a high-temperature source. A source like this can be characterized by its spectral radiance (brightness) as

$$B(\omega, \mathbf{r}, \hat{e}_k) \equiv d^3P/d\omega\,dA_\perp\,d\Omega, \tag{169}$$

which is the power emitted per unit angular frequency per unit perpendicular-source area element per unit solid angle. Such a source will usually be of large size, large angular spread, and wide frequency spectrum. In considering how much of this source power can be efficiently coupled into the FEL cavity mode, we point out that an ideal optical system could at best keep the phase-space volume of a radiation beamlet $\Delta A_\perp \, \Delta\Omega$ unchanged. Thus the phase-space volume by which we have to multiply the source brightness [Eq. (169)] is the mode phase-space volume $A_{em} \times \Delta\Omega_{em}$. This is equal to $\frac{1}{4}\lambda^2$ for the open resonator Gaussian mode [Eq. (164)] and is roughly the same for the waveguide resonator modes. Considering the finite gain linewidth $\Delta\omega_N$ [Eq. (159)], the effective input power in this case is

$$P_{in} = \tfrac{1}{4} B\lambda^2 \, \Delta\omega_N. \tag{170}$$

This can be used in Eq. (167) to calculate the output power with the incoherent continuous-source radiation gain curve $G_N(\bar\theta)$ [Eq. (153)].

If the radiation source can be characterized by a temperature T, then from Planck's law it follows that

$$B(\omega) = \frac{1}{\pi}\frac{1}{c^2 h^2}\frac{(\hbar\omega)^3}{\exp(\hbar\omega/k_B T) - 1}, \tag{171}$$

where h is Planck's constant. In the limit $\hbar\omega \ll k_B T$ we obtain

$$B(\omega) = (1/\pi)(k_B T/\lambda^2), \tag{172}$$

and Eq. (170) assumes the particularly simple form

$$P_{in} = (k_B T \, \Delta\omega_N)/(4\pi). \tag{173}$$

To conclude this section we give here the approximate expression for the FEL power saturation level (Gover and Sprangle, 1981; Sprangle and Smith, 1980; Sprangle *et al.*, 1979, 1980). The beam power-extraction efficiency in the tenuous-beam, low-gain regime is given approximately by

$$\eta = 1/2N_w, \tag{174}$$

and hence the saturation power generation is

$$\Delta P_{sat} = V I_0 \eta = (\gamma_0 - 1)(mc^2/e)I_0(1/2N_w). \tag{175}$$

In Gover and Sprangle (1981) and Sprangle *et al.* (1980), expressions for the electron-beam extraction efficiency were given also in other gain regimes. For example, in the cold-beam gain, strong pump regime ($\bar{Q} \gg \pi, \bar\theta_p, \bar\theta_{th}$) the efficiency is

$$\eta = (1/2N_w)(\bar{Q}^{1/3}/2\pi). \tag{176}$$

This will be substantially higher than the low-gain efficiency only when $\bar{Q}$ is very large, which is when the exponential folding length of the electromagnetic-wave growth is much shorter than the interaction length L. In intermediate regimes and when accurate estimates are desirable, a numerical calculation of the saturation level is required (Hull and Iles, 1980). However, for the sake of preliminary design of rf acceleration FELs it is usually sufficient to use Eqs. (174), (175) as estimates for efficiency and saturation power. With practical FEL parameters, the gain parameter will be in the regime $\bar{Q} \lesssim \pi$ and examination of Eq. (176) suggests that, even in the intermediate regime $\bar{Q} \simeq \pi$, Eq. (175) is a good estimate.

The output power expression [Eq. (167)] is valid only as long as $P_{out} < P_{sat}$. If the rf accelerator macropulse is long enough to reach saturation, then the approximate number of round-trip traversals necessary for the oscillation to build up to saturation is calculated approximately by equating the output-power generation in the Nth traversal to the saturation single-path power extraction from the electron beam:

$$[G(\bar{\theta}_m) - 1] \times \begin{Bmatrix} G_{N-1}(\bar{\theta}_m) \\ \tilde{G}_{N-1}(\bar{\theta}_m) \end{Bmatrix} \times P_{in} = \Delta P_{sat}. \tag{177}$$

Using Eqs. (151), (153), or (155) for the multipath gain function, we obtain

$$N_{sat} = \frac{\log(\Delta P_{sat}/P_{in}) - \log[G(\bar{\theta}_m) - 1]}{\log G_{net}(\bar{\theta}_m)} + 1, \tag{178}$$

or

$$N_{sat} = \frac{\log(\Delta P_{sat}/P_{in}) - \log\{[G(\bar{\theta}_m) - 1]/[G_{net}(\bar{\theta}_m) - 1]\}}{\log G_{net}(\bar{\theta}_m)} + 1, \tag{179}$$

$$N_{sat} = \frac{\log(\Delta P_{sat}/P_{in}) - \log\{[G(\bar{\theta}_m) - 1]/[G_{net}^{1/2}(\bar{\theta}_m) - 1]^2\}}{\log G_{net}(\bar{\theta}_m)} + 1, \tag{180}$$

corresponding to the cases of instantaneous, continuous-incoherent, or continuous-coherent input power sources, respectively. In the specific case of operation in the low-gain regime [Eq. (12)], we can substitute in Eqs. (178)–(180) $G(\bar{\theta}_m) \simeq 1 + 0.27\bar{Q}$, $G_{net}(\bar{\theta}_m) \simeq 1 + 0.27\bar{Q} - (1 - R)$, where $0.27\bar{Q}, 1 - R \ll 1$.

It should be pointed out that the expression for the extraction efficiency [Eq. (174)] was derived with the assumption that saturation (electron trapping) takes place at the end of the interaction region of length L. Evidently, saturation will occur at an earlier point whenever the single-path small-signal gain is larger than the round-trip losses in the cavity (positive net gain). To accelerate the power-buildup process we always try to maximize the round-trip small-signal net gain; therefore, the oscillation power growth in a cavity with passive output coupling will not stop at the saturation start point when beam-extraction saturation (trapping) is reached right at the end of the interaction region. Rather, the spatial point of the beam saturation (trapping) will move backward to some point before the end of the inter-action length. When this saturation evolution takes place, a reabsorption process develops in the end of the interaction length, and a full nonlinear analysis is required (Colson, 1982; Kroll *et al.*, 1981; Liewer *et al.*, 1981; Prosnitz *et al.*, 1981; Sprangle *et al.*, 1980; Tang and Sprangle, 1982). This of course limits the validity of the approximation [Eq. (174). Furthermore, contrary to the warm-beam gain regime, the saturation process in the cold-beam gain regimes is homogeneously broadened (Gover, 1980; Yariv, 1976). This means that if the input source of radiation is not monochromatic (e.g., spontaneous emission), then a longitudinal-mode competition process will turn on once saturation is reached by a range of frequencies (longitudinal modes). Then, much more elaborate nonlinear theory is required to derive the radiation power development in the cavity. We conclude that the use of Eqs. (177) through (179) with the saturation-power generation expression (175) is valid only for determining at what time saturation of the peak gain frequency is attained right at the end of the interaction length. Only at this time is the radiative power generation given by Eq. (175). At later times, the calculation of the radiative power generation requires a full nonlinear analysis.

VI.   Seed Radiation Injection

Seed radiation injection provides an input radiation field in the oscillator cavity at the arrival time of the electron beam at a power level higher than the spontaneous emission or noise radiation. In this case, the free-electron laser operates as a regenerative amplifier until the saturation point is reached. Seed injection reduces the time required for the radiation to build up to interesting power levels. This is important in accelerators whose electron beams can only run for a few microseconds. The major problem with the idea is that the tunability of the free-electron laser is reduced to that of the laser providing the seed. By using sets of tunable lasers, e.g., lead-salt diode lasers (for operation in the infrared regime), the tuning range can still be sufficiencietly broad.

Most seed-injection schemes affect the cavity losses and the cavity dead time. Dead time is the time a photon spends outside the gain region of the laser in one round trip through the cavity. The losses and dead time are the major considerations in the design of injection systems for free-electron lasers with electron-beam macropulse times that are short compared with the time to saturate the gain.

Several seed-injection schemes are illustrated in Fig. 11a–e. In Fig. 11a, the seed radiation is brought in through a leaky mirror. This system has high

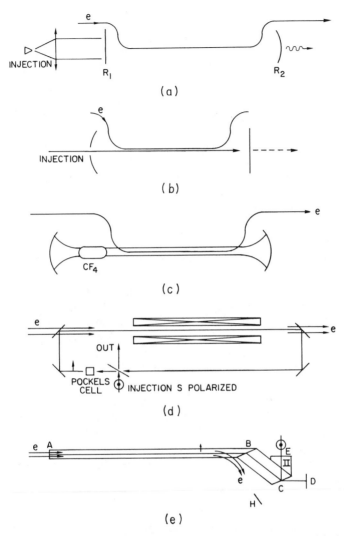

FIG. 11   Various schemes of seed radiation injection into an FEL cavity: (a) leaky mirror; (b) hole in mirror; (c) intracavity injection; (d) electrooptic switch in a ring laser; (e) electrooptic switch in a straight laser.

losses and permits only a small fraction of the light from the seed laser to enter the cavity.

Figure 11b shows an unstable resonator configuration similar to the CO_2 regenerative laser experiment of Bélanger and Boivin (1976). Seed radiation enters through a hole in one mirror and makes a number of passes before exiting the cavity. With this system it would be difficult to obtain a good fill factor.

Figure 11c shows an intracavity injection scheme. Here a CF_4 laser amplifier is placed inside the free-electron laser cavity. It is made to lase in the cavity just prior to the arrival of the electron pulse or throughout the pulse. The CF_4 lasers are line tunable and by replacing the gases they can cover wide infrared–far-infrared ranges (Tee and Wittig, 1977). Some thought would have to be applied to the problem of light absorption in the gas and dead time added by the length of the CF_4 laser (Tee and Wittig, 1977).

Figures 11d and 11e illustrate two variations of a class of injection schemes involving an electrooptic switch. Typically, an electrooptic switch consists of two components. One component rotates the polarization through 90 degrees, and the second transmits one polarization and reflects the other. Figure 11d illustrates the simplest application of this technique in a ring laser. Here the injection optics, consisting of a Brewster plate and a Pockels cell, are in the return path. Injection light is S polarized, so some of it reflects off the Brewster plate into the return path toward the pockels cell. The Pockels cell rotates the polarization through 90° so that it will couple with the electron beam in the wiggler and also pass through the Brewster plate without reflective losses on the return trip. When injection is complete, the Pockels cell is turned off and therefore transmits light without affecting the polarization. The light is trapped in the resonator and grows, because of the laser gain, until the cell is again energized. The trapped polarized light is then rotated to S polarization, which reflects off the Brewster plate to the output. The Brewster plate can be replaced with a number of alternate devices. For example, it could be a wire-grid polarizer (now made to work well on 10.6 μm light), a set of prisms arranged for interference-enhanced frustrated total internal reflection (Hull and Iles, 1980), or a semiconductor laser-actuated mirror (Alcock *et al.*, 1976). In this last instance, the Pockels cell would be superfluous.

A practical modification of the Brewster plate is shown in Fig. 11e in a straight laser configuration. Here the Pockels cell surfaces are cut at Brewster's angle. Mirrors A and D form the resonator cavity. S-polarized light is injected into the Pockels cell by the coupling prism II. Some of the light reflects off the Brewster surface C and into the laser, just as described for the ring laser. Output radiation follows the reverse path. Some improvement may accrue by adding a mirror H to capture some of the energy lost at the Brewster surface during injection and ejection.

VII. Short-Pulse Propagation Effects

Because the electron beam emitted from an rf accelerator consists of short microbunch pulses in the picosecond range, the continuous-electron-beam model used up to this point may be limited in describing the FEL operation in various conceivable conditions of experimental design. Recent experimental results of the Stanford University FEL oscillator (Eckstein *et al.*, 1982) with ultrashort electron-beam pulses (1 mm) showed many interesting short-pulse effects. Those experimental results stimulated many theoretical research works (Al-Agawi *et al.*, 1979; Colson, 1981; Dattoli *et al.*, 1981; Sprangle *et al.*, 1983), in this area. For practical design considerations, we are mainly interested in the total power gain per pass including the short-pulse effects, such as pulse slippage, laser lethargy, and electron-pulse length and shape.

The first effect of short-pulse propagation is a simple kinematic effect that arises from the different velocity of the electron beam and the electromagnetic wave. The radiation pulse slips by the electron pulse by the amount of

$$\Delta l_{\mathrm{R}} = (c - v_z)(L/v_z) \simeq L/2\gamma_z^2, \tag{181}$$

where the last part of the equality corresponds to the extreme relativistic limit. Using the radiation condition (10), this can also be written as

$$\Delta l_{\mathrm{R}} = \lambda N_{\mathrm{w}}, \tag{182}$$

which corresponds to slippage of one optical wavelength for each wiggler period.

The laser lethargy effect, i.e., the fact that the group velocity of the radiation pulse is less than the speed of light, is another short-pulse effect. A descriptive explanation (Colson, 1982) of this effect is presented using a simplified picture with a long radiation pulse and a step electron pulse, i.e., the density of the electrons is constant within the electron-pulse length (see Fig. 12). After one traversal through the interaction length L, the electromagnetic-beam pulse that was generated by stimulated emission attains a roughly

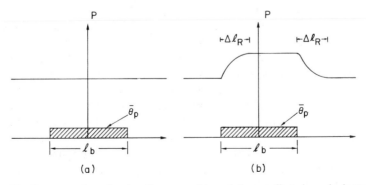

FIG. 12   Demonstration of pulse slippage and laser lethargy effects in a single-traversal FEL amplifier structure with a short electron bunch and a continuous radiation wave: (a) entering the wiggler; (b) leaving the wiggler.

trapezoidal shape with a full width equal to $l_b + \Delta l_R$ (where l_b is the electron-beam bunch length) and a top width of

$$l_R = l_b - \Delta_R. \tag{183}$$

The radiation-pulse fall-off sections of width Δl_R at the front and the back of the radiation pulse is the result of the pulse-slippage effect. The radiation gain in the leading edge of the radiation pulse is produced by the electrons in the leading edge of the electron pulse, whereas the radiation gain in the trailing edge is produced by the electrons at the trailing edge. There is a larger gain at the trailing edge of the radiation pulse than the leading edge. This is because the radiation at the trailing edge is mostly generated near the end of the traversal cycle when the electrons are bunched. The bunched electrons radiate more than unbunched electrons. The dynamic pulse effect involving a short radiation pulse and an arbitrary electron-beam density profile is still governed by the same physics. This difference in the gain causes the trailing edge of the pulse to build up more than the leading edge. This is manifested by the reduced group velocity of the radiation pulse and is the fundamental source of the lethargy effect. The lethargy effect, however, is not devastating on the gain because it can be compensated by reducing the mirror separation from the exact synchronous condition, $m t_{rf} = 2l/c$, where $t_{rf} = 1/f_{rf}$ is the period of the electron pulses, f_{rf} is the rf frequency of the accelerator and is an integer.

From the experiment-design point of view, the electron-pulse length and shape seem to be more fundamental. We are interested in the criterion that the gain is not affected by the electron-pulse length. Sprangle *et al.* (1983) have shown that the gain approaches the gain for the continuous, long electron beam when the pulse slippage becomes unimportant, i.e.,

$$l_R = l_b - \Delta l_R \gg \Delta l_R. \tag{184}$$

Figure 13 is a plot of the normalized optimal gain, i.e., the ratio of the gain of the short electron pulse (g) versus the gain of the continuous electron beam g_0, as a function of the electron-beam length for a step electron beam using the Stanford experimental parameters: $\lambda_w = 3.3$ cm, $B_w = 2.3$ kG, $L = 528$ cm, $\gamma_0 = 85$, $\gamma_{0z} = 69$, $l_b = 1$ mm, $I_0 = 1.3$ A, and filling factor $= 0.017$. (Here $g \equiv \Delta P/P$.) For an electron pulse with a more realistic profile, the optimal gain will vary slightly from the constant-density electron pulse.

The parameter values for the slippage [Eq. (182)] and the single-path optical-pulse top [Eq. (183)] are, in the case of the Stanford experiment, $\Delta l_R = 530$ μm and $l_R = 470$ μm, respectively. The strong inequality (184) is not only dissatisfied with these parameters but even reversed. Strong short-pulse gain-reduction effects should be expected in this experiment, and indeed single-path gain of about 5% was measured instead of the calculated

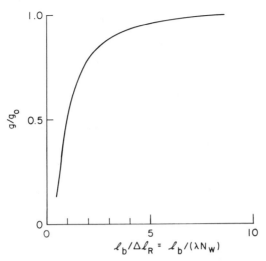

FIG. 13 Ratio of the gain g of the short electron pulse to the gain g_0 of the continuous electron beam bersus the electron beam length.

70% (Eckstein *et al.*, 1982). We conclude that short-pulse gain-reduction effects can be quite devastating when condition (184) is strongly dissatisfied. It is physically understandable that the gain would be drastically reduced when the inequality (184) is reversed. It can be appreciated from Fig. 12 that when this happens, there will be no electrons in the electron bunch that can overlap the flat top section of the pulse (of width l_R) along the entire interaction lenght. In a uniform optical field, electrons radiate in cubic proportion to their effective interaction length; therefore, gain is strongly diminished. Also, weak satisfaction of Eq. (184) will present substantial improvement over the reversed conditions that were present in the Stanford experiment.

To avoid short-pulse effects altogether, we need to satisfy condition (184). Using Eq. (182) we can write this in the form

$$N_w \ll l_b/2\lambda. \tag{185}$$

It is thus desirable to produce long microbunch pulses if we mean to use long wigglers at short operating wavelengths.

It is worth pointing out that in rf accelerators the microbunch length l_b is related to the energy spread of the beam because of the acceleration process. If we assume that a single electron keeps a constant phase value ϕ relative to the rf wave throughout the entire interaction process in the acceleration gaps, then the electron energy will be given approximately by $E = E_{max} \cos \phi$, where E_{max} is the maximum acceleration energy that is obtained when the electron is synchronous with the maximum electric-field point of the rf wave ($\phi = 0$). If the microbunch-beam phase has a full spread $\Delta\phi_b$ around a central phase ϕ_b, then the full-width energy spread is

$$\Delta E/E \simeq \tan \phi_b \, \Delta\phi_b. \tag{186}$$

Around the maximum acceleration phase $\phi_b = 0$ the energy spread is minimal, and one needs to expand the electron energy $E_{max} \cos \phi$ to second order about $\phi_b = 0$. Instead of Eq. (186) we then have

$$\Delta E/E \simeq \tfrac{1}{2}(\tfrac{1}{2}\Delta\phi_b)^2. \tag{187}$$

The microbunch-beam phase spread is related to its physical length by the following:

$$\Delta\phi_b = 2\pi(l_b/\lambda_{rf}) = (2\pi/c)f_{rf}l_b, \tag{188}$$

and this relates l_b to the beam energy spread $\Delta E/E$ using Eq. (186) or (187). For the case of minimum energy spread ($\phi_b = 0$), we obtain from Eqs. (187) and (188):

$$l_b = (\lambda_{rf}/\pi)\sqrt{\Delta E/E}. \tag{189}$$

When l_b is given by Eq. (189), the condition for neglecting short-pulse effects [Eq. (185)] is given by

$$N_w \ll (1/2\pi)(\lambda_{rf}/\lambda)\sqrt{2(\Delta E/E)}. \tag{190}$$

If we also want to keep the beam coldness condition given by Eq. (78), condition (190) reduces to the simple relation

$$N_w \ll [(1/2\pi)(\lambda_{rf}/\lambda)]^{2/3}, \tag{191}$$

which sets the ultimate restriction on the FEL wavelength λ and number of wiggles for a given accelerator when we desire to satisfy both the beam coldness condition and the condition for neglect of short-pulse effects. This basic limitation can be relaxed only if ways are devised to expand the beam bunches after the accerlation stage.

It should be pointed out that Eq. (187) is a somewhat optimistic estimate of the energy spread, assuming that the beam can really be kept at constant zero-acceleration phase along all the acceleration stages and assuming that the dominant source of energy spread is in these stages. Nevertheless, with careful control of the accelerator parameters it is possible to get close to the estimate [Eq. (187)] when operating at high γ values. Even though the zero phase-acceleration point is slightly phase unstable, operation at this point is realizable (Smith, 1982).

VIII.   FEL Design Procedure

In this section we will summarize the equations and inequalities derived in the previous sections and formulate a logical design procedure. We find it most convenient to present the design procedure in the form of two flow-charts. The first is useful for calculating the single-path gain of the FEL with either a linear or a helical wiggler and either an open-cavity or a waveguide-cavity optical resonator. The results of the single-path gain calculation are then used in the second flowchart for calculating the power buildup in the resonator from either the spontaneous emission or an externally injected seed radiation.

The input parameters needed for calculating the single-path gain are, in general, the beam parameters, the wiggler parameters, and the optical cavity parameters, which are listed in the "data in" table of Appendix A. Among the four dependent parameters λ_w, λ, γ_0, and $\bar{a}_w$ that are related through Eqs. (7), (10), we choose to keep γ_0 as the dependent (implicit) parameter in some of the expressions to emphasize the realistic scaling-law dependencies. The beam-energy parameter γ_0 can usually be varied in rf accelerators over a wide range of values. Nevertheless, after a choice of λ_w, λ, $\bar{a}_w$ one should verify that γ_0 falls within the range of the accelerator operating limits.

The chart presents expressions of optimal design parameters that were derived in the Section VII. However, it leaves the freedom to not choose the optimal values in calculating other parameters. This is because optimal parameter values are sometimes not obtainable for technical or practical reasons. The parameters for which an optimal value is calculated in the chart are marked with an asterisk. If choice of the optimal value is practical, these parameters need not be inserted with the input data but are calculated from the other parameters.

The expressions in the chart may be used with any consistent sets of units except when specific units are indicated. The chart can be used as a hand-calculation design guide and can easily be put into the form of a computer program.

A.   IMPLEMENTATION OF THE GAIN CALCULATION CHART

The beam current I_0 (A1) is the microbunch peak current of the rf acceler-ator-generated electron beam. The normalized emittance is defined by $\varepsilon_n = \gamma\varepsilon$, where the beam emittance ε is defined in Eq. (58) and it is assumed that $\varepsilon_x = \varepsilon_y = \varepsilon$. It is useful to express all the parameters in terms of ε_n rather than ε, because ε_n is a beam property that is independent of γ or the focusing parameters of the beam (Lawson, 1977).

The value of the normalized emittance is fundamentally limited by the properties of the injector cathode in which the accelerator current was gen-erated. This sets a low-limit value on ε_n, given by the Lawson–Penner law, which is an empirical relation that states a proportionality relation between the normalized emittance and the square root of the total current. It holds well for many rf accelerators at high beam energy $\gamma_0 > 30$. The physical source of this relation is the fact that the normalized emittance is ideally conserved throughout the acceleration stages and has the same value right at the injector cathode. At this point the normalized emittance is proportional to the cathode radius, which is, of course, proportional to the square root of the cathode current. The proportionality coefficient of the Lawson–Penner law will be limited ideally only by the current density that can be emitted by the cathode and the transverse velocity spread at this point (determined by the cathode temperature). Because most accelerators use electron injectors with quite similar thermionic cathodes, the proportionality coefficient of the Lawson–Penner law is roughly the same for all of them. In this chapter we use for the Lawson–Penner proportionality coefficient a value in amperes from Smith and Madey (1982) and Sprangle *et al.* (1983):

$$\varepsilon_n[\text{cm}\cdot\text{rad}] = 3 \times 10^{-2} \times I_{\text{AV}}^{1/2}, \tag{192}$$

where I_{AV} is the average current in the macropulse. In the case of a linear wiggler, we permit different focusing in the x and y dimensions.

The parameter y_{b0} (A4) is the half-width of the beam waist in the y direction, which is the direction of the magnetic field for the linear wiggler. It is assumed that in the y dimension the beam is focused to its waist at the wiggler entrance.

The parameter x_{b0} (A5) is the half-width of the beam waist in the x direction, which is the direction perpendicular to the magnetic field for the linear wiggler. It is assumed that the x-dimension beam waist is at the wiggler center. In the helical wiggler, the parameter x_{b0} is ignored. In the case of a helical wiggler, the electron beam is assumed to be circularly symmetric and focused to a waist of radius r_{b0} (A6) right at the wiggler entrance.

As explained previously, the input electron-beam parameters do not include the energy parameter γ_0. This is calculated in the chart from the wiggler parameters (A9) to (A11). Note that the parameter $\bar{B}_w$ is the rms value of the periodic wiggler magnetic field. In the helical wiggler $\bar{B}_w = B_w$, where B_w is the amplitude of the periodic magnetic field; but in the linear wiggler, $\bar{B}_w = B_w/\sqrt{2}$. The parameters c_1, c_2, c_3, c_4 (A12) are arbitrary design parameters that determine how strongly we wish to satisfy various beam-quality conditions.

The Gaussian-radiation beam-waist radius w_0 (A14) is determined from the open-resonator design of mirror curvature and mirror spacing (Yariv, 1976). It is assumed here that the beam waist is at the center of the wiggler length. In the case of a waveguide-resonator design, one should insert for the input parameter a (A15) the radius of the waveguide in the case of a circular waveguide, or its half-thickness (in the y dimension) in the case of a rectangular or parallel-plate waveguide. In the case of a rectangular waveguide, one should insert also the parameter b (A16) that is half the width of the waveguide in the x dimension. For a parallel-plate waveguide, one inserts w_{0x} (A17) that is the half-width of the guided-wave waist in the x dimension. The parameter A is the waveguide attenuation dimensionless parameter that is defined in Eq. (100). Its numerical values are given later in the chart in the specific cases of copper waveguides with a radiation wavelength around $\lambda = 10\ \mu m$ for mode EH_{11} in the case of a circular waveguide, and TE_{01} in the case of a rectangular or parallel-plate waveguide.

The first computation block is used to compute the basic FEL parameters common to both helical and linear wigglers. The beam energy parameter γ_0 (A24) is calculated from the synchronization condition [Eqs. (7), (8), and (10). The chart produces the value of the optimal rms periodic magnetic field $\bar{B}_w^{opt}$ (A23) that corresponds to $\bar{a}_w = \sqrt{2}$ and maximum gain, but subsequent expressions will use any value $\bar{a}_w$ (A21) that is calculated from the input $\bar{B}_w$ if for any reason operation at optimal conditions is not feasible. The maximum microbunch length l_b^{max} consistent with the beam energy spread [Eq. (189)] and the minimum beam emittance ε_n^{min} consistent with the Lawson–Penner Law [Eq. (192)] are given in (A25) and (A26), respectively. These parameters can be used as optimal design parameters when there is no definite knowledge of the rf accelerator parameters l_b and ε_n.

Using the input parameters and the parameters calculated in the first box, we can now calculate the dimensions of the electron beam separately for the cases of linear and helical wigglers. In the linear wiggler case, the optimal and actual beam thickness and width (A28–A31) are calculated on the basis of Eqs. (38), (39), (48), (49). The parameter x_b (A31) is the maximum width of the beam at the wiggler ends, assuming focusing at the wiggler center. If, instead, the beam is focused to its waist at the wiggler entrance point, one should rather use Eqs. (45), (46) to calculate x_{b0}^{opt} and x_b, respectively. The parameters r_b^{max} (A32) and $\bar{r}_b$ (A33) are the maximum and the average beam radius, respectively.

In the case of a helical wiggler, the optimal beam radius r_{b0}^{opt} and actual (maximum amplitude) beam radius r_b are given in (A35, A36) based on the expressions (62), (59), respectively. Again it is assumed that the beam is focused to its waist at the wiggler entrance. The parameters r_b^{max}, $\bar{r}_b$, y_b, and x_b are identical in this case with r_b (A37).

In the next box of the chart, we calculate the maximum values of the wiggler lengths that are consistent with the various beam-quality conditions derived earlier. L_1^{max} (A38) is limited by the beam-energy spread constraint (78). L_2^{max} (A39) is limited by the beam-emittance constraint for either a linear wiggler [Eq. (80)] or a helical wiggler [Eq. (83)]. L_3^{max} (A40) is limited by the short-pulse effect neglect condition (185). L_4^{max} (A41) is limited by the space-charge effect neglect condition (85).

The condition for neglecting space-charge effect $\bar{\theta}_p \ll \pi$ [Eq. (85)] was derived originally with the assumption of a circular-beam cross section. Its use in (A41) with an average beam radius $\bar{r}_b$ is an approximation. If the wiggler length L is smaller than L_4^{max} with a linear wiggler, and $\bar{\theta}_p > \pi$, then from the discussion in Section II [Eqs. (68), (87)], we conclude that, with these design parameters, significant space-charge expansion of the beam is expected in the x dimension, and redesign of the FEL is required. In the case of a helical wiggler, it is possible to operate in the collective regime $\bar{\theta}_p > \pi$ without space-charge-beam expansion as long as condition (88) is satisfied. If this condition is not satisfied, redesign of the FEL parameters is again required. If it is satisfied, program COLD (Appendix C) can be used to calculate the FEL gain in the collective regime. Also, when L is larger than either of L_1^{max}, L_2^{max}, L_3^{max}, redesign of the FEL parameters is required to assure satisfaction of all inequalities. If the wiggler length is a free design parameter, choice of $L \doteq L^{max}$ (A42) guarantees satisfaction of all beam-quality constraints.

We will now consider the FEL resonator parameters. In the case of a circular waveguide resonator, we calculate the optimum waveguide radius a_{opt} (A45) that corresponds to maximum gain [based on Eq. (103)]. If a is a free parameter, it may be desirable to choose $a = a_{opt}$. But whether this is the case or not, it should always be verified that the waveguide radius is large enough to contain the beam (filling factor $A_b/A_{em} < 1$). When this condition is not satisfied, we choose for the practical waveguide radius a' (A46) the smallest radius for which $A_b/A_{em} = 1$ [based on Eq. (101)] and continue the design with this new optimal value.

In the case of rectangular and parallel-plate waveguides, we calculate the effective electromagnetic-mode width x_{em} separately for a rectangular waveguide (A54) and parallel-plate waveguide (A55). The optimum waveguide thickness a_{opt} is given in both cases by (A54), [based on Eq. (115)]. Also, in the present case it should be checked that the electron beam is contained within the radiation-mode cross section in both the x (A58) and the y (A59) dimensions. If this is not the case, then [based on Eqs. (118)–(120)] new optimum waveguide dimensions $a' = 2y_b$ (A58) and $b' = 2x_b$, $w'_{0x} = 2\sqrt{2/\pi}\, x_b$ (A59) are chosen.

In the case of an open resonator, we calculate the optimal beam-waist radius w_0^{opt} (A49). Whether we choose $w_0 = w_0^{opt}$ or w_0 as an input parameter, we again need to check that the electromagnetic-mode cross section contains the electron beam completely [Eq. (96)]. If this is not the case, we choose a new Gaussian beam waist w'_0 (A50) that corresponds to a unity filling factor [Eq. (97)].

The effective electromagnetic-wave cross-section area A_{em} that was calculated separately for each kind of laser resonator is now used in (A62) to calculate the normalized gain parameter $\bar{Q}$ (A18). If $\bar{Q} < 1$ (small-gain regime), the net single-path gain is calculated in (A63). If $\bar{Q} \gtrsim \pi$, program COLD can be used to calculate the gain. It can be also estimated from the curve given in Fig. 4.

B. IMPLEMENTATION OF THE POWER BUILDUP CALCULATION CHART

The flowchart of Appendix B is useful for calculating the power-buildup parameters in the FEL oscillator resonator. When the single-pass gain and round-trip transmission are known, the flowchart produces the multipath regenerative power gain in the resonator $[G(\bar{\theta}_m), \tilde{G}(\bar{\theta}_m)]$ during the period of the electron-beam macropulse t_{MP}. If saturation is not achieved during this period the output power can be calculated for three initial-radiation sources: spontaneous emission, incoherent-radiation injection, and coherent-radiation injection. If saturation is achieved during the macropulse period, then we can estimate the oscillation buildup time, up to the saturation point, and the saturated output power.

The parameters in row (B1) were defined or calculated in the flowchart of Appendix A. The cavity length L_c $(L_c \geq L)$ is the total resonator optical-path length, including dead-time regions. In the case of a ring laser configuration (Fig. 11d), L_c is half the round-trip optical path.

The lossless FEL maximum-gain factor $G(\bar{\theta}_m)$ (B4) is the single-path output-to-input power ratio [Eq. (11)] at the maximum-gain point. In the limit $\bar{\theta}_p$, $\bar{Q} \gtrsim \pi$ (low-gain, tenuous cold-beam limit) this parameter need not be inserted. It is calculated in the flow chart based on Eq. (I-14): $G(\bar{\theta}_m) = 1 + 0.27\,\bar{Q}$ (Eq. (39)], where $\bar{Q}$ was calculated in Appendix A. In other gain regimes, one should use program COLD (Appendix C) to cacluate $G(\bar{\theta}_m)$. In the tenuous-beam limit $\bar{\theta}_p \simeq 0$, the gain factor [Eq. (18)] is given as a function of $\bar{Q}$ in Fig. 4a, which was calculated from program COLD. We see that the use of the low-gain approximation is quite accurate up to gain parameter value $\bar{Q} \simeq 3$. The second derivative of the gain curve at the maximum-gain point $G''(\bar{\theta}_m)$ (B5) is also drawn in Fig. 4b as a function of $\bar{Q}$ in the limit $\bar{\theta}_p = 0$. In the low-gain limit $\bar{Q} < \pi$, the analytic approximation for $G''(\bar{\theta}_m)$ is used in the flowchart (B13) to calculate this parameter.

The parameter R (B6) is the total round-trip transmission factor including mirror-reflection coefficients, diffraction-loss coefficient, transmission coefficients of all optical components placed in the optical path (e.g., an electro-optic switch), and, in the case of a waveguide resonator, also the waveguide-transmission factor and the coupling-efficiency coefficients into and out of the waveguide (when external mirrors are used). In terms of the waveguide-loss parameter L_G that was calculated in Appendix A (A61), the waveguide attenuation coefficient is $R_{WG} = 1 - L_G$ in the low-loss limit $L_G \ll 1$, and in the general case $R_{WG} = \exp(-L_G)$.

The parameter T (B7) is the resonator output transmission factor. When the power is coupled out actively via an optical switch, as described in Section VI (Fig. 11d, e), T is the out-coupling efficiency of the switch. In passive-output coupling through a front mirror, we always have $T \le 1 - R$. The equality takes place only in a lossless system.

The parameters of the radiation source (B8)–(B10) should be inserted according to the kind of input radiation source used to build-up the power in the resonator. No additional data is needed when the radiation source is spontaneous emission. When the input radiation is injected from an incoherent radiation source, we need to insert in the source spectral bright-ness radiance B (B8) that is evaluated at the peak gain frequency and in the direction that excites the relevant resonator mode. In the case of a co-herent-radiation injection source, we need to insert the source power P_{inj} (B9) and frequency spread $\Delta\omega_{inj}$ (B10).

In the second block of the flowchart, we calculate the number of round-trip traversals N (B11) within the electron-beam macropulse duration t_{MP}. We then calculate the net single-path round-trip gain (B12) and the total multipath gain after N round-trip traversals, assuming saturation is not obtained (B14). This is calculated for the case of instantaneous radiation injection during a period shorter than a round-trip time $2L_c/c$.

In (B13) we calculate $G''(\bar{\theta}_m)$ in the regime of cold, tenuous beam and low gain. In other regimes, this parameter is inserted as input data. When $\bar{\theta}_p = 0$, this parameter is as shown in Fig. 4b in the gain-parameter regime $0 < \bar{Q} < 15$. The parameter $G''(\bar{\theta}_m)$ is used to calculate the full-width, half-maximum (FWHM) detuning parameter linewidth $\Delta\bar{\theta}$ (B16) of the total gain curve for N traversals. This consequently yields the relative (B17) and absolute FWHM frequency linewidth of the total gain curve.

The expression for the approximate saturation-power extraction from the electron beam ΔP_{sat} (B19) is calculated to determine whether saturation was reached during the macropulse duration. The limited validity of this expression is discussed in Section V in relation to Eq. (175).

In (B20)–(B23) we calculate the solid angle occupied by the relevant reso-nator mode for four different cases: open resonator (fundamental mode), circular-waveguide resonator (EH_{11} mode), parallel-plate and rectangular resonators (TE_{10} modes). This is used in (B24)–(B29) to calculate the effective input power that is coupled into the FEL resonator mode in three different cases: spontaneous emission, incoherent injection source, and coherent injection source. These expressions are based on the derivation in Section V where the reservation about the validity of (B24) in the case of a waveguide resonator is also discussed. Note that $I(\omega_m, \hat{e}_k = \hat{e}_z)$ [see (B25)] is slightly smaller by a factor 0.55 than $I(\omega_0, \hat{e}_k = \hat{e}_z)$ (ω_0 is the peak spontaneous-emission frequency and ω_w the slightly smaller peak gain frequency). The numerical factor is obtained by substituting in the spontaneous-emission curve $\sin^2(\frac{1}{2}\theta)/(\frac{1}{2}\theta)^2$ [Eq. (140)] $\bar{\theta} = \theta_m$, where we assumed $\bar{\theta}_m = -2.6$, as in the small-gain regime (B13).

To find out whether saturation is attained during the macropulse-duration period t_{MP}, we calculate the multipath maximum gain $\tilde{G}(\bar{\theta}_m)$ for the cases of continuous incoherent-radiation source (B30) and continuous coherent-radiation injected source (B32). The instantaneous-injection multipath total gain has been calculated (B15). Based on Eqs. (178)–(180), the number of round-trip traversals N_{sat} is calculated for three different cases, (B31), (B33), (B34). This is used in (B35) to find the start saturation time t_{sat}, which is the time when the saturation-power extraction ΔP_{sat} is extracted out of the electron beam in the last traversal. The validity of these expressions is discussed in the end of Section V. If $t_{sat} > t_{MP}$, then saturation is not reached during the macropulse period, and the output power P_{out} is given by (B35). In the opposite case, saturation is reached within the macropulse duration, and the saturated power output is given by (B37).

Equation (B37) was derived from the equality $\Delta P_{sat} = P_{sat}(L) - P_{sat}(0) = P_{sat}(L)(1 - 1/G_{sat})$, where G_{sat} is the lossless FEL gain at saturation. By definition, $G_{sat} R = 1$, and consequently $P_{sat}(0) = \Delta P_{sat}/(1 - R)$. The output power (B37) is obtained by multiplying this expression by the output-transmission coefficient T. In the case of passive output coupling, $T \leq 1 - R$ and equality corresponds to the case of zero losses; when all the generated power is coupled out through the front mirror, $P_{out} = \Delta P_{sat}$. In the case of active coupling (optical switch), it is possible to have $1 \gtrsim T \gg 1 - R$ and get a power output higher by a factor $T/(1 - R)$ than the single-path saturation-power generation ΔP_{sat}. The reason is that in this case we couple out in a short period not only the power generated in one traversal but also most of the stored energy. The FEL operates then as a Q-switched laser in the cavity-dumping mode of operation. As discussed at the end of Section V, the validity of (B19) and consequently of (B37) is limited to the time when the peak gain frequency starts reaching saturation right at the end of the interaction region. At a later time when the saturation process proceeds, a full nonlinear analysis is required to calculate the saturated output power.

In reference to FEL operation after start-saturation time is reached, we point out that although maximum round-trip transmission factor $R \to 1$ is desirable to reach saturation as early as possible, it may be preferable, once saturation time is reached, to reduce R down to some finite *nonzero* optimal value, to obtain maximum power extraction from the electron beam during the rest of the macropulse duration. For example, if at the start-saturation time we set $R = 1/G(\theta_m)$, the FEL will keep saturating right at the end of the interaction length L and also after the start-saturation time. If intracavity losses are small, then the output power during the rest of the macropulse period will stay $P_{out} = \Delta P_{sat}$.

The possibility of controlling the round-trip transmission factor R is in fact available in active output-coupling configurations, as shown in Fig. 11 d, e. Rather than applying the full switching voltage across the Pockels cell at the start-saturation time and then, with $T \simeq 1$, obtain maximum momentary output power, it may be preferable to apply a lower voltage that produces a lower output transmission factor T. This will keep $R > 0$, and sufficient stored energy will stay in the cavity to provide efficient power extraction from the electron beam during the rest of the macropulse duration. Higher average power generation is obtained this way.

C. NUMERICAL EXAMPLES

We shall now demonstrate the use of the FEL design flowcharts as well as the prospects for useful rf accelerator FELs by considering simple numerical examples.

We choose for the input parameters of the gain-calculation flowchart the following values: $I_0 = 3$ A, $\Delta E_k/E_k = 0.5\%$, $y_{b0} = 0.65$ mm, $x_{b0} = 1.3$ mm, $f_{rf} = 3$ GHz, $\lambda_w = 4.4$ cm, $L = 1.91$ m, $c_1 = c_2 = c_3 = c_4 = 0.5$, $\lambda = 16 \mu$m, $a = 1.4$ mm, and $b = 2.8$ mm.

We thus choose a rectangular waveguide and left some of the design parameters to be determined via the flowchart calculation as the optimal design parameters (marked with an asterisk in Appendix A). For a linear wiggler we obtain $\gamma_{0z} = 37.09$, $\bar{a}_w = \bar{a}_w^{opt} = \sqrt{2}$, $\bar{B}_w = \bar{B}_w^{opt} = 3.44$ kG, $\gamma_0 = 64.24$, $l_b = l_b^{max} = 3.2$ mm, $\varepsilon_n = \varepsilon_n^{min} = 9.3 \times 10^{-3}$ cm·rad, $a_w = 2$, $y_{b0}^{opt} = 0.38$ mm, $y_b = y_{b0} = 0.65$ mm, $x_{b0}^{opt} = 0.662$ mm, $x_b = x_b(0) = x_b(L) = 1.34$ mm, $r_b^{max} = 1.34$ mm, $\bar{r}_b = 0.93$ mm, $L_1^{max} = 2.2$ m, $L_2^{max} = 1.91$ m, $L_3^{max} = 2.2$ m, $L_4^{max} = 16.4$ m, $L^{max} = 1.91$ m, $x_{em} = 1.4$ mm, $x_{em}^{opt} = 1.33$ mm, $a^{opt} = 14 \mu$m, $a' = a = 1.4$ mm, $b' = b = 2.8$ mm, $A_{em} = 3.92$ mm^2, $L_G = 7 \times 10^{-5}$, $\bar{Q} = 2.12$, and $\Delta P/P = 57.2\%$.

For comparison we consider an open-resonator configuration with the same parameters and optimal Gaussian-beam waist: $w_0' = w_0 = w_0^{opt} = 2.2$ mm, $A_{em} = 7.64$ mm^2, $\bar{Q} = 1.087$, and $\Delta P/P = 29.3\%$.

The higher gain that was obtained in the waveguide-resonator configuration is owing to the smaller effective electromagnetic beam area (larger filling factor) that was obtained in the first case without excessive waveguide losses.

Note that the wiggler length L was chosen at maximum value $L = L^{max} = L_2^{max}$, and is limited by the beam emittance and by insufficient focusing in the y dimension at the wiggler entrance ($y_{b0} > y_{b0}^{opt}$). If stronger focusing power is applied on the electron beam by the input-coupling electron lens, we can design an FEL with a longer interaction length $L = L_1^{max} = L_3^{max} = 2.2$ m and, consequently, with higher gain. For a given λ_w, λ, and f_{rf}, it is not possible to further increase the interaction length because any increase of L_1^{max} (A38) by decreasing the energy spread $\Delta E_k/E_k$ involves decrease of the bunch length l_b (A25) and, consequently, decrease in L_3^{max} (A40). This is, of course, assuming that l_b is determined by (A25).

We shall now use the FEL design parameters that we have just calculated in the power buildup calculation flowchart of Appendix B. We add the following input-parameter data: $t_{MP} = 1$ μsec, $L_c = 2.11$ m ($L_D \equiv L_c - L = 20$ cm), $R = 87\%$, $T = 60\%$, $P_{inj} = 10$ mW, and $\Delta\omega_{inj}/\omega \ll 3 \times 10^{-3}$.

A lead-salt injection laser is proposed as the injection source. Although commercial lasers of this kind operate cw at power levels less than 1 mW with spectroscopic-grade linewidths, it is believed that 10 mW useful power can be coupled into the cavity within the broad linewidth requirement during a short injection time $2L_c/c = 14$ nsec. The following parameter values are found for the rectangular waveguide cavity configuration: $N = 71$, $G(\bar{\theta}_m) = 1.572$, $G_{net}(\bar{\theta}_m) = 1.368$, $G_N(\bar{\theta}_m) = 4.5 \times 10^9$, $G''(\bar{\theta}_m) = -0.1875$, $\Delta\bar{\theta} = 0.809$, $\Delta\omega_N/\omega = 3 \times 10^{-3}$, $\Delta P_{sat} = 1.12$ MW, $\Delta\Omega_{em} = 8.16 \times 10^{-6}$ sr, $P_{in} = P_{inj} = 10$ mW, $N_{sat} = 62$, $t_{sat} = 0.86$ μsec, $P_{out} = 5.17$ MW.

Considering the case of spontaneous-emission input-radiation source with the same parameters values, we obtain: $\tilde{G}_N(\bar{\theta}_m) = 1.22 \times 10^{10}$, $I(\omega_m, \hat{e}_k = \hat{e}_z) = 2.5 \times 10^{-11}$ w/(s-rad/sec), $P_{in} = 7.86 \times 10^{-5}$ W, $N_{sat} = 73$, $t_{sat} = 1$ μsec, and $P_{opt} = 575$ kW. We see that FEL saturation is almost obtained at the end of the macropulse period even without input-radiation injection. However, as discussed in Sections V, VI, the calculations for this case are not very reliable.

For comparison, we also calculate the power-buildup parameters for the example of an open resonator, as follows: $G_{net}(\bar{\theta}_m) = 1.125$, $G_N(\bar{\theta}_m) = 4.3 \times 10^3$, $\Delta\bar{\theta} = 1.026$, $\Delta\omega_N/\omega = 3.77 \times 10^{-3}$, $\Delta\Omega_{em} = 8.4 \times 10^{-6}$ sr, $P_{in} = P_{inj} = 10$ mW, $N_{sat} = 169$, $t_{sat} = 1.6$ μsec, and $P_{opt} = 25.6$ W. With a spontaneous-emission radiation source, we have $\tilde{G}_N(\bar{\theta}_m) = 3.4 \times 10^4$, $P_{in} = 1 \times 10^{-4}$ W, $N_{sat} = 184$, $t_{sat} = 2.6$ μsec, and $P_{opt} = 2.1$ W.

The striking differences in the output power of the different examples are directly traceable to the exponential nature of the total gain formula [Eqs. (13), (14)]. For the same reason, the FEL power will be highly sensitive to optimization of the FEL gain parameters, because even small changes in the single-path gain parameter leads to great changes in the total gain. In these examples ($N = 71$), any 10% increase in the net single-path gain corresponds to three orders of magnitude increase in the total gain and power (before saturation).

IX. Conclusion

In this chapter, we have derived a design procedure and design criteria for free-electron lasers driven by electron beams from rf accelerators. We have also suggested a number of ways to reduce the oscillation-buildup time to reach saturation, or at least to reach a substantially high power level, as long as the short electron-beam pulse still lasts. One way to reduce this buildup time is to increase the single-path gain by using a waveguide for propagating the electromagnetic wave. This helps to reduce the FEL filling factor in cases when it would otherwise be small because of the wave diffraction. Another way is to inject a seed radiation wave into the cavity to provide an initial source radiation at a power level higher than the spontaneous emission.

The design procedure and design criteria that were derived here permit an estimate of the oscillation-buildup time and the laser output power of various FEL schemes: with waveguide resonator or open resonator, with initial seed-radiation injection or with spontaneous-emission radiation source, with a linear wiggler or with a helical wiggler. The expressions that were derived for computing the various FEL parameters allow for the design and optimization of the FEL operational characteristics under ideal conditions or with nonideal design parameters that may be limited by technological or practical constraints. In either case, the design procedure enables one to derive engineering curves and scaling laws for the FEL operating parameters. This can most conveniently be done with a computer program based on the flowcharts in Appendixes A and B.

It is out of the scope of this chapter to derive the complete scaling laws of the FEL operating parameters and their optimization. The need to keep various inequalities for proper operation of the FEL, and various technological and practical limitations that appear in specific design problems, make it quite difficult to derive practical analytitic scaling laws for all cases. We will brifly discuss some of the more important dependencies of the FEL operating parameters on the design parameters. The interested reader is also referred to Renieri (1979) and Smith and Madey (1982), where some basic scaling laws of the open-resonator FEL power generation and single-path gain are discussed assuming ideal design conditions.

The basic scaling law of the FEL is based on Eq. (20), predicting that the small-signal single-path gain [Eq. (14)] is proportional to I_0, $\lambda^{3/2}$, L^3 and inversely proportional to $\lambda_w^{5/2}$. This is a representative scaling law as long as we assume that the parameter γ_0, which is indirectly determined by λ and λ_w from the synchronization condition (10), is a free design parameter that can be chosen at will. It is also assumed that the parameter A_{em} in Eq. (20) is independent of the other parameters. This last condition is satisfied in an open-resonator configuration only if diffraction effects are small enough to keep $r_{em} \simeq r_b$ [Eq. (97)] along the whole interaction length L. If λ or L are large enough, then the minimum radiation-beam radius $r_{em} = [\lambda L/(4\pi)]^{1/2}$ becomes larger than r_b [Eq. (96)]. In this case, the scaling law of the gain is described by Eq. (26), indicating proportionality to $I_0 \lambda^{1/2} L^2 \lambda_w^{-5/2}$.

The gain parameter $\bar{Q}$ grows monotonically with λ and L in both cases of Eqs. (14) and (95). We conclude that FEL design in the infrared regime should be easier than at short wavelength, and indeed design papers of FEL systems in the far-infrared regime predict high gain and favorable parameters (Bizzarri et al., 1982; Liewer et al., 1981; Shaw and Patel, 1980). However, at longer wavelengths and interaction lengths, when diffraction becomes more significant, we always reach regime (27), where the growth of the gain parameter (95) with λ and L is smaller than in the negligible diffraction regime (20). It is exactly in this case that waveguide resonators can help to increase the gain. The effective electromagnetic wave area in the waveguide can be made smaller than the diffraction-limited area [Eq. (94)], and the filling factor A_{em}/A_b can be made close to 1. The improvement is especially apparent in the case of a linear wiggler and a rectangular waveguide. In this case, the optimal electron beam is usually ribbon shaped and a rectangular waveguide can match well its cross-sectional area. When losses in the waveguide are small, the FEL-gain scaling law is again described by Eq. (20). As λ and L are further increased, losses in the waveguide can become significant, and the scaling laws change, as discussed in Section III.

The practical gain scaling laws may also deviate from the ones discussed here when the inequalities that should be satisfied depend on the design parameters. If the beam radius r_b that appears in inequality (27) is an emittance-limited free parameter, it can be chosen r_{b0}^{opt}, which is proportional to $\lambda_w^{1/2}$ [Eq. (A35) in Appendix A]. Thus the open-resonator helical-wiggler FEL gain scaling with λ_w will be modified. As long as $[\lambda L/(4\pi)]^{1/2} < r_{b0}^{opt}(\lambda_w)$, we can substitute in Eq. (20) $A_{em} = A_b = \pi(r_{b0}^{opt})^2$ and the gain scales with λ_w such as $\lambda_w^{-7/2}$. Also, the beam-quality conditions (A39)–(A41) can modify the gain scaling laws whenever they are relevant. If L is a free design parameter and is not limited by practical constraints, we will choose $L = L^{max}$ [Eq. (A42)]. The dependence of L^{max} on the other parameters is described by (A38)–(A41), and it will change the gain scaling law dependence on these parameters in a complex way when substituted in the gain expression.

In all cases discussed, increasing the interaction length L has a significant effect on augmenting the gain. The interaction length is usually limited by the beam-quality conditions (A39)–(A41). The space-charge-neglect condition is usually well satisfied with practical rf accelerator parameters and $L < L_4^{max}$ is satisfied. The angular-spread (emittance) limited-length L_2^{max} is often an important limiting parameter (as in the example considered in

Section VIII). When optimal beamwidth [Eq. (38)] or radius [Eq. (62)] is practical, the limitation on L_2^{max} is given by Eq. (81) or (83). L_2^{max} is then proportional to $\lambda_w^2/\varepsilon_n$. Notice that because high current is desirable, the Lawson–Penner law [Eq. (192)] limits increase of L_2^{max} because high current involves increase in ε_n.

The energy-spread length L_1^{max} and the length L_3^{max} corresponding to short-pulse-effect neglect are interrelated through the energy-spread parameter $\Delta E_k/E_k$. This is correct as long as the beam microbunch length l_b is limited by Eq. (189). Increasing $\Delta E_k/E_k$ reduces L_1^{max} and increases L_3^{max}, and vice versa. Equating $L_1^{max} = L_3^{max}$ leads to the optimal parameters choice (when L_1^{max} or L_3^{max} are the length-limiting parameters and l_b is a controllable parameter):

$$L_{1,3}^{max} = \tfrac{1}{2}c_1\lambda_w\left[(c_3/c_1)(\sqrt{2}/\pi)(\lambda_{rf}/\lambda)\right]^{2/3}, \tag{193}$$

$$(\Delta E_k/E_k)^{opt} = \left[(c_1/c_3)(\pi/\sqrt{2})(\lambda/\lambda_{rf})\right]^{2/3}, \tag{194}$$

$$l_b^{opt} = \left[(c_1/c_3)(2/\pi^2)\lambda\lambda_{rf}^2\right]^{1/3}, \tag{195}$$

where $\lambda_{rf} = c/f_{rf}$. Normally the rf frequency of the accelerator is a fixed design parameter. We comment that if rf accelerators were designed specifically for FEL application, then Eq. (194) and also (A21) suggest that low rf frequency is advantageous for obtaining higher-quality beam parameters. Nevertheless, we should never increase the rf wavelength beyond the limit $\lambda_{rf} < 2 L_c$, to keep at least one microbunch electron-beam pulse coming into the cavity after each round-trip traversal of the radiation pulse. Furthermore, it should be realized that decreasing f_{rf} will involve a penalty of decreased FEL average power owing to a reduced duty cycle within the electron-beam macropulse.

In regard to gain scaling with length L, it is pointed out that while the FEL single-path gain grows strongly with increasing L in all regimes, the total multipath gain $G_N(\theta)$ or $\tilde{G}_N(\theta)$ increases more moderately with L because the number of cavity bounces during the macropulse duration $N = t_{MP}c/(2L_c)$ falls down with increasing L. Increasing L affects the total multipath gain significantly only if the single-path net gain is small, $G_{net}(\theta_m) = G(\bar{\theta}_m)R \gtrsim 1$, or when $L \lesssim L_D$.

In considering the multipath gain process, we should point out that a proper FEL design must satisfy the beam-quality condition (77), $\Delta E/E \ll 1/(2N_w)$, not only for each microbunch but also for the whole macropulse duration. This is not a trivial condition because it requires a "flat top" rf-power macropulse in the accelerator cavities. If this power (and consequently the acceleration energy) is varying temporally during the macropulse duration beyond the limit set by Eq. (77), then the FEL radiation pulse can be amplified only during part of the macropulse duration, or, worse than that, it can be reabsorbed during the periods in the macropulse when the accelerating voltage is slightly reduced relative to the synchronous beam-velocity voltage.

As we explained here and in Section III, the use of leaky waveguides for FEL resonators in the infrared regime may provide a significant increase in the single-path FEL gain that is so important in short-pulse-duration rf accelerations. In addition, a waveguide resonator can be used with planar electron-beam-transparent mirrors (Fig. 1b). The short cavity length obtained this way provides the additional benefit of an increased number of cavity bounces, which enhances the multipath total FEL gain and reduces the oscillation-buildup time. However, both the techniques of waveguide cavities and electron transparent mirrors need to be demonstrated experimentally for practical application in FELs construction.

The seed radiation-injection scheme and in particular the active input–output cavity-coupling configuration that are discussed in Section VI are very promising approaches in the effort to decrease the oscillation-buildup time in rf linacs. There is still a need to identify and improve appropriate seed-radiation sources that operate at the wavelength desirable for FEL operation. They should preferably be tunable, to keep the tunability advantage of FELs. In the near-infrared-wavelengths regime (3 μm–30 μm), we have identified lead-salt injection lasers as an appropriate injection source. Though commercial lasers of this kind operate cw at power levels smaller than 1 mW, it is conceivable that power levels higher than 10 mW can be obtained with very short pulses. In an instantaneous injection scheme with an optical switch, the injection duration is smaller than a round-trip traversal time $2 L_c/c$. This is in the 10-nsec range and is very short indeed. The obtainable injected-input radiation power can be, thus, larger than the spontaneous-emission power by many orders of magnitude, and can provide short oscillation-buildup time and efficient extraction of the macropulse beam energy. Furthermore, if the saturation-start point is reached within the macropulse duration, then the active output-coupling scheme has an additional advantage, as was discussed in Section VIII. By controlling the output-coupling coefficient, we can either optimize the power extraction during the saturation time and obtain high average power output, or dump all the cavity power at one time and obtain high peak power when the radiative power stored in the cavity is maximal.

In considering seed radiation injection sources, high spectral brightness of the source is essential to couple a significant amount of power into the cavity-mode phase-space volume and the frequency linewidth. Therefore, incoherent sources such as flash tubes or thermal sources will not be good candidates for this purpose even if they deliver considerably higher total power. For example, even a thermal source with effective temperature of

$T = 20,000$ K would produce, according to Eq. (173), a miniscule effective input power of 7.4 $\times$ 10^{-9} W into a frequency linewidth of $\Delta\omega_N = 3.5 \times 10^{11}$ rad/sec. Atomic discharge tubes may have larger spectral brightness $B(\omega)$ at the emission-line frequencies, but because of their high angular spread they will also produce low effective input power, according to Eq. (170). They also deprive the FEL of its tunability feature.

The conclusion of this chapter and the examples given in Section VIII show that FEL operation with electron beams from short-pulse rf accelerators is conceivable, and the slow oscillation-buildup problem may be surmountable by optimizing the design parameters and using various techniques suggested here. The examples given in Section VIII do not necessarily correspond to an optimal design. To realize infrared FEL oscillators operating with rf accelerators, the design parameters should be optimized, and an experimental effort should be carried out to demonstrate that the calculated design parameters are realizable and that the waveguide cavity and seed-radiation-injection techniques that were suggested here can be successfully applied to FELs. Finally, we point out that quite straightforward technological improvements are expected to make rf accelerator parameters more favorable for FEL applications and will relax FEL design constraints (Madey, 1983). Use of high-power long-pulse klystrons for driving the rf accelerator will allow production of electron beams with macropulse duration of tens of microseconds and more. Electron-energy-compression techniques are known in the art of accelerator technology. Such schemes can be employed after the accelerator unit, reducing the beam-energy spread ΔE and at the same time elongating the microbunch length l_b (by reducing the peak current). This alleviates the fundamental restriction [Eq. (191)] on the number of wiggler periods. Recent developments of high-quality rf accelerator cavities permit high-acceleration gradients and show promise for development of rf linacs considerably smaller than presently existing ones.

Appendix A: Flowchart for FEL Gain Computation†

DATA IN

Beam parameters:

	I_0 [Amp]	— beam peak current		(A1)
*	ϵ_n [length-rad]	— normalized beam emittance		(A2)
	$\Delta E_k/E_k$	— FWHM beam energy spread		(A3)
*	y_{b0}	— e-beam waist half-width in y direction at the wiggler entrance.	(lin. wigg.)	(A4)
*	x_{b0}	— e-beam waist half-width in x direction at the wiggler center.	(lin. wigg.)	(A5)
*	r_{b0}	— e-beam waist radius at the wiggler entrance.	(hel. wigg.)	(A6)
	l_b	— e-beam microbunch length		(A7)
	f_{RF}	— acceleration RF frequency		(A8)

Wiggler parameters:

	λ_w	— wiggler period	(A9)
*	L	— wiggler length	(A10)
*	$\bar{B}_w$ [kG]	— wiggler magnetic field RMS value	(A11)
	$c_1, c_2, c_3, c_4 < 1$	— inequalities satisfaction design parameters	(A12)

Optical parameters:

	λ	— radiation wavelength		(A13)
*	w_0	— Gaussian beam waist radius	(open resonator)	(A14)
*	a	— circular W.G. radius or rectangular W.G. half-thickness	(W.G. resonator)	(A15)
*	b	— rectangular W.G. half-width	(rect. W.G. resonator)	(A16)
*	w_{0x}	— parallel-plate W.G. half-waist width	(p.p. W.G. resonator)	(A17)
*	A	— waveguide attenuation parameter	(W.G. resonator)	(A18)

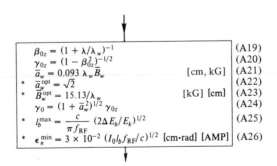

$$\beta_{0z} = (1 + \lambda/\lambda_w)^{-1} \qquad \text{(A19)}$$
$$\gamma_{0z} = (1 - \beta_{0z}^2)^{-1/2} \qquad \text{(A20)}$$
$$\bar{a}_w = 0.093\,\lambda_w\bar{B}_w \qquad \text{[cm, kG]} \quad \text{(A21)}$$
$$*\quad \bar{a}_w^{\text{opt}} = \sqrt{2} \qquad \text{(A22)}$$
$$*\quad \bar{B}_w^{\text{opt}} = 15.13/\lambda_w \qquad \text{[kG] [cm]} \quad \text{(A23)}$$
$$\gamma_0 = (1 + \bar{a}_w^2)^{1/2}\,\gamma_{0z} \qquad \text{(A24)}$$
$$*\quad l_b^{\max} = \frac{c}{\pi f_{RF}}\,(2\Delta E_k/E_k)^{1/2} \qquad \text{(A25)}$$
$$*\quad \epsilon_n^{\min} = 3 \times 10^{-2}\,(I_0 l_b f_{RF}/c)^{1/2} \quad \text{[cm·rad] [AMP]} \quad \text{(A26)}$$

† Parameters for which an optimal value is calculated are marked with an asterisk. W. G., waveguide; lin. wigg., linear wiggler; hel. wigg., helical wiggler; p.p., parallel plate.

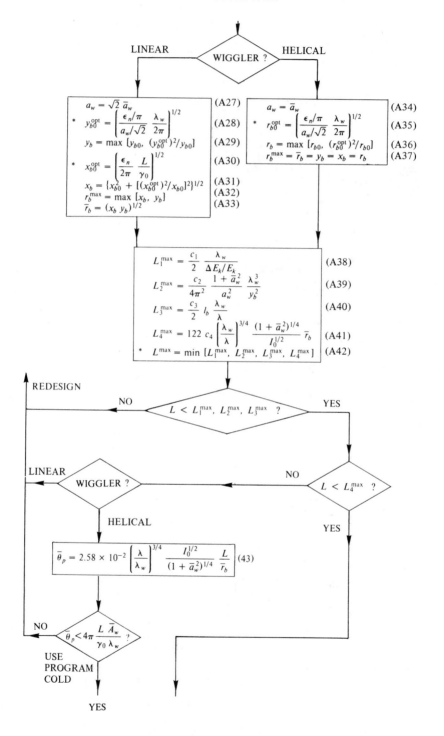

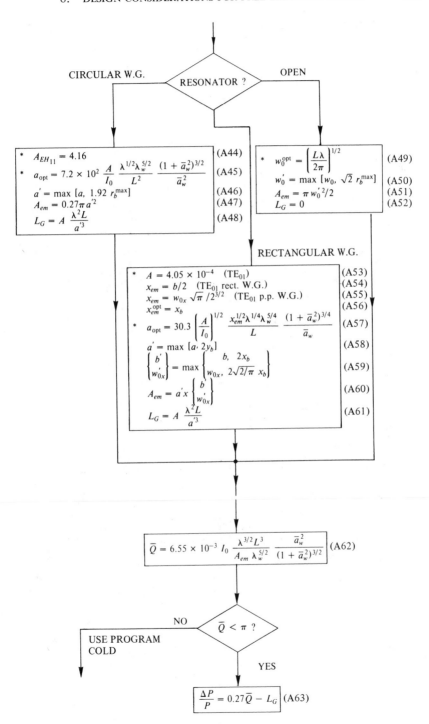

Appendix B: Flowchart for Computation of FEL Oscillator Power Buildup

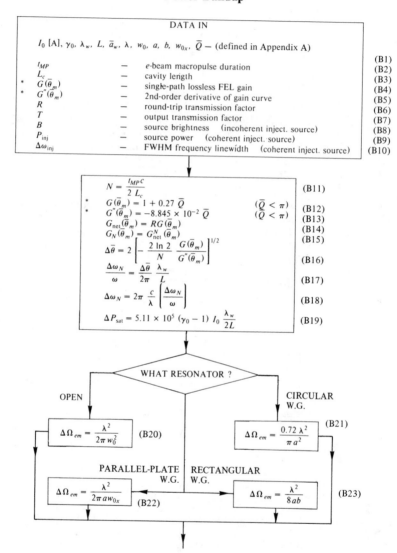

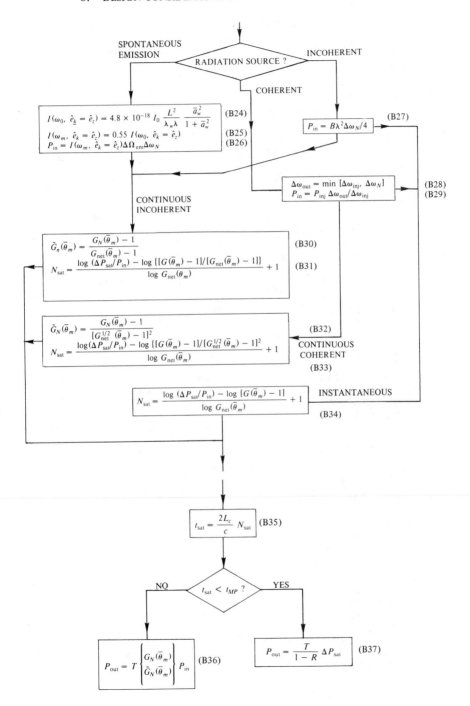

$$I(\omega_0, \hat{e}_k = \hat{e}_z) = 4.8 \times 10^{-18} \, I_0 \, \frac{L^2}{\lambda_w \lambda} \, \frac{\bar{a}_w^2}{1 + \bar{a}_w^2} \qquad \text{(B24)}$$

$$I(\omega_m, \hat{e}_k = \hat{e}_z) = 0.55 \, I(\omega_0, \hat{e}_k = \hat{e}_z) \qquad \text{(B25)}$$
$$P_{in} = I(\omega_m, \hat{e}_k = \hat{e}_z)\Delta\Omega_{em}\Delta\omega_N \qquad \text{(B26)}$$

SPONTANEOUS EMISSION

RADIATION SOURCE ?   INCOHERENT

COHERENT

$$P_{in} = B\lambda^2\Delta\omega_N/4 \qquad \text{(B27)}$$

$$\Delta\omega_{out} = \min\,[\Delta\omega_{inj}, \ \Delta\omega_N] \qquad \text{(B28)}$$
$$P_{in} = P_{inj}\,\Delta\omega_{out}/\Delta\omega_{inj} \qquad \text{(B29)}$$

CONTINUOUS INCOHERENT

$$\tilde{G}_n(\bar{\theta}_m) = \frac{G_N(\bar{\theta}_m) - 1}{G_{net}(\bar{\theta}_m) - 1} \qquad \text{(B30)}$$

$$N_{sat} = \frac{\log(\Delta P_{sat}/P_{in}) - \log\{[G(\bar{\theta}_m) - 1]/[G_{net}(\bar{\theta}_m) - 1]\}}{\log G_{net}(\theta_m)} + 1 \qquad \text{(B31)}$$

$$\tilde{G}_N(\bar{\theta}_m) = \frac{G_N(\bar{\theta}_m) - 1}{[G_{net}^{1/2}(\bar{\theta}_m) - 1]^2} \qquad \text{(B32)}$$

$$N_{sat} = \frac{\log(\Delta P_{sat}/P_{in}) - \log\{[G(\bar{\theta}_m) - 1]/[G_{net}^{1/2}(\bar{\theta}_m) - 1]^2\}}{\log G_{net}(\bar{\theta}_m)} + 1 \qquad \text{(B33)}$$

CONTINUOUS COHERENT

$$N_{sat} = \frac{\log(\Delta P_{sat}/P_{in}) - \log[G(\bar{\theta}_m) - 1]}{\log G_{net}(\bar{\theta}_m)} + 1 \qquad \text{(B34)}$$

INSTANTANEOUS

$$t_{sat} = \frac{2L_c}{c} \, N_{sat} \qquad \text{(B35)}$$

NO $t_{sat} < t_{MP}$?   YES

$$P_{out} = T \left\{ \frac{G_N(\bar{\theta}_m)}{\tilde{G}_N(\bar{\theta}_m)} \right\} P_{in} \qquad \text{(B36)}$$

$$P_{out} = \frac{T}{1 - R} \, \Delta P_{sat} \qquad \text{(B37)}$$

Appendix C: Program Cold

Program COLD computes the FEL small-signal gain in any gain regime in the cold-beam limit (low- or high-gain tenuous beam or collective regime). The program variables are defined in terms of the FEL parameters used in this paper in the following way: ETP $\equiv \bar{\theta}_p$, ETA $= \bar{\theta}$, XKPA $= \bar{Q}/\bar{\theta}_p^2$. In the tenuous beam regime, substitute $\bar{\theta}_p = 0.01$.

```
        PROGRAM COLD (INPUT, OUTPUT)
C       THIS PROGRAM COMPUTES THE FRACTIONAL INCREASE IN
            POWER OF AV
C       ELECTROMAGNETIC WAVE THAT INTERACTS WITH A COLD
            (MONOENERGETIC)
C       ELECTRON BEAM.
C       THE COMPUTATION IS PERFORMED BY INVERTING LAPLACE
            TRANSFORMATION
C       USING PARTIAL FRACTIONS EXPANSION.
C       ETP = PLASMA FREQUENCY * INTERACTION LENGTH/BEAM
            VELOCITY
C       ETA = NATURAL WAVEGUIDE WAVE NUMBER — RADIAN
            FREQUENCY/BEAM
C       VELOCITY)* INTERACTION LENGTH
C       XKPA = NORMALIZED INTERACTION PARAMETER
C       XL1 = MIN VALUE OF ETA
C       XL2 = MAX VALUE OF ETA
C       DELTA = INCREMENT OF ETA
        DIMENSION X (303), Y (303), KK (6), LL (6)
        COMPLEX F, ALFA1, ALFA, BETA, FI, W, S1, S2, AA, S3, M1, M2, M3
        CALL NAMPLT
        READ 10, XLONG, YLONG
10      FORMAT (2F10, 3)
        READ 20, DELTA, XL1, XL2, MISP
20      FORMAT (3F10.3, I3)
        AA = (0., 0.)
        DO 56 I = 1, 5
        KK(I) = 10H
        LL(I) = 10H
56      CONTINUE
        KK(6) = 8H   DP/D
        LL(6) = 7H   ETA
        N = (XL2 − XL1)/DELTA + 1
        DO 200 JJ = 1, MISP
        ETA = XL1
        READ 17, ETP, XKPA
17      FORMAT (2F10.3)
        PRINT 22
22      FORMAT (1X, *ETP =*, 10X, *XKPA =*)
        PRINT 24, ETP, XKPA
24      FORMAT (1X, 2F10.5)
        PRINT 15
```

```
15        FORMAT (10X, *ETA*, 16X, *G*, 22X, *S1*, 30X, *S2*, 30X, *S3*)
          A0 = XKPA*ETP**2
          OO 100 K = 1, N
30        A1 = ETA**2 - ETP**2
          A2 = 2*ETA
          O1 = A2**2/9. - A1/3.
          O2 = A1*A2/6. - A2**3/27. - A0/2.
          AA = O2**2 - O1**3
          F = CSQRT(AA)
          ALFA1 = (CLOG(O2 - F))/3.
          ALFA = CEXP(ALFA1)
          BETA = O1/ALFA
          FI = (0., 1.)
          W = (-1.+FI*SQRT(3.))/2.
          S2 = A*ALFA + W**2*BETA - A2/3.
          S1 = ALFA + BETA - A2/3.
          S3 = W**2*ALFA + W*BETA - A2/3.
          M1 = (S1 + ETA + ETPA)*(S1 + ETA - ETP)/((S1 - S2)*(S1 - S3))
          M2 = (S2 + ETA + ETP)*(S2 + ETA - ETP)/((S2 - S1)*(S2 - S3))
          M3 = (S3 + ETA + ETP)*(S3 + ETA - ETP)/((S3 - S1)*(S3 - S2))
          G = CABS(M1*CEXP(FI*S1) + M2*CEXP(FI*S2) + M3*CEXP(FI*S3))**2 - 1.
          PRINT 40, ETA, G, S1, S2, S3
40        FORMAT (1X, 8E15.5)
          X(K) = ETA
          Y(K) = G*10000.0
          ETA = ETA + DELTA
100       CONTINUE
          IF(JJ.NE.1)GO TO 150
          CALL SCALE(Y, YLONG, N, 1)
          CALL SCALE(X, XLONG, N, 1)
          XS = -X(N + 1)/X(N + 2)
          YS = -Y(N + 1)/Y(N + 2)
          A = Y(N + 1)
          B = Y(N + 2)
          CALL AXIS(XS, 0.0, KK, 58, YLONG, 90.0, Y(N + 1), Y(N + 2), -1, 1)
          CALL AXIS(0.0, YS, LL, -57, XLONG, 0.0, X(N + 1), X(N + 2), 0, 1)
150       Y(N + 1) = A
          Y(N + 2) = B
          CALL LINE (X, Y, N, 1, 0, 1)
200       CONTINUE
          CALL PLOT (15., 0., 3)
          CALL ENOPLT
          END
```

REFERENCES

Abrams, R. L. (1972). *IEEE J. Quantum Electron.* **QE-8**, 838–843.

Adam, B., and Kreubühl, F. (1975). *Appl. Phys.* **8**, 281–291.

Al-Abawi, H., Hopf, F. A., Moore, G. T., and Scully, M. O. (1979). *Opt. Commun.* **30**, 235–238.

Alcock, A. J., Corkum, P. B., James, D. J., Leopold, K. E., and Samson, J. C. (1976). *Opt. Commun.* **18**, 543–545.

Bélanger, P.-A., and Boivin, J. (1976). *Can. J. Phys.* **54**, 720–727.

Bizarri, U., *et al.* (1982). *Phys. Quantum Electron.* **9**, 667–696.

Blewett, J. P., and Chasman, R. (1977). *J. Appl. Phys.* **48**, 2692–2968.

Brau, C. A., and Cooper, R. K. (1980). *Quantum Electron.* **7**, 647–664.

Colson, W. B. (1977). "Free-Electron Laser Theory." Ph.D. Thesis, Department of Physics, Stanford University, Stanford, California.

Colson, W. B. (1982). *Phys. Quantum Electron.* **8**, 457–488.

Dattoli, G., Mario, A., Renieri, A., and Romanelli, F. (1981). *IEEE J. Quantum Electron.* **QE-17**, 1371–1387.

Deacon, D. A. G., *et al.* (1977). *Phys. Rev. Lett.* **38**, 892–894.

Eckstein, J. N. (1982). Personal Communication, Department of Physics, Stanford University, Stanford, California.

Eckstein, J. N., *et al.* (1982). *Phys. Quantum Electron.* **8**, 49–76.

Elias, L. R. (1982). Personal communication, Quantum Institute, University of California at Santa Barbara, Santa Barbara, California.

Freund, H. (1981). *Phys. Rev. A.* **24**, 1967–1979.

Garmire, E., McMahon, T., and Bass, M. (1976). *Appl. Opt.* **15**, 145–150.

Gover, A. (1980). *Opt. Lett.* **5**, 525–527.

Gover, A., and Sprangle, P. (1981). *IEEE J. Quantum Electron.* **QE-17**, 1196–1215.

Hull, J. R., and Iles, M. K. (1980). *J. Opt. Soc. Am.* **70**, 17–28.

Jarby, E., and Gover, A. (1984). To be published.

Kapitza, S. P., and Melekhin, V. N. (1978). "The Microtron." Harwood Academic Publishers, London.

Kroll, N. M., and McMullin, W. A. (1978). *Phys. Rev. A* **17**, 300–308.

Kroll, N. M., Morton, P. L., and Rosenbluth, M. N. (1981). *IEEE J. Quantum Electron.* **QE-17**, 1436–1468.

Lawson, J. D. (1977). "The Physics of Charged Particles." Clarendon Press, Oxford.

Lenham, A. P., and Terhune, D. M. (1966). *J. Opt. Soc. Am.* **56**, 683–685.

Liewer, P. C., Lin, A. T., Dawson, J. M., and Capoui, M. Z. (1981). *Phys. Fluids* **24**, 1364–1372.

Livni, Z., and Gover, A. (1979). Quantum Electron Laboratory Scientific Report AFOSR 77-3445, School of Engineering, Tel Aviv University, Tel Aviv, Israel.

Madey, J. (1983). Personal communication, High Energy Physics Laboratory, Stanford University, Stanford, California.

Marcatili, E. A. J., and Schmeltzer, R. A. (1964). *Bell Syst. Tech. J.* **43**, 1783–1809.

Moltz, H. (1951). *J. Appl. Phys.* **22**, 527–535.

Nakahara, T., and Kurauchi, N. (1967). *IEEE Trans. Microwave Theory Tech.* **MTT-15**, 66–71.

Neil, V. K. (1979). JASON Technical Report JSR-79-10, SRI International, Arlington, Virginia.

Nishihara, H., Inone, T., and Koyama, J. (1974). *Appl. Phys. Lett.* **25**, 391–393.

Parker, R. K., *et al.* (1982). *Phys. Rev. Lett.* **48**, 238–242.

Prosnitz, D., Szoke, A., and Neil, V. K. (1981). *Phys. Rev. A* **24**, 1436–1451.

Renieri, A. (1979). "Feasibility and Performance of the Free Electron Laser," Frascati Report 79/30. ENEA, Dipartimento Tecnologie Intersettorial: d: Base Centro Ricerche Energia Frascati, Rome, Italy.

Shaw, E., and Patel, C. K. N. (1980). *Phys. Quantum Electron.* **7**, 665–669.

Smith, P. W., Wood, O. R., III, Maloney, P. J., and Adams, C. R. (1981). *IEEE J. Quantum Electron.* **QE-17**, 1166–1181.

Smith, T. (1982). Personal communication, Hansen Laboratory, Stanford University, Stanford, California.

Smith, T. I., and Madey, J. M. J. (1982). *Appl. Phys. B* **27**, 195–199.

Sprangle, P., and Smith, R. (1980). *Phys. Rev. A* **21**, 293–301.

Sprangle, P., and Tang, C. M. (1981). "The Three-Dimensional Nonlinear Theory of the Free-Electron Laser in the Amplifying Configuration." NRL Memorandum Report 4663, Naval Research Laboratory, Washington, D.C.

Sprangle, P., Smith, R., and Granatstein, V. L. (1979). *In* "Infrared and Millimeter Waves," Vol. 1 (K. J. Button, ed.), pp. 279–327. Academic Press, New York.

Sprangle P., Tang, C. M., and Manheimer, W. M. (1980). *Phys. Rev. A.* **21**, 302–318.

Sprangle, P., Tang, C. M., and Bernstein, I. B. (1983). "Start-Up of a Pulsed Beam Free Electron Laser (FEL) Oscillator." NRL Memorandum Report 5011, Naval Research Laboratory, Washington, D.C.

Tang, C. M. (1982). "The Effects of Transverse Gradient of Static Magnetic Wigglers on the Free-Electron Laser." NRL Memorandum Report 4820, Naval Research Laboratory, Washington, D.C.

Tang, C. M., and Sprangle, P. (1982). *J. Appl. Phys.* **53**, 831–839.

Tee, J. J., and Wittig, C. (1977). *Appl. Phys. Lett.* **30**, 420–422.

Yamanaka, M. (1977). *J. Opt. Soc. Am.* **67**, 952–958.

Yariv, A. (1976). "Introduction to Optical Electronics." Holt, Rinehart, and Winston, New York.

INDEX